NETWORK
INFRASTRUCTURE
AND ARCHITECTURE

NETWORK INFRASTRUCTURE AND ARCHITECTURE

Designing High-Availability Networks

KRZYSZTOF INIEWSKI
CARL McCROSKY
DANIEL MINOLI

WILEY-
INTERSCIENCE

A JOHN WILEY & SONS, INC., PUBLICATION

Published by John Wiley & Sons, Inc., Hoboken, New Jersey.
Published simultaneously in Canada.

For general information on our other products and services or for technical support, please
contact our Customer Care Department within the United States at (800) 762-2974, outside
the United States at (317) 572-3993 or fax (317) 572-4002.

Wiley also publishers its books in a variety of electronic formats. Some content that appears in
print may not be available in electronic formats. For more information about Wiley products,
visit our web site at www.wiley.cm.

Library of Congress Cataloging-in-Publication Data:

Iniewski, Krzysztof.
   Network infrastructure and architecture: designing high-availability networks /
Krzysztof Iniewski, Carl McCrosky, Daniel Minoli.
       p. cm.
   Includes index.
   ISBN 978-0-471-74906-6 (cloth)
   1. Optical communications. 2. Integrated circuits—Very large scale integration.
3. Data transmission systems—Design and construction. I. McCrosky, Carl, 1948–
II. Minoli, Daniel, 1952– III. Title.
   TK5103.59.I49 2008
   621.382′7—dc22

                                                                    2007034273

10 9 8 7 6 5 4 3 2 1

For
Ferdynand Iniewski
Judy Berlyne McCrosky
Anna
with affection and thanks

CONTENTS

PREFACE

Modern communications networks consist of collections of *network elements* connected by *communications paths*. Network elements and communications paths depend on, or are developed as, three fundamental layers: (1) optical transmission facilities, which carry information between network elements; (2) protocol engines, which enforce the constructive discipline of communications protocols and provide many protocol supported capabilities; and (3) switching and queuing components, which direct information to its intended destination. Protocol engines, switching/queuing components, and optical interfaces are used to construct network elements; network elements are then connected by optical transmission facilities to form communications networks. This book provides an introduction to these building blocks and to the networks formed by their aggregation for advanced undergraduate students in electrical engineering, computer engineering, and computer science.

We take a unique approach to this material by covering the ideas underlying these networks, the architecture of the network elements, and the implementation of these network elements in optical and VLSI technologies. The authors believe the nature of these networks to be in large part determined by these twin underlying implementation technologies, and that it is useful for the student to study networks in this broader, implementation-oriented context. Consequently, in this book we show how communications systems are implemented with these two fundamental technologies, and how the choice of these two technologies affects the design of communications systems.

Communications systems have been undergoing dramatic change for several decades. It is reasonable to expect this period of dramatic change to continue. In times of dramatic change, it is difficult to predict the future. Consequently, selecting material to be included in the book has been challenging. Our choices have been based on several principles: (1) networks will continue to incorporate multiple protocols in order to support a variety of applications to maintain smooth transitions from legacy systems; (2) although there will be a highly visible move toward packet-based services, there are strong technical reasons why underlying networks will continue to have circuit and TDM layers and components; and (3) storage networks will continue to grow beyond the data center and occupy increasing capacity within wide area networks.

It is abundantly clear that wireless protocols, networks, and applications will be a highly visible growth area. This is a vital area of study for future telecommunications workers, so vital that the area requires a full, separate treatment instead of a few chapters in a larger text. At the same time, wireless networks will connect to the underlying WAN network elements, and regardless of wireless developments, optical WANs will continue to be of crucial importance. Consequently, we see the wireless network connection to the WAN as a natural boundary for this book.

Why write another book about networking? For various reasons. First, almost all networking texts focus on TCP/IP routing. Although IP routing is a very important subject, treatment of physical transport and switching is equally important and deserves better coverage. In that respect the book fills a hole in the marketplace and offers a complementary view of optical networking from a bits and bytes perspective.

Second, many existing treatments of optical networking lack coverage of hardware issues. Hardware is as important as software, and we believe that electrical engineers need to develop some understanding of how networking hardware is built. The book offers a unique perspective on VLSI and its linkages to advances in networks.

Finally, optical networking is a rapidly evolving theme. Many new technologies are being deployed in the marketplace, such as metro wave-division multiplexing, resilient packet rings, optical Ethernet, and multiprotocol label switching, to name a few. The complex picture of broadband networking is constantly evolving and as a result, needs clear delineation as to what is important and what are basic principles behind its evolution. The book provides an understanding that helps answer these questions.

Acknowledgments

The authors would like to acknowledge the valuable contributions of Jeff Bull, Hiroshi Kato, and Benjamin Tsou to Chapter 2 and of Jerzy Świć to Chapter 7.

KRZYSZTOF INIEWSKI
CARL MCCROSKY
DANIEL MINOLI

1

INTRODUCTION TO NETWORKING

1.1   INTRODUCTION

We all use the Internet. However, few of us are aware of exactly what is happening once we click a mouse after typing a key word in a search engine. How is our question traveling to Google computers? What type of hardware

Network Infrastructure and Architecture: Designing High-Availability Networks,
By Krzysztof Iniewski, Carl McCrosky, and Daniel Minoli
Copyright © 2008 John Wiley & Sons, Inc.

1

is involved in this process? How are bits and bytes moved from point A to point B? How do they know which way to go? Who takes care of the process of data delivery?

In this chapter we provide a gentle start on our journey of discovering how data networks, which are responsible for the delivery of information across the Internet, work in today's world. The text serves as an introduction to broadband networking and deals with basic concepts in communications. We start by discussing the various transmission media used and show that optical fiber is a much better medium than copper wire or coaxial cable. This is followed by a description of the classes of various networks present today. We briefly mention the basic characteristics of access, local area, storage, and metropolitan and wide area networks. In the following section we go over possible network topologies. We show differences between point to point, hub, ring, and mesh architectures. Our discussion of network topologies is followed by a review of an open system interconnect (OSI) model. Following the OSI model we discuss three methods of multiplexing data: wavelength division, time domain, and statistical multiplexing. At the end of the chapter we describe various classes of networking equipment. We examine briefly the functionality of networking devices such as regenerator, hub, switch, or router.

1.2 TRANSMISSION MEDIA

Four types of media can be used to transmit information in the world of communications: copper wire, coaxial cable, optical fiber, and air (wireless communication).

We discuss all four options briefly to illustrate that optical fiber is by far the best option available when information is to be transmitted over long distances.

1.2.1 Copper Wire

In the old days, copper wire was the only means of transmitting information. Technically known as unshielded twisted pair (UTP), the connection consists of a number of pairs, typically two or four, of copper wires. The wire pairs are called "twisted" because they are physically twisted. This helps reduce both crosstalk between wires and noise susceptibility.

UTP cable does not have a shield, and therefore a high-frequency part of the signal can "leak out." Also, the twisting on the copper pair can be quite casual, designed as much to identify which wires belong to a pair as to handle transmission problems. Although not perfect, this cabling technology was for many years quite satisfactory for voice communication purposes. Consequently, there are millions of kilometers of copper wires in the public switched telephone network. However, not only did the copper wire itself have limitations, but things were done to the wiring to make it even more unsuitable for high-speed data transmission. These actions took many forms, although we only mention one here—load coils—as an example.

Load coils were frequently added to wiring loops longer than a few kilometers. The coils were essentially low-pass filters, which meant that they passed without a loss of the low frequencies that correspond to your voice, but blocked higher frequencies. High-frequency blocking is disastrous for data communications, as we actually rely on high frequencies to achieve the desired speed of data transmission.

In conclusion, what was once a good system for voice transmission has become a big problem for data transmission. In fact, out of all possible wires, UTP copper wire is probably the least desired, as it has the worst transmission properties. However, it is present virtually everywhere, as most households, at least in developed countries, have copper wire phone connections already in place.

For networking applications, the term UTP generally refers to 100-Ω category 5 cables, the characteristics of which are tightly specified in data networking standards. Techniques have been developed to provide means of having a phone conversation and high-speed data transfer using the same twisted-pair copper wire. These techniques, called DSL (digital subscriber loop) technology, come in many different flavors, asymmetrical DSL being the most popular for home Internet connectivity. However, even with state-of-the-art hardware, the highest bandwidth of the DSL connection might get to a level of 50 million bits per second (Mb/s). In practice, a more typical number would be around 1 Mb/s. As we will see in a moment, this is a relatively low number compared to that of optical fiber. The difference results from the poor transmission characteristics of unshielded copper wires compared with the excellent transmission characteristics of optical fibers.

The low bandwidth of UTP cabling is only one of its problems. The second problem is its signal attenuation, which is very high in this medium. For example, a DSL signal virtually disappears after transmission over a few kilometers. It needs to be recovered, amplified, and retransmitted by sophisticated electronic equipment. You can imagine that it would have taken many UTP wires and retransmission points to send a stream of 10 Gb/s from New York to San Francisco. So although UTP cable is useful at home, clearly we need something better to send significant amounts of data over long distances. Let us look next at coaxial cable.

1.2.2 Coaxial Cable

Coaxial cable consists of a single strand of copper running down the axis of the cable. This strand is separated from the outer shielding by an insulator made of a dielectric material. A conductive shield covers the cable. Usually, an outer insulating cover is applied to the overall cable. Because of the coaxial construction of the cable and the outer shielding, it is possible to send quite high frequencies. For example, in cable television systems in North America, 20 TV channels, each with 6 MHz of bandwidth, can be carried on a single coaxial cable.

Coaxial cable access was used originally for the purpose of broadcast video. As a result, the system is inherently well suited for high-speed data transfer. Techniques have been developed to provide broadcast video and high-speed Internet access on the same coaxial cable. We refer to these techniques collectively as *high-speed cable modem technology*. However, this term involves a bit of cheating. The term *coaxial cable* notes, in practice, a system that is really a hybrid of optical fibers and a coaxial cable with the optical fiber portion of the cable network being hidden from the end user. The reason for this "hybridization" of the coaxial cable network is that although coaxial cables have better signal transmission properties than UTP cables, they are inferior to optical fibers. As a result, your cable company, which provides you with TV programming, only needs to install optical fibers in its infrastructure in the last few kilometers adjacent to you when using coaxial cables.

Another issue with coaxial cable is the fact that cable access is not available universally. From this brief description, the conclusion seems to be that although coaxial cable is somewhat better than UTP wires, it is still not good enough for the purpose of transmitting large amounts of information over long distances. Let us look next at optical fiber.

1.2.3 Optical Fiber

Fiber is the third transmission medium we discuss. We will say immediately that it is unquestionably the transmission medium of choice. Whereas transmission over copper utilizes frequencies in the megahertz range in the best of cases, transmission over fiber utilizes frequencies approximately 1 million times higher, in the terahertz range. Terahertz frequencies translate into terabytes per second (TB/s) of bandwidth; this is very large indeed. For example, the most congested Internet connection today still requires lower bandwidth than 1 TB/s. To say this another way, a single strand of optical fiber is at present sufficient to carry all Internet traffic between North America and Europe. Not bad.

Fiber-optic technology, offering virtually unlimited bandwidth potential, is the transmission medium of choice for long-haul networks. It is also widely considered to be the ultimate solution to the delivery of broadband access to the end user, called the *last mile*, the network space between the carrier's central office and the user location. The last-mile piece of the network is typically the area where the majority of bottlenecks occur that slow the delivery of data services. Right now, UTP wire and coaxial cable are still used in most cases for last-mile data delivery.

Beyond its enormous transmission bandwidth, optical fiber has another great advantage: its very low attenuation loss. Optical signals travel for kilometers without losing virtually any of their strength. Recall for comparison that electrical signals traveling over copper wires basically disappear after a few kilometers! If you think of the optical fiber as a piece of glass, you would be able to see objects several kilometers away when looking into it. This is tremendously impressive and useful. There is no need to regenerate optical signals for several kilometers of its transmission along the optical fiber.

Besides the available bandwidth, which is enormous compared to copper wire or coaxial cable, and the very low signal loss, which is again much lower, optical fiber has other advantages. For example, optical signal propagating in the fiber is insensitive to external noise. Chapter 2 is devoted to issues of optical transmission, as they lay the foundation for optical networking, so we defer a full discussion until then. For now, let us just summarize by saying that the data signal in an optical fiber has a very high bandwidth, a very low attenuation coefficient, and is very robust (it is quite difficult to disturb).

It appears as though we are done with the choices for suitable transmission media: Optical fiber wins hands down. But wait a moment, what about wireless? Everything is wireless these days: your CDMA or GSM cellular phone, your WiFi 802.11 connection in your laptop, and even your garage door opener. Could it be that wireless is more suitable than optical fiber? Not really, but for the sake of completeness we discuss wireless next.

1.2.4 Wireless Communication

Wireless communications is our final option for a transmission medium. In this case the medium is simply air. Wireless transmission can take several forms: microwave, low-Earth-orbit satellites, cellular, or wireless local area networks. In every case, however, a wireless system seems to obviate the need for a complex wired infrastructure. There are also other advantages. In the case of satellites, transmission can take place across any two points on the globe. With microwave communications there is no need to install cabling between microwave towers. Cellular phones, for example, afford significant mobility. Similarly, your laptop WiFi wireless card lets you surf the Net in any Starbucks with hotspot availability.

These advantages come at a price. Wireless communications is significantly less efficient than optical communication systems in terms of bandwidth available. The most advanced, third-generation cellular phone systems provide bandwidth of hundreds of kilobytes per second at best, whereas standard optical transmission runs routinely at rates of several gigabyte per second. Using advanced optical technology called wave-division multiplexing, which we discuss in detail in Chapter 3, transmissions of terabits per second in one optical fiber are possible.

At this point, it is difficult not to stop and wonder over the fact that there is a difference of a few orders of magnitude in terms of available signal bandwidth between wireless and optical networks. Looking at this comparison another way, it would take about 1 million times longer to transfer a large file using a wireless connection than to send it using optical fiber. Clearly, you would not want to send a 1-gigabyte file using air as the medium!

The reasons for such lower bandwidths in wireless systems are fairly obvious. Air is not a very suitable medium for transmission, as signals propagate in all directions. There are many signal distortions as signals bounce from numerous objects and interfere with hundreds of other signals present simultaneously. Most important, unlike in an optical fiber case, where the signal is confined to a very small physical space, a wireless signal loses it strength very quickly as its energy propagates in three-dimensional sphere. Finally, mobility of the user places an additional burden on the wireless communication system.

Therefore, the conclusion seems to be that although wireless systems are very useful in many applications, they are not suitable for sending large amounts of data over long distances. Optical fiber is a clear winner for this application, as we anticipated from the start. For this reason, for the remainder of the book, unless specified otherwise, we assume that data transmission takes place over optical fibers.

1.3   BASIC NETWORKING CONCEPTS

The global broadband network is a connection of thousands of different networks implemented in various countries and cities by companies using a variety of technologies and protocols, as shown schematically in Figure 1.1. For a number of historical reasons, the global broadband network is more complicated than it could have been if it were build today from scratch, but that is a nature of many technologies. We hope to shed some light on this complex jungle of networking hardware technologies by explaining the basic concepts. Armed with this knowledge, you should be able to navigate Internet infrastructure with better understanding, and you should be able to find additional details, if needed, in the appropriate reference sources.

1.3.1   LAN, SAN, MAN, and WAN

To start our discovery process in network complexities, we will divide the global network into five classes of networks that can be identified based on their geographical span and functionality:

- *Access networks*: networks that connect consumers and corporate users with Internet infrastructure. DSL and cable modems are examples of access technologies.
- *Local area networks* (LANs): networks that connect multiple users in a contained environment such as a corporate building or a campus. Typically, Ethernet is used as the protocol of choice.

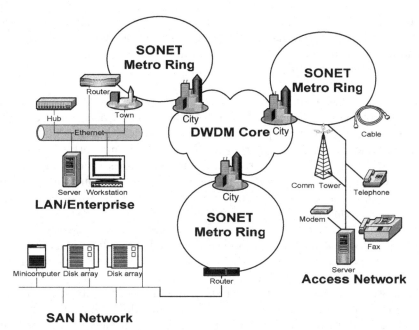

FIGURE 1.1 Broadband network elements consisting of a network core, synchronous optical network (SONET) rings, a local area network, storage, and access networks.

- *Storage area networks* (SANs): corporate data storage networks that connect backend storage disks via high-speed interfaces using primarily a fiber channel protocol.
- *Metropolitan area networks* (MANs): networks that connect data and voice traffic at the city level, typically using synchronous optical network (SONET) rings.
- *Wide area networks* (WANs): networks that connect multiple corporate locations or cities across long distances, also known as *core* or *long-haul networks*. WANs use optical fiber infrastructure exclusively.

Access Networks Access networks are used to gain access to broadband Internet infrastructure. At home we might use a conventional, old-fashioned telephone dial-up, a DSL, or a cable modem technology to make that connection. Some of us might use wireless via advanced cellular phones or by hooking up our laptop using a WiFi 802.11 wireless card. Only a lucky few have access to dedicated optical fiber connections with practically unlimited bandwidth. An *access network* is defined somewhat more precisely as the portion of a public switched network that connects central office equipment to individual subscribers. All these access connections, coming from multiple users, somehow have to be groomed and delivered to Internet service providers (ISPs), companies that provide Internet access service in terms of content. In general, access networks provide data, video, and voice connectivity to all required locations for both consumers and corporate customers using a number of protocols and technologies.

Local and Storage Area Networks   University campus networks or networks in large companies are examples of LAN networks. Local area networks connect PCs, workstations, printers, and other devices inside a building or campus, typically using an Ethernet protocol. A LAN network starts with a single user in a corporate environment, but where does it end? Typically, a local area network is connected to the public network via a firewall. The firewall provides data and security protection for a business. Firewalls also provide a convenient demarcation point between LAN and WAN/MAN infrastructure. LAN networks typically use copper wires and coaxial connections, as these are more readily available than optical fiber, and the distances involved are short, hundreds of meters at most. As a result, the use of optical fiber in LAN networks is quite low, nonexistent in most cases.

Storage area networks (SANs) are specialized local area networks that deal exclusively with storage devices such as tape drives, disks, and storage servers (Figure 1.2). Storage networks that connect backend storage devices via high-speed interfaces use a protocol called a *fiber channel*. We do not discuss LAN or SAN networks much in this book, however, as our focus is on core data networks.

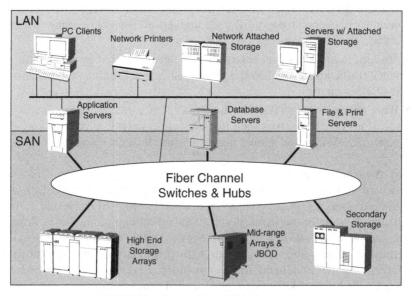

FIGURE 1.2   LAN and SAN networks.

Wide and Metropolitan Area Networks   Wide area networks can be divided into metropolitan area networks (MANs) and long-haul networks. Long-haul networks provide transmission services over long distances, typically over hundreds of kilometers. Consequently, they use big "pipes" to carry the traffic and their main service is to deliver from point A to point B. Heavy trucks on a transcontinental highway system are a good example of a long-haul network. MANs encompass large metropolitan areas and therefore cover distances of about 80 to 120 km. The size of the data "pipe" is smaller than in a long-haul network, and services become more varied. Small distribution trucks in a city would be a good analogy for metropolitan area networks.

What is required to build economical wide area networks? First, a network should be flexible, which can be accomplished by providing a large number of interchanges to offload traffic at various points in the network. Second, means of going long distances need to be provided. With optical transmission this is not a problem under certain conditions. Finally, if many "cars" could use the same "highway," the networking system would be able to ensure large throughput. Fortunately, a technology that does precisely that wavelength-division multiplexing, has been developed and is discussed in detail in Chapter 3.

Optical Links   Optical links are deployed in each segment (i.e., access, SAN, LAN, MAN, and WAN) of the global networking infrastructure. Networks such as long-haul have lots of optical fiber, whereas some, such as the LAN of a small company, have little or none. Very few networks will be purely optical, but more often than not, networks will involve both optical and electrical components.

1.3.2   Network Topologies

Many network topologies exist in optical data networks, from point to point, to hub, to ring, to fully meshed networks. Each network has advantages and disadvantages. Schematic representations of all of these topologies are shown in Figure 1.3.

Point-to-Point Topology   The simplest network topology is point to point, shown schematically in Figure 1.3(a). In this configuration network management is straightforward, but the link has low reliability. If the fiber is cut between points A and B, there is no way to recover. Point-to-point links are sometimes used in long-haul networks, such as in cabling under oceans.

Hub Topology   The hub network architecture accommodates unexpected growth and change more easily than do simple point-to-point networks. A hub concentrates traffic at a central site and allows easy reprovisioning of the

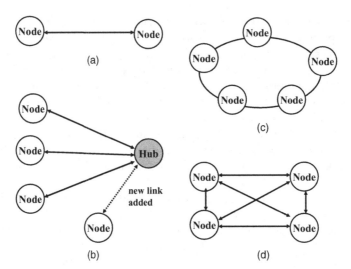

FIGURE 1.3 (a) Point-to-point, (b) hub (star), (c) ring, and (d) mesh network topologies.

circuits. A process of adding one more node in the hub configuration is simple and is shown in Figure 1.3(b). The hub configuration, however, still has the "single point of failure" problem. Ethernet hubs are frequently used in LANs.

Ring Topology   Ring topology relies on a connectivity of the network nodes through a circular type of arrangement, as shown in Figure 1.3(c). The main advantage of the ring topology is its survivability: If a fiber cable is cut, the nodes have the intelligence to send the services affected via an alternative path. SONET/SDH uses ring topology predominantly.

Mesh Topology   A meshed network consists of any number of sites connected together arbitrarily with at least one loop. A fully meshed topology is shown in Figure 1.3(d) for a simple network of four nodes. A typical characteristic of a meshed network is that nodes within the network can be reached from other nodes through at least two distinct routes. In practice, meshed networks are often constructed using large rings and numerous subrings. Meshed topology can be very expensive in terms of the hardware used. WDM technology frequently has a mesh infrastructure.

Network Topology Comparison   All four topologies have been summarized in Table 1.1. It should be noted that other intermediate forms of network topologies exist. Star or bus topologies are very similar to the hub topology, tree topology is a combination of the hub and point-to-point configurations, meshed networks can be only partially meshed, and so on. These intermediate

TABLE 1.1   Network Topology Comparison

Topology	Benefits	Shortcomings	Example
Point-to-point	Very simple	Single point of failure	Long-haul links
Hub	Simple	Single point of failure	Ethernet LAN
Ring	Some redundancy	Scalability problems	Metropolitan SONET/SDH
Mesh	Full redundancy	Complex, hardware intensive	WDM core

forms retain, to a large extent, the benefits and shortcomings presented in Table 1.1.

1.3.3   Circuit vs. Packet Switching

Transmission in telecommunications networks is digital by nature, and the transmission medium of choice is fiber. But how are the ones and zeros to be arranged? At what speed are they to travel? What route should they take? Answers to questions such as these have taken many forms and have made for the most complicated aspect of telecommunications networks. In this section we introduce some basic networking concepts. We discuss circuit and packet switching and describe ways of multiplexing the data. To start, then, consider that points A and B in Figure 1.4 want to exchange information and that the network that will enable the desired connection is as shown. How can the connection get established? There are two ways in principle: circuit switching and packet switching.

Circuit Switching   In circuit switching, a dedicated connection is established for the time of the data transfer between points A and B. This process seem to be straightforward (Figure 1.5), although in practice, finding an available

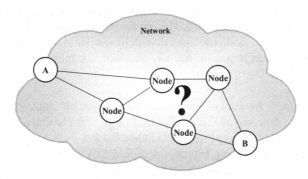

FIGURE 1.4   Network connection problem between points A and B.

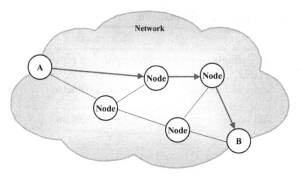

FIGURE 1.5 Circuit-switching principle. A permanent connection is established between points A and B.

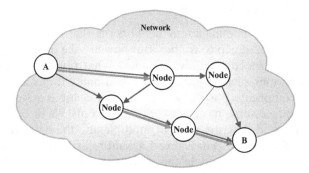

FIGURE 1.6 Packet-switching principle. Each packet travels using a different route from paint A to paint B, as indicated by arrows.

connection in a congested network is often difficult. Telephone services use circuit-switching connections. Once the connection is established, you can talk as long as you want. During your talk the network circuits (i.e., in practice, the wires and switch ports) are open, while other hardware pieces remain dedicated to your phone conversation. This means that nobody else can use them at the same time. As a result, the circuit-switching system is simple and reliable, but not efficient. Even if you do not say anything for minutes, the connection is reserved for you.

Packet Switching   In packet switching, each piece of data, or *packet*, can travel from A to B using a different path. The process is illustrated in Figure 1.6 for three packets taking different routes. Packet switching is an example of connectionless technology. Internet protocol (IP) is a routing protocol used to find suitable routing paths in Internet networks. IP uses a packet-switching principle.

Switching Technology Comparison   What are the benefits and shortcomings of packet-switching concept compared to circuit switching? Packet switching is more complex, as each piece of data is switched differently. It is also less reliable, as the packets reach destination point B at different times. In fact, because a network has to ensure that each packet reaches its destination in a reasonable time, looping around in large networks has to be avoided. Packet switching has two huge advantages, though: resource sharing and flexibility. Network resources are used only when needed, and as soon as a resource becomes available, it can be used to serve another connection. In this sense, resources are shared between various connections. In addition, this model is flexible, as data packets can be sent using the routes that are available at that particular moment in time.

The battle between circuit switching and packet switching has been one of the most interesting technology battles to watch. On one side you have public switched telephone network and traditional telephone service. On the other side, you have IP and related voice over IP (VoIP) technologies. In a general sense, circuit switching is a centralized model. To establish a global path between A and B, some authority, in the form of a central management system, has to decide where that dedicated path is established. Packet switching, on the other hand, is a decentralized system. Local paths between network elements can be established locally only by viewing large networks in close proximity. In this way, there is less need for centralized management, and large portions of system intelligence can be distributed through the network.

1.3.4   Wavelength vs. Time vs. Statistical Multiplexing

There are several way of multiplexing communications signals. Consider your cable television system. To get multiple TV channels, each channel is broadcast on a different frequency. All signals are "mixed" or, to use a more technical term, *multiplexed*, using frequency-division multiplexing (FDM). FDM is rarely used in optical networks but has a close relative called wavelength-division multiplexing (WDM).

Wavelength-Division Multiplexing   WDM works on a principle similar to that of FDM. It is a technique of "combining" or multiplexing multiple wavelengths in an optical fiber. The basic concept is illustrated in Figure 1.7. WDM takes optical signals, each carrying information at a certain bit rate, gives them a specific wavelength, and then sends them down the same fiber. Each input optical signal has the illusion of possessing its own fiber. WDM and FDM are conceptually similar, as there is a direct fundamental relationship between frequency and wavelength. On the other hand, FDM equipment for multiplex electrical signals and WDM equipment for multiplex optical signals are very different.

In this chapter we have started to use a highway analogy for communication along optical fiber. In this context, WDM gets more cars to travel. It

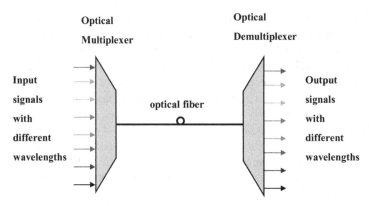

FIGURE 1.7   Principle of wavelength-division multiplexing.

does this not by increasing their speed but by making them travel in parallel in their own dedicated lanes. Traffic in each lane can travel at different speeds, as each lane is independent. The wavelengths used for WDM are chosen in a certain range of frequencies, and details of this selection are discussed in Chapter 3.

Synchronous Time-Division Multiplexing   Another method of multiplexing signals is called time-division multiplexing (TDM). The basic concept is illustrated in Figure 1.8. In this example, three data streams (A, B, and C) are multiplexed in a time domain in the following way: Each stream is assigned a fixed time slot in a multiplexed output stream. In time-synchronized TDM, it does not matter whether the stream has any data to send. If it does not, the

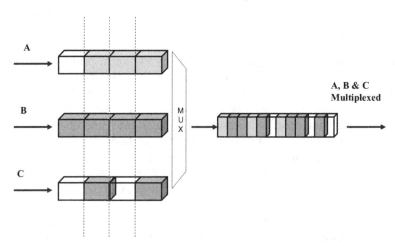

FIGURE 1.8   Principle of synchronous time-division multiplexing.

time slot is simply wasted. As a result, the multiplexed system retains all empty slots of the input data stream, so the TDM system is not efficient in this regard.

Note that for the system shown in Figure 1.8, the bit period of the output stream is three times smaller than the bit period of the input stream. To put this differently, the output bandwidth is three times greater than the bandwidth of an individual input stream. Therefore, in the example given, the networking clock for the output network needs to be three times as fast as the networking clock for the input network. As an example of networking application, TDM has been used in the multiplexing of voice signals. In the early 1960s, Bell Labs engineers created a voice multiplexing system that digitized a voice sample into a 64-kB/s data stream. They organized this data stream into 24-element frames with special conventions for determining where all the bit slots were positioned. The frame was 193 bits long and created an equivalent data rate of 1.544 Mb/s. This rate is referred to as *T1*. European public telephone networks modified the Bell Lab approach and created *E1*, a multiplexing system for 30 voice channels running at 2.048 Mb/s.

Based on these T1 and E1 concepts, the entire hierarchy of T- and E-type signals was created. For example, T3 was created to have a bandwidth three times greater than that of T1. Figure 1.8 can serve as a conceptual representation of multiplexing three T1 signals into a T3 data stream. Multiples of T1 were used to create the SONET multiplexing hierarchy. For example, 84 T1 signals create STS-1, which serves as a building block for SONET and has the rate of 155 Mb/s (84 times 1.544 Mb/s). Similarly, E1 was used to create the synchronous digital hierarchy (SDH) system in Europe. We talk about SONET and SDH in much more detail in Chapter 4.

Asynchronous Time-Domain Multiplexing   A different technique for multiplexing relies on an asynchronous principle and, as it turns out, is well suited for packet switching. Asynchronous multiplexing takes packets from each data stream and allocates them in order in the output queue. The process shown conceptually in Figure 1.9 looks deceivingly simple. In

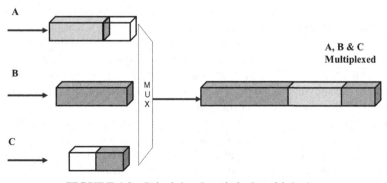

FIGURE 1.9   Principle of statistical multiplexing.

practice, one has to realize that packets arrive at different times and have different lengths, so some form of packet buffering is required. In addition, a recipe for packet ordering is required. Overall, the process of packet switching can be a very complex, and we discuss it later in the book. The big advantage of asynchronous multiplexing is that empty slots in input data streams are effectively eliminated. Only "real" data packets are included in the output queue. As a result, this technique has the potential for a statistical throughput gain.

1.4 OPEN SYSTEM INTERCONNECTION MODEL

1.4.1 Basic Concept

The open system interconnect (OSI) model is somewhat abstract, but is nevertheless a very useful tool in understanding networking concepts. To describe it clearly, we will first use a loose analogy: Imagine that two ships on the Atlantic Ocean are passing each other. Assume that there is a Chinese cook on the first ship and a French cook on the second ship, and they want to exchange recipes. There are a few problems they have to overcome. First, the French cook does not speak Mandarin or Cantonese. In addition, the Chinese cook does not speak French. Fortunately, English translators are available on both ships, so both cooks get their recipes translated into English.

Although both sets of recipes are translated into a common language, there is still a problem. Due to the distance involved, neither party can talk to the other. Fortunately, the ships are close enough that people can see each other. Both have specialists in semaphore on board who can communicate using flags. The translation process goes like this. First, the Chinese recipe is translated from English into semaphore and signaled by the flag so the other party can read the message. On the other side the semaphore signal is received and decoded into English. Subsequently, the English text is translated into French so that the French cook can read it and cook a Chinese dinner for his crew. The process can be repeated as many times as needed as long as visual flag connection is maintained between the two parties (and obviously, this process works in reverse as well).

As we can see from this simple example, the communications problem can be viewed hierarchically. The first level, layer 1, is the physical layer at which the communication has occurred. In our example this is the semaphore signaling layer using flags. At this level the signals are physical in nature. The flag is up or down, and it would be very difficult for an untrained observer to understand what the message is saying. In digital communication, the corresponding physical signal is a stream of zeros and ones.

The second layer, layer 2, is the English language layer. Anyone who speaks English can participate in communications at this level. But people who use different language, or in a digital communication sense, different protocols, cannot participate effectively unless English text is translated into their languages. Obviously, people can speak two or more languages, but that is another story.

Finally, the third layer, layer 3, is the cooking layer. Only cooks or people skilled in cooking can participate in communications at this level. But people who do not cook will not know what a phrase such as "make French sauce" means and will therefore be excluded from communication.

1.4.2 OSI Model and Data Encapsulation

The OSI model presented in Figure 1.10 is built on the layering principles outlined in our example. *Layer 1* is a *physical* or *transport layer.* This layer is concerned with electrical and optical signals, their amplitude, jitter, frequency of operation, and so on. An example of the transport layer is the SONET protocol, discussed in detail in Chapter 4.

Layer 2 is a *data link layer.* The data link is concerned with moving data across the physical links in the network. Layer 2 is responsible for switching data, typically using some form of data address to determine where to direct the message. The data link layer ensures that the connection has been set up, formats data into chunks called *frames*, and handles acknowledgments from a receiver that the data have arrived successfully.

Layer 3 is a *routing layer.* The routing layer is concerned with finding the next routing point in the network to which a message should be forwarded toward its end destination. It decides which way to send the data based on the

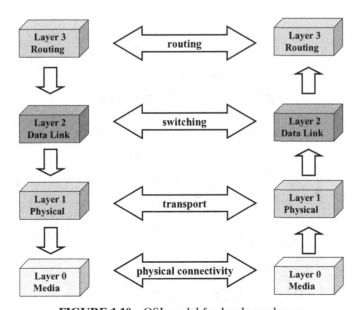

FIGURE 1.10 OSI model for hardware layers.

device's understanding of the state of the network to which the routing device is connected. An example of the routing protocol is IP, which uses a number of storage mechanisms to create and maintain tables of available routes and their conditions. The routing table information, coupled with distance and cost algorithms, is used to determine the best route for a given packet. IP routing is a complex technology that we discuss in later chapters.

There is one additional layer indicated in Figure 1.10 which has not been described so far. It is *layer 0*, a *media layer*. This additional layer represents media and technologies that operate at the media level. As discussed earlier, the media might be UTP copper wire, coaxial cable, or optical fiber, with optical fiber being the medium of choice in this book. Wave-division multiplexing is an example of a layer 0 technology, as it effectively "multiplies" the fiber capacity, as explained in Chapter 3.

One word about the naming convention is in order before we continue. Depending on where we are in the OSI hierarchy, we might use different names for the communication signal. At layer 0 we will probably talk about *optical pulses* or *electrical currents*. At layer 1 we will probably use the term *frame*: for example, *SONET frame*. At layer 2 we might also use the term *frame*, or if the frame is of constant length, we will call it a *cell*, as in asynchronous transfer made (ATM) *cell*. Finally, at layer 3 we usually refer to a data signal as a *packet*.

It is important to realize that frames, cells, and packets have generally similar structures, as shown in Figure 1.11. They all consist of the real data they are carrying and some additional overhead information. The real data are the

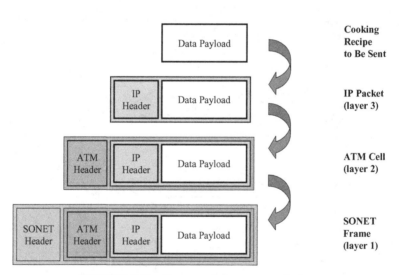

FIGURE 1.11 Process of data encapsulation.

messages to be transferred across the network, referred to as a *data payload*. Most of the data are in the data payload section, although the amount of data overhead will vary depending on the particular protocol implementation. The overhead information can either be appended before the payload in the form of a header or after the payload in the form of a trailer, and is frequently appended at both ends. For simplicity of illustration we assume in this chapter that only the header is attached.

As illustrated in Figure 1.11, when the data travel in the OSI hierarchy, its overhead information changes. Assume that the aforementioned cooking recipe is the data payload to be sent across the network. At the IP routing layer the packet consists of the data payload (the recipe) and IP header. When sent to a data link layer, a frame appends its own header (in the example shown, it is the ATM header). Finally, when sent down to layer 1, the transport protocol, in this case SONET, appends its own header. The entire process is reversed on the opposite end. First, the transport header is stripped off, and the remaining information is sent to the data link layer. At layer 2 the switching header is removed and the remaining information is sent to the routing layer. Finally, at the routing layer the layer 3 header is removed and only the data payload remains intact.

1.4.3 Network Overlay Hierarchy

As a result of the OSI layer structure, the networks are frequently overlaid one on top of the other. A typical overlay situation is shown in Figure 1.12. IP networks are overlaid on top of layer 2 networks such as ATM and Ethernet

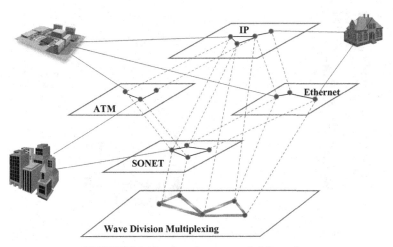

FIGURE 1.12 Overlay network hierarchy.

networks. These in turn can be overlaid on top of layer 1 networks such as
SONET networks. Finally, all of the networks above can be overlaid on top
layer 0 WDM networks. The degree of overlay can vary and in numerous cases,
some layers are omitted. As you can imagine, this fact creates a large number
of possible scenarios. One popular example of this networking overlay is the
following combination: IP on top of ATM on top of SONET on top of
WDM.

1.5   NETWORKING EQUIPMENT

Networking hardware equipment consists of various networking "boxes" that
are used to build broadband data networks. To connect your computing
resources and enjoy the benefits of network computing, you need these "boxes"
filled with processors, switches, and cabling interfaces. One might argue here
that the term *boxes* is somewhat of an understatement, as the box can be quite
powerful, as illustrated in Figure 1.13, which uses a core router as an example.
We might also want to mention that a core router can easily cost over $1
million. After this example, hopefully, networking boxes have gained some
respect in your eyes. Corresponding to their new, more appreciated status for
networking boxes, we instead use the terms *network elements* or *networking
equipment* for the remainder of the book.

   We can divide networking elements according to the OSI layer number at
which they operate, as illustrated in Figure 1.14. For example, IP routers

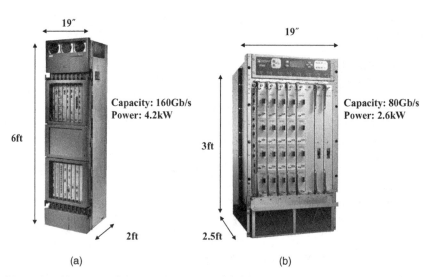

Capacity: 160Gb/s
Power: 4.2kW

Capacity: 80Gb/s
Power: 2.6kW

(a)   (b)

FIGURE 1.13   Examples of core routers: (a) Cisco GSR 12416; (b) Juniper M160.
(Courtesy of Cisco and Juniper Corporations.)

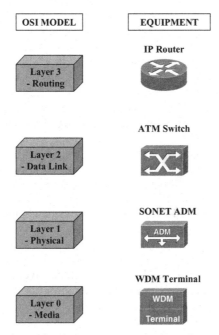

FIGURE 1.14 Networking elements and corresponding OSI layer hierarchy.

operate at routing layer 3, while switches operate at switching layer 2. Regenerators, modems, hubs, and add/drop multiplexers operate at layer 1.

1.5.1 Regenerators, Modems, Hubs, and Add–Drop Multiplexers

Regenerators A *regenerator* is a simple networking element operating on electrical bits. It does not add functionality in terms of traffic handling but is present in the network for transmission purposes. The regenerator regenerates a signal back to its original shape by processes of amplification and/or regeneration. In optical networks, a popular method of optical signal regeneration is to convert optical pulses (O) to electrical signals (E) and then back again to the optical domain (O). We discuss this O-E-O conversion process later in the book.

Modems *Modems* convert digital data into an analog signal and then back again. The conversion process involves *modulation* of the digital signal and, in the reverse direction, *demodulation* of the analog signal into the digital domain: hence the name *Modem*. A typical example of the modem is the device in your PC used to connect to the Internet for dial-up connections capable of 56-kb/s bandwidth. Similar devices for DSL networking are called *DSL modems* and for broadband cable connection are called *cable modems*. All modems use some form of compression and error correction. Compression algorithms

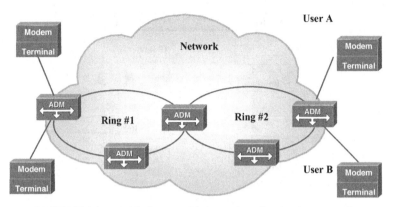

FIGURE 1.15   Add–drop multiplexer function in the network.

enable throughput to be enhanced two to four times over normal transmission. Error correction examines incoming data for integrity and requests retransmission of a packet when it detects a problem.

Hubs   A *hub* is a device that aggregates data traffic from many low-bandwidth connections. In local area networks hubs are connected to PCs, workstations, and servers. *Passive hubs* simply provide cable connectivity, while *active hubs* also provide management capabilities.

Add–Drop Multiplexers   A device that resembles a hub, called an *add–drop multiplexer* (ADM), is shown in Figure 1.15. ADMs are typically connected to form rings, but each device can add or drop some traffic coming from network points outside the ring. ADMs are used in very large quantities in metropolitan SONET/SDH networks.

1.5.2   Switches

A *switch*, as the name implies, switches frames between various ports. It assigns dedicated bandwidth to be designated to each device on the network connected to the switch. Switches split large networks into smaller segments, decreasing a number of users sharing the same resources. Typical examples of switches include ATM switches in wide area networks, Ethernet switches in local area networks, and fiber channel switches in storage area networks. A switch function is illustrated in Figure 1.16.

1.5.3   Routers

Routers direct data packets, following IP rules, from one network to another. To accomplish this task they examine the content of data packets flowing

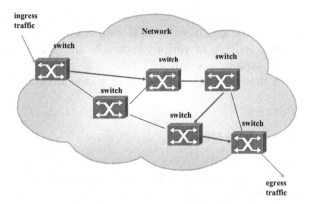

FIGURE 1.16   Switch function in the network.

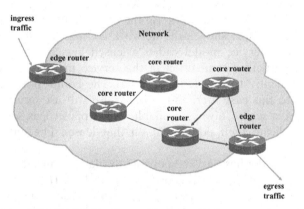

FIGURE 1.17   Router function in the network.

through them. Routers have to determine the most efficient path through the network using complex routing algorithms. After finding the most efficient path, routers switch frames between various ports. In this sense, routers perform the same switching function as that of switches. Routers are much more complex devices, however, as they have to deal with large routing tables for the global Internet and find appropriate routing addresses. Switches, on the other hand, merely switch frames and are aware only of their close network proximity. In terms of packet processing functions, routers perform an order of magnitude of many operations compared to switches.

Router equipment can be classified as core routers, and edge routers as shown in Figure 1.17. *Core routers* reside in the core of the network, so they typically have more capacity than edge routers. *Edge routers*, on the other hand, have more diversified functions than core routers, as they reside on the edge of the network and have to deal with traffic coming from access and MAN networks arriving in various protocols and speeds. Core routers have "big pipe" interfaces operating at Gb/s rates, whereas edge routers have to deal with many "smaller pipes" with bandwidths as low as a few Mb/s.

1.5.4 Networking Service Models

Having described briefly various pieces of networking equipment, we can now start putting together some simple networks and see what services can be offered to network customers. Let us utilize layer 1 equipment to form a transport network as shown in Figure 1.18. Customers with locations A and B will need to have some routers, as routers are always needed to find a path through the network. We can assume that routers belong to a customer and are installed on customer premises. As such, they belong to a customer premise equipment class. A network, on the other hand, belongs to a service provider. In the example shown, the network consists of SONET add–drop multiplexers. In a service model called a leased line, the service provider can lease one particular connection in the service provider network to be assigned permanently to locations A and B. The connection will be dedicated entirely to this customer, and nobody else will use any portion of it. The connection is secure, always available, but expensive, as service providers have to dedicate their equipment on that leased line entirely to one customer.

If a service provider has a layer 2 network available, which it normally does, it can offer a different service, as shown in Figure 1.19. This type of service is called a *virtual private network* (VPN). User traffic from point A is sent to a layer 2 network. After traversing through this switching network it reaches

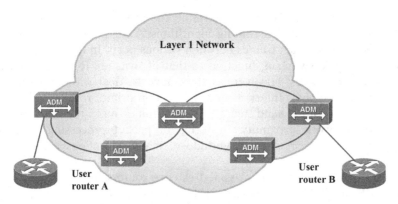

FIGURE 1.18   Leased line layer 1 transport network.

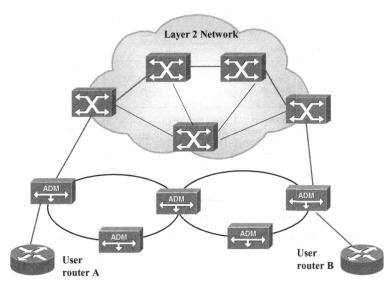

FIGURE 1.19 Virtual private network.

destination B. It is worthwhile to note that in this example, somewhere on the edges of the layer 2 network, layer 1 equipment is involved in the data transport as well, though only to a limited extent.

VPN service has some interesting advantages over leased lines. It uses the network resources much more efficiently, as resources can be shared across the entire customer base, and layer 2 networks can take advantage of statistical multiplexing. Some customers might have some apprehension regarding the security of the VPN connection, as the packets from one customer "mix" somewhere in the system with packets from another customer. In a properly engineered VPN system, though, security should not be a concern.

It is important to realize that from a customer's point of view, the statistical nature of a layer 2 network is not visible. From his or her point of view, the connection between A and B is his or her private connection, hence the term *virtual private network*. Finally, it has to be pointed out that to create a VPN in Figure 1.19, we used a layer 2 switched network. It is equally possible to create a layer 3 VPN, where the routing network is used to achieve the same function. VPN layers 2 and 3 are both offered commercially by service providers, in addition to traditional leased-line models based on layer 1 networks.

In closing, this chapter has shown for us that networking equipment comes in various sizes, "flavors," and functionality subsets. Figure 1.20 lists a number of network elements and their corresponding product naming, depending on whether they are deployed at LAN, MAN, or WAN networks.

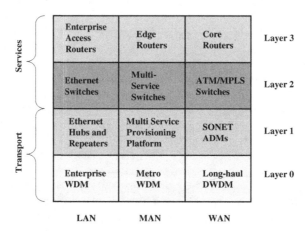

FIGURE 1.20   Various examples of networking equipment.

We discuss network elements' functionality in great detail in Chapters 12 and 13.

KEY POINTS

Transmission media:

- Four transmission media are available for data communication: copper wire, coaxial cable, optical fiber, and air (wireless connection).
- Optical fiber is the most efficient medium, as its has the lowest signal attenuation and is insensitive to electrical noise.

Network classification:

- Five distinct classes of networks are present in global Internet infrastructure: access networks, local area networks, storage area networks, metropolitan area networks, and wide area networks.
  - Access networks connect consumers and corporate users with Internet infrastructure.
  - Local area networks connect multiple users in a contained environment like a corporate building or a campus. Typically, Ethernet is used as the protocol of choice.
  - Storage area networks are corporate data storage networks that connect backend storage disks via high-speed interfaces using primarily fiber channel protocol.

- Metropolitan area networks connect data and voice traffic at the city level typically using SONET rings.
- Wide area networks connect multiple corporate locations or cities across long distances, also known as core or long-haul networks.

Network topologies:

- Various network topologies are used in different networks: point to point, hub, ring, or mesh. Each has advantages, disadvantages, and a particular level of complexity.
  - Point-to-point links are the simplest form of network. Typically, they are used in long-haul connections: for example, under the Atlantic Ocean.
  - Hub networks aggregate multiple connections into one higher-throughput connection. Ethernet hub topology is frequently employed in local area networks.
  - Ring topology introduces the possibility of higher reliability at the cost of more hardware. SONET/SDH rings are frequently used in metropolitan area networks.
  - Mesh topology introduces the possibility of large redundancy, but it is more expensive in terms of hardware complexity. WDM links are used to create mesh in some core networks.

Switching concepts:

- To establish a connection between two points in the network, two techniques can be used: circuit switching and packet switching.
  - In circuit switching, a dedicated connection is established for the time needed. Telephone service is an example of the circuit-switching technique.
  - In packet switching, each packet travels across the network using a different path. Internet protocol routing is an example of the packet-switching technique.

Multiplexing concepts:

- There are three ways to multiplex data in optical networks: wavelength-division multiplexing (WDM), synchronous time-domain multiplexing (TDM), and asynchronous TDM.
  - WDM takes optical signals, each carrying information at a certain bit rate, gives them a specific wavelength, and sends them down the same fiber. As a result, each optical signal has the illusion of having its own fiber.

- Synchronous TDM takes multiple synchronized streams of data, each carrying information at a certain bit rate, and assigns each piece of data a precise time slot in the output stream.
- Asynchronous TDM takes multiple asynchronous variable-size packets and assigns them in the output stream. The length of the time slot is determined based on the relative needs of input data streams. Asynchronous TDM is more efficient but much more complex than synchronous TDM. This technique requires the use of packet buffering and queuing.

OSI model:

- An open system interconnect (OSI) model is a very useful conceptual model used to classify various networking functions.
- OSI has the following hardware layers: network (layer 3), data link (layer 2), transport (layer 1), and media (layer 0).
- The data encapsulation at layer 3 is usually referred to as a packet, at layer 2 as a frame or cell, and at layer 1 as a frame or electrical/optical signal.
- The typical protocol stack in today's optical networks is IP over ATM over SONET over WDM.

Networking equipment:

- Modems, regenerators, and hubs belong to the layer 1 class of networking equipment. They perform physical functions such as sending or receiving data at the terminal nodes (modem), regenerating signals on their way (regenerator) or combining signals into a larger data stream (hub). DSL modems, SONET regenerators, and Ethernet hubs are examples of this class of equipment.
- Switches belong to the layer 2 class of networking equipment. Switches perform switching functions by sending frames from inputs to desired output ports at the switch I/Os. Core ATM switches, enterprise Ethernet switches, and director fiber channel switches are examples of this class of equipment.
- Routers belong to the layer 3 class of the networking equipment. Routers perform the complex task of finding the most effective way of sending data packets through the global network. Core IP routers and edge routers are examples of this class of equipment.

REFERENCES

Agarwal, G. P., *Fiber-Optic Communication Systems*, Wiley, Hoboken, NJ, 1997.

Freeman, R., *Fiber-Optic Systems for Telecommunications*, Wiley, Hoboken, 2002.

Goralski, W., *SONET: A Guide to Synchronous Optical Networks*, McGraw-Hill, New York, 1997.

Kartalopoulos, S., *Next Generation Sonet/SDH*, IEEE Press, Piscataway, NJ, 2004.

Mukherjee, B., *Optical Communication Networks*, McGraw-Hill, New York, 1997.

Ramaswami, R., and K. Sivarajan, *Optical Networks: A Practical Perspective*, Academic Press, San Diego, CA, 1998.

Tomsu, P., and C. Schmutzer, *Next Generation Optical Networks*, Prentice Hall, Upper Saddle River, NJ, 2002.

2

FIBER-OPTIC TRANSMISSION

2.1 INTRODUCTION

Consider the task of transmitting data over a great distance, say from the east to west coasts in North America or from New York to London. If a reach of hundreds or even thousands of kilometers is required, the question arises:

Network Infrastructure and Architecture: Designing High-Availability Networks,
By Krzysztof Iniewski, Carl McCrosky, and Daniel Minoli
Copyright © 2008 John Wiley & Sons, Inc.

What is the best physical medium to use to send information rapidly? It turns out that sending pulses of light down an optical fiber is the fastest and most cost-effective means for high-capacity data transmission over large distances. Light can travel in glass fibers with significantly lower attenuation levels than high-frequency electrical signals can be transported using conductors. Optical signals can thus travel much farther before any need for amplification or regeneration is required.

Propagation of light over great distances has been made possible by the development of low-loss single-mode fiber which has achieved a phenomenal loss of 0.2 dB/km at a wavelength of 1550 nm (Miya et al., 1979). Imagine, if our eyes could see 1550-nm light, looking through a 1-km-thick window of glass and seeing no noticeable attenuation! Another key development that has revolutionized long-haul fiber transmission is the optical amplifier. Optical amplification has allowed transmission distances on the order of 10,000 km without the need for electrical signal regeneration, enabling all-optical inter-continental links (Agrawal, 2002).

As we will see later in this chapter, light also has much more capacity than electrical carriers for sending information. Optical fiber bandwidth is on the order on tens of terahertz, while that of the highest-speed coaxial cable available commercially is on the order of 100 GHz. The loss of coaxial cables is, however, prohibitive for high-speed transmission over any appreciable distance, being on the order of 1000 dB/km at 10 GHz.

In this chapter we deal with the fundamentals of fiber-optic communications, including fiber propagation characteristics, light generation, amplification and detection, and optical modulation. Many books are dedicated to each of these topics, so in one chapter we are only able to touch briefly on each area. We hope that the reader will acquire a general overview of the topics and will seek out the references contained herein for elaboration on particular areas of interest.

2.2 FIBER OPTIC COMMUNICATION

2.2.1 Why Optical Fiber?

Optical fiber is one of several types of transmission media used in communication systems, including twisted-pair wire and coaxial cable (both typically of copper construction) and air (wireless transmission, including satellites). Optical fiber is unquestionably the transmission medium of choice for sending large bandwidth signals over long terrestrial or undersea distances. As an example of the challenges with electrical signal transmission, consider the example of the TAT-6, transatlantic coaxial cable installed between France and Rhode Island in 1976 (Paul et al., 1984). This cable, 6300 km in length and 53 mm in diameter, had 50,000 dB of attenuation while supporting 4200 voice circuits (~30 MHz bandwidth). Repeaters were placed every 9.5 km, which gives about 75 dB of attenuation between repeaters. The signal power transmitted through each span was therefore only 30 parts in 10^9. In comparison, an optical fiber with 0.2 dB/km, transmits 65% of the power over the same distance. We can extend the fiber distance to 100 km and still deliver 1% of the transmitting power using fiber, whereas the power transmitted with coaxial cable would be unimaginably close to zero. If the fiber transmitting power is 1 mW, we will detect 10 μW, which although small, is still a detectable level.

Low attenuation is not the only advantage that fiber-optic transmission has over copper transmission. The maximum possible frequency that can be transmitted is dramatically higher in optical fibers. Coaxial cables have a maximum usable frequency that is related to the physical dimensions of the cable's cross section. When the electrical wavelength becomes smaller than the cable's radial dimensions, multiple modes can propagate, causing increased loss and signal distortion. For 50-GHz cables, the inner diameter of the ground (shield) is 2.4 mm. Although this is still large compared to a fiber's cross section, the cable's resistance increases as the dimensions are reduced, imposing a practical limit on scaling coaxial cables to much higher frequencies. Furthermore, bandwidth is limited by the resistive loss in coaxial cables, which increases with increasing frequency (as $\sqrt{f}$) due to the skin effect (Ramo et al., 1994).

To understand the origin of the phenomenal bandwidth of optical fibers, recall that light is simply electromagnetic radiation, fundamentally no different than an FM radio station's broadcast. The difference is simply one of frequency: in this case, ~100 MHz vs. ~200 of THz. The photon, which is the smallest packet of electromagnetic energy for any particular frequency, is an equally valid description of radio waves as it is for light, although the photon is not commonly used in the analysis of megahertz or gigahertz types of type frequencies. If light waves are the same phenomenon as radio waves, aside from frequency, why can't we transmit light waves along tiny copper wires? In addition to the frequency-scaling problems mentioned above for coaxial cables, the conductivity of metals is quite poor at optical frequencies.

The purpose of both metallic cables and optical fibers is to contain the electromagnetic energy in transverse directions while the signal travels in a line from the transmitter to the receiver. In metallic cables, this transverse energy containment is provided by the free electrons in the metal. Optical fibers, however, do not rely on conductivity at all, but instead, rely on the principle of total internal reflection (TIR). Although electrons are still integral to TIR, in dielectrics such as glass their motions are restricted to small displacements about the atomic nuclei, resulting in a dielectric polarization. In high-purity glass, this polarization can oscillate at 200 THz with extremely low power dissipation, allowing the low-propagation losses that we exploit for fiber-optic telecommunications.

The fact that we can use a 200-THz carrier frequency in optical fibers provides a huge potential for transmission bandwidth. A relatively narrow bandwidth of 1% of the carrier frequency still provides a 2-THz passband! The approximate usable bandwidth for a single fiber, exploiting the 1450- to 1650-nm low-loss window in silica, is an astounding 25 THz. Depending on how efficiently this bandwidth is used, 25 THz allows roughly 25 Tb/s of data to be transmitted; that's the equivalent of 25,000 × 1 Gb/s lines in a single fiber.

Another advantage of optical transmission is immunity to electromagnetic interference (EMI). Electrical signals carried by conductors are very sensitive to interference from many electrical sources, and a large part of electronic system design involves minimizing these effects. In fiber-optic transmission, the fields that typically cause EMI are at such low frequencies compared to the optical wave that the fiber will simply not transmit them. Furthermore, even if there were EMI sources present at 200 THz, the interfering radiation would simply pass through one side of the fiber and exit through the other as a consequence of exceeding the critical angle for TIR. Finally, the most important reason that optical fibers provide EMI immunity is that they are nonconductive. Lack of conductivity eliminates capacitive and inductive coupling mechanisms from EMI sources to either the transmitter or receiver.

If optical fiber has so many advantages over electrical wires, why is fiber not used everywhere? There are two main difficulties using fiber. First, most information signals are generated by electronic devices. Computers and most storage devices are electronic in nature. An electrical-to-optical conversion process is therefore required at the transmitter side, whereas optical-to-electrical conversion is required at the receiver. Conversion adds complexity and cost and is warranted only if the data rates and/or transmission distances are high. Second, fiber is often more expensive to install than unshielded copper wires or coaxial cables.

Laying fiber on a per-kilometer basis costs somewhat more than laying copper. However, on a per-circuit basis, there is no contest; fiber wins hands down. At the time of the market explosion for fiber in the late 1990s, people anticipated connecting houses with optical fibers, and the acronym FTTH [*fiber to the home*; also called *fiber to the premise* (FTTP)] was born. Shortly after the first FTTH trials, it was realized that there was little need to install fiber for the final few meters of the link. The industry then moved to *fiber to the curb* (FTTC). In such a system, fiber would carry a plurality of channels to the "curb," whereupon they would be split and applied to copper wire leading to the home. In many cases even this technology is overkill, and *fiber to the neighborhood* (FTTN) became the preferred approach. There are, however, regional preferences for the various fiber stop locations, which are referred to collectively as *FTTx*. Japan has made significant FTTH deployments. Nippon Telephony and Telegraphy has reported that their growth in FTTH connections is growing at a faster rate than ADSL, with a targeted 30 million FTTH connections by 2010. The bottom line is that fiber deployment makes sense when it is economical to do so.

Another network approach is a combination of optical fiber and coaxial cable, usually referred to as *hybrid fiber–coax* (HFC). Coaxial cable has a greater bandwidth than twisted-pair copper wire but is obviously still much smaller than fiber. HFC is commonly used in the cable TV industry to carry a variety of analog and digital services. Whether alone or combined with twisted pair or coaxial cable, optical fibers are being deployed everywhere. The tremendous capacity of fiber has permitted dramatic growth in data traffic; however, placing so much traffic on a single fiber strand makes for greater vulnerability. Most disruptions in long-distance networks are a result of physical interruption, such as accidentally cut fibers. As we will see in later chapters, communication protocols introduce mechanisms to detect and recover from network failures such as fiber cuts.

2.2.2   Propagation: Single- and Multimode Fibers

Optical fibers exploit the remarkable material properties that are possible with high-purity glass. The basic material requirements needed for optical fibers are the ability to control the index of refraction, the ability to form long strands with a precisely controlled cross section, and of course, high transparency. Fused silica (SiO_2) combined with dopants such as germanium provides these features. A preform is first constructed with scaled-up dimensions of the desired cross section and refractive index profile. The preform is then heated and *pulled* into a long fiber (on the order of tens of kilometers) that has a diameter typically of $125\,\mu m$.

To laterally confine, or to *guide*, light inside an optical fiber, the index of refraction is generally highest at the center of the fiber (the *core*) and decreases toward the perimeter (the *cladding*) (see Figure 2.1). Although knowing the exact propagation characteristics of these structures requires solving Maxwell's equations (Snyder and Love, 1983), an intuitive understanding is provided by the total internal reflection. Light propagating in the fiber's core will be guided along the fiber if the angle φ between the propagation direction and the core–cladding interface normal is larger than the *critical angle* predicted by *Snell's law* (see Figure 2.2):

$$\phi_{critical} = \sin^{-1}\frac{n_2}{n_1}$$

The diameter of the core also greatly affects the propagation characteristics. A fiber *mode* is a particular cross-section distribution of the optical fields (electric and magnetic) that travels along the length of the fiber unchanged. The mode is a unique transverse resonance which, loosely speaking, means that light can bounce from one side of the fiber core to the other and back again, arriving with the same phase that the light started with. Light traveling at the greatest angle of incidence to the core–cladding interface (smallest angle with respect to the fiber axis) at which a resonance occurs is known as the

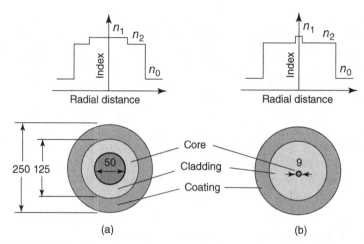

FIGURE 2.1 Refractive index profiles and cross sections for (a) multimode and (b) single-mode fiber. The coating, typically a polymer, protects the fiber without influencing the propagation characteristics.

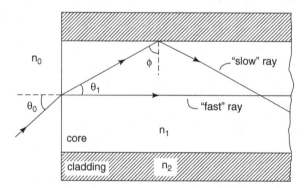

FIGURE 2.2 Ray propagation through a step-index multimode fiber. "Fast" and "slow" refer to the group velocities (i.e., the velocities of energy transport along the fiber axis).

fundamental mode. If only one angle exists that provides lateral resonance while maintaining internal reflection conditions, we have a *single-mode fiber* (SMF). If more than one angle can exist, we have a *multimode fiber* (MMF).

Figure 2.1 shows the typical fiber cross sections for both multimode and single-mode *step-index fibers,* where the transition from core to cladding is an abrupt change in refractive index. The large difference in core diameter between an SMF (~9 μm) and an MMF (~60 μm) affects not only the propagation characteristics, but also the ease of manufacture and use. The considerably larger core of the MMF makes it much less sensitive to fiber-to-fiber misalignment at fiber interconnects, as well as less susceptible to contaminated connections. It is also easier to inject light into an MMF with good coupling efficiency, so manufacturing tolerances for MMFs as well as components employing MMFs are looser and costs are lower. As a result, MMF is used extensively in local area networks and access networks, where the bandwidth and distance limitations of multimode propagation are permissible. A short pulse entering an MMF will result in a distribution of energy among many modes in the fiber. At the fiber end, each mode will emerge at slightly different times, producing a stretching of the pulse in time. This phenomenon, known as *multimode dispersion*, gives rise to an *effective modal bandwidth*, which limits the modulation frequency that the fiber can be used for at a given distance.

The modal dispersion limitations in MMFs can readily be analyzed by considering the difference in travel times for the slowest and fastest rays through the fiber. Figure 2.2 shows a cutaway of an MMF. We know from Snell's law that the steepest ray that will be guided in the fiber core has a critical angle given by $\sin \varphi_c = n_2/n_1$, where n_1 and n_2 are the core and cladding refractive indices, respectively. A ray traveling at the critical angle will have the longest path through the fiber, while the shortest path will be a straight trajectory along the fiber axis. The time difference between these two rays over a fiber of length L is given by (Agrawal, 2002)

$$\Delta t = \frac{n_1 L}{c \sin \phi_c} - \frac{n_1 L}{c} = \frac{L n_1^2}{c n_2} \Delta$$

where c is the speed of light in vacuum and $\Delta = (n_1 - n_2)/n_1$ is the normalized index difference between the core and the cladding. A rough estimate of maximum tolerable value of Δt for digital links is bit period T_b, which is the inverse of the bit rate B. We can therefore estimate the maximum bit rate–distance product as

$$BL < \frac{n_2 c}{n_1^2 \Delta}$$

For a typical value of $\Delta = 5 \times 10^{-3}$ and $n_1 = 1.5$, we find that the bit rate–distance product, a convenient figure of merit for digital links, is $BL < 40 \ (\text{Mb/s}) \cdot \text{km}$. Clearly, the step index MMF is not suitable for high-capacity long-distance links.

A significantly better BL product can be achieved in MMFs by employing a graded-index profile instead of using a step-index change. In a *graded-index fiber*, the refractive index varies smoothly from a high point in the center of the fiber to a lower value at the perimeter. Rather than light reflecting off of a sharp core–cladding interface, the light is refracted gradually, following a curved trajectory. One particular graded-index profile that is of interest is the parabolic profile, which has the unique property that the time taken for a ray to emerge at the output is independent of the angle of the ray. This behavior results from the increase in the speed of light (decrease in refractive index) when rays are farther away from the fiber center, which compensates for the difference in path length. Graded-index MMFs provide a dramatic improvement over step-index MMFs in the bit rate–distance product. For example, the Gigabit Ethernet standard 10GBASE-SR supports 10 Gb/s up to 300 m using 850-nm light sources and 50-μm-core MMFs with a modal bandwidth of 2 GHz·km (IEEE, 2005). The corresponding *BL* product is 3 (Gb/s)·km.

For wide area networks, single-mode fibers are used to achieve the required transmission capacity and distance. SMFs eliminate multimode dispersion by employing a sufficiently small core to support only one propagating mode (ignoring polarization effects). As a result, the bandwidth and distance achievable with SMFs are several orders of magnitude higher than for those MMFs. Ultimately, other types of dispersion, as well as nonlinear interactions, limit SMF links. We discuss these issues in Section 2.6.

Unlike MMFs, SMFs have a single mode of propagation that travels at a single *phase velocity* for a given optical frequency. The phase velocity is typically expressed in terms of an *effective refractive index*, or *effective index*, n_{eff}, that takes into account the actual refractive index of the fiber as well as the impact that the waveguide structure (core and cladding) has on the modes velocity. In step-index SMF, the effective index satisfies $n_2 < n_{eff} < n_1$ as a consequence of the conditions to maintain internal reflection.

The attenuation and dispersion properties of fibers are strongly wavelength dependent. At short wavelengths (800 nm), Rayleigh scattering dominates the loss spectrum, whereas at longer wavelengths (1800 nm), infrared absorption due to molecular vibrational resonances dominates the loss (Miya et al., 1979). Between these two effects, fibers are suitably transparent for fiber-optic communications. Figure 2.3 shows the attenuation vs. wavelength of a typical single-mode fiber in the region 1250 to 1650 nm. An additional feature of attenuation, known as a *water peak*, which is the result of resonances of unintentional OH impurities, is seen around 1380 nm. More recent *full-spectrum fibers* such as AllWave from Furukawa, and SMF-28e from Corning, are now available with negligible water peaks.

The *first window*, named for its early use in fiber-optic transmission, is around 850 nm. Although the losses are considerably higher (~3 dB/km) at 850 nm than at longer wavelengths, the 850-nm region was attractive historically because of the availability of light sources and detectors at this wavelength. Both light-emitting diodes, and more recently, vertical-cavity surface-emitting lasers at 850 nm exploit the relatively low-cost AlGaAs/GaAs material system. As a result, 850 nm is still used for short-reach links.

The *second window* to become commonly used for optical transmission is centered near 1310 nm (the O-band in Figure 2.3). This window offers lower

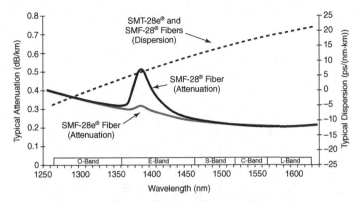

FIGURE 2.3 Attenuation spectrum of a typical single-mode fiber (SMF-28) and the reduced water peak fiber SMF-28e. Group velocity dispersion is also shown. (Courtesy of Corning Corporation.)

loss than the first window, being between 0.3 and 0.4 dB/km on modern fibers but with increased cost over 850-nm light sources. One of the initial advantages of the second window was that group velocity dispersion is nearly zero at 1310 nm. The low group velocity dispersion limits pulse distortion; however, the full impact of group velocity dispersion on a fiber link depends on many parameters and can be either beneficial or detrimental. Dispersion is covered in more detail in subsequent sections.

The *third window*, and now most exploited wavelength region for long-haul fiber transmission, is the C-band, which includes approximately 1530 to 1570 nm. C-band not only provides the lowest loss, at approximately 0.2 dB/km, but also has the advantage that fiber-optic amplifiers are readily available for this part of the spectrum. The disadvantage of using the C-band is that light sources near 1550 nm, at least historically, are more expensive than sources at either 850 nm or 1310 nm.

Ultimately, the choice of optical wavelength for any given link is governed by the permissible attenuation, amplification requirements, group velocity dispersion requirements, and the availability/cost of optical components.

2.3 LIGHT EMISSION AND DETECTION

2.3.1 Light Sources

The two commonly used light-sources in optical communication systems are light-emitting diodes (LEDs) and laser diodes (LDs). LEDs are inexpensive but have relatively low output power and broad spectral width (*linewidth*). LDs, on the other hand, are more expensive but provide higher output power with significantly narrower linewidth. As we will see in subsequent sections,

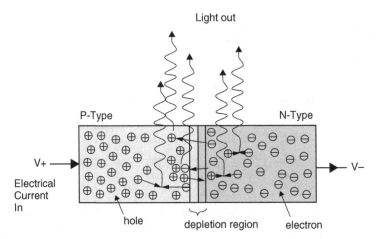

FIGURE 2.4 Basic operating principle of an LED. The radiation is omnidirectional but is shown to emit from one edge for illustrative clarity. The charge symbols represent free carriers; the fixed charges that keep the bulk p- and n-type regions charge neutral are not shown.

linewidth can be the limiting factor in transmission speed and distance as a result of group velocity dispersion in the fiber.

The operating principle of an LED is illustrated in Figure 2.4. The key part of the LED is the interface formed between the p- and n-type doped semiconductor layers (the PN junction). A depletion region (void of free carriers) is formed at the junction as carriers diffuse across the boundary (Pulfrey and Tarr, 1989). When the diode is forward biased, electrons and holes are injected across the depletion region, where they become minority carriers in a sea of oppositely charged majority carriers. The minority carriers diffuse away from the depletion region and ultimately recombine with the majority carriers. In direct-bandgap materials such as GaAs and InP (referred to as *III–V materials*), where the minimum energies of electrons and holes are at the same momentum, a recombination event can release energy in the form of a photon (radiative recombination). The energy of the photon E is related to the optical frequency ν by

$$E = h\nu \qquad (2.1)$$

where h is Planck's constant ($4.1357 \times 10^{-15}\,\text{eV·s}$). Since $\nu = c/\lambda_0$, where c is the speed of light in vacuum and λ_0 is the wavelength in vacuum, we can use energy, wavelength, or frequency as we like, to describe essentially one property of the photon. When analyzing absorption or generation of light, it is convenient to use the photon energy; when analyzing interactions of light geometric structures such as waveguides, it is generally more convenient to use the wavelength.

For materials with an indirect bandgap (e.g., silicon), recombination events have a much higher probability of generating *phonons*, which are quantized lattice vibrations dissipated as heat. To promote the radiative recombination process, and to confine the minority carriers near the junction, more sophisticated LEDs use a double heterostructure (junctions made from different materials with different bandgaps) (Liu, 1996). In addition, the double heterostructure can function as a waveguide for the light that is generated, allowing more efficient coupling of light into a fiber. LEDs can be optimized for either edge- or surface-emitting configurations.

The radiation from LEDs is dominated by spontaneous emission; each radiative recombination event occurs independent of other photons in the semiconductor. The spontaneous process results in a broad emission spectrum (tens of nanometers) and hence incoherent light. Rate equations relating drive current, carrier densities, and photon density can be used to model the physical processes. The recombination time is on the order of 1 ns, and the rate equations predict the maximum useful modulation speed to be less than 1 GHz (Kazovsky et al., 1996). LEDs tend to have a wide cone of emission as a result of the spontaneous emission being omnidirectional, making coupling efficiency into SMFs poor. The broad emission spectrum and low output power means that LEDs are deployed primarily in low-speed short-distance links with multimode fibers. Common wavelengths are 850 nm and 1300 nm.

For multigigabit long-haul optical networks, semiconductor laser diodes are used (Agrawal and Datta, 1993; Coldren and Corzine, 1995). *Laser* is an acronym for light amplification by stimulated emission of radiation. The structure of a semiconductor laser is similar to that of an LED, where an active region is formed between n- and p-type layers (Figure 2.5). A key difference is that optical feedback is introduced to the structure by partially reflective facets, giving some photons the possibility to be amplified. Another key difference between an LED and a laser is that the active region of the laser

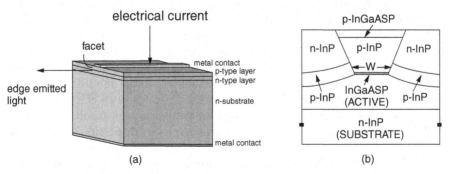

FIGURE 2.5 (a) Basic structure of an edge-emitting laser diode; (b) cross section of an etched mesa buried heterostructure laser. (From Agrawal and Datta, 1993.)

requires high doping and strong current injection to achieve *population inversion*. Under these conditions, the active region begins to produce gain: a photon in the active region becomes more likely to produce another photon through *stimulated* emission than it is likely to be absorbed.

At low forward bias current, emission from a laser is dominated by spontaneous emission and the device functions essentially as an LED. When the drive current is increased to obtain population inversion, the active region provides gain through stimulated emission. When the gain is sufficiently high that it exceeds the reflection, absorption, and scattering loss accumulated on the round trip from one facet to the other and back again, the laser reaches threshold and coherent light with narrow spectral width is emitted. Laser output power typically increases linearly with the drive current once a threshold is reached. As with an LED, rate equations can be used to model a laser operating above and below the threshold. The recombination time associated with stimulated emission is much shorter than the spontaneous emission recombination time, giving lasers useful modulation bandwidths into the tens of gigahertz.

Lasers are structurally more complicated than LEDs, and costs are higher accordingly. To increase the efficiency, low-dimensional structures such as multiple quantum wells are commonly incorporated into the active region (Zory, 1993). Since quantum wells have insufficient thickness to guide light (~5 nm), additional cladding regions are included above and below the active region, forming many layers for design and manufacturing. The laser's facets perform a critical role by providing directional and wavelength-selective optical feedback. Directional selectivity results because only light propagating perpendicular to the reflective facets can make multiple passes through the active region, experiencing significant gain. The facets also provide wavelength selectivity since the cavity will resonate only at wavelengths that satisfy $\lambda/2 = L/N$, where L is the cavity length and N is an integer.

The structure just described, with partially reflective facets, is known as a *Fabry–Perot* (FP) *laser*. FP lasers generally suffer large chirp (frequency variation under amplitude modulation) when the drive current is modulated as a result of the refractive index of the cavity varying with the drive current. In addition to producing undesirable chirp, FP lasers operate in multiple longitudinal modes, corresponding to the multiple resonances of the cavity length (the laser is still single mode in cross section, analogous to SMF). An FP laser's linewidth is about an order of magnitude narrower than an LED's (a few nanometers vs. tens of nanometers). Unfortunately, under modulation conditions the FP laser's linewidth increases further as a result of chirp. FP lasers are suitable for short- to medium-range transmission as a result of the fact that their broad linewidth interacts with group velocity dispersion (see Section 2.6.1).

For long-haul and high-bit-rate transmission (10 Gb/s and up), *distributed feedback* (DFB) *lasers* are used. In DFB lasers, the functions provided by the mirrors of the FP laser are accomplished with a reflective grating incorporated into the active region. The key feature of a DFB laser is that the grating can be made to filter out all but one longitudinal mode, making the spectral width very narrow (~10 MHz). The DFB structure is more complicated to manufacture, and costs are higher than for FP lasers. To enhance performance further, a DFB laser package may contain an optical isolator, power monitoring diode, thermoelectric cooler/heater, and electronic feedback circuit to maintain constant output power and wavelength. Commercially available DFB lasers are available in many wavelengths between 1530 and 1612 nm on a grid set out by the ITU (International Telecommunication Union). Output powers can exceed 50 mW.

Both FP and DFB lasers are classified as edge-emitting lasers since the light is emitted from the edge of the substrate. This configuration makes it impossible to test each laser until it is cleaved from the wafer. A structure that allows light to exit vertically from the wafer surface has been developed and is known as the vertical-cavity surface-emitting laser (VCSEL) (Yu, 2003). In this structure, the mirror function is performed by distributed Bragg reflectors (DBRs). DBRs in VCSELs are multilayer structures (tens of layers with alternating high and low refractive indices) capable of providing high reflectivity (>99%). The quantum well active layer is grown between upper and lower DBRs, which provide optical feedback in the vertical direction. Optical confinement is achieved by etching circular mesas on the wafer. Electrical contacts are provided by electrodes deposited at the top and bottom of the structure, as shown in Figure 2.6. Currently, VCSELs work well between 850 and 980 nm and have been deployed in short links (a few hundred meters) operating up to 10 Gb/s. Devices working at 1310 and 1550 nm are being introduced, but material issues are still being pursued in research laboratories.

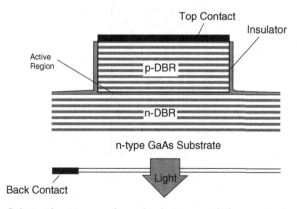

FIGURE 2.6 Schematic cross section of a bottom-emitting vertical-cavity surface-emitting laser. (After Peters et al., 1993.)

2.3.2 Photodetectors

Photodetectors are used to convert optical signals into electrical signals to recover optically transmitted information. Photodetectors exploit the photoelectric effect: generation of an electron–hole pair in response to an absorbed photon. As with optical sources, semiconductor diodes are the devices of choice for detectors in optical communications. The two types of photodetectors commonly used are p-type/intrinsic/n-type (PIN) diodes and avalanche photodiodes (APDs). Both generate a current (photocurrent) that is proportional to the envelope of the optical power received and characterized by a responsivity in amperes and watts. Mathematically, the photodetector's current is proportional to a low-pass-filtered version of the optical power. A fast photodetector's bandwidth is in the range 10 to 100 GHz and is thus much too slow to respond to anything but the envelope of the ~200-THz optical carrier. Because the optical power is proportional to the square of the optical electric field, the photodetector is a square-law device.

The structure of a PIN diode (Figure 2.7) is similar to that of a pn junction except for the addition of either a very lightly doped or undoped (intrinsic) layer grown between the p- and n-type regions to increase the frequency response (Sze, 1981). The intrinsic layer increases the width of the depletion region, allowing most of the light to be absorbed in the depletion region, where

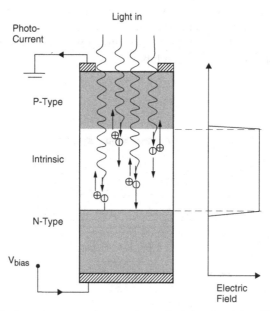

FIGURE 2.7 Basic structure and operating principle of a PIN diode (only newly generated carriers are illustrated, for clarity). The electric field distribution is shown at the right and is high in the depleted intrinsic layer.

the carriers are quickly swept out by the electric field. The PIN diode operates under reverse bias to strengthen the electric field in the depletion region. Light absorbed in the depletion region generates electron–hole pairs, which are then separated by the electric field. Carriers drifting across the depletion region produce a current at the diode's terminals. A thinner intrinsic layer yields a higher-frequency response because the carriers have less distance to travel. The penalty is a lower quantum efficiency: Fewer photons result in electron–hole pair generation.

Unlike laser diodes and LEDs, photodetectors do not require direct-bandgap semiconductors, so silicon and germanium are used in low-speed, low-cost applications. Direct-bandgap semiconductors (often, InGaAs) are used for high-speed applications because the intrinsic layer can be made thinner while maintaining good quantum efficiency. APDs are similar in structure to PIN diodes (see Figure 2.8) but with an additional doped *multiplication layer* at one end of the intrinsic layer (Sze, 1981). Like the PIN diode, absorbed photons generate electron–hole pairs in either a depleted lightly doped or intrinsic layer. With a large reverse bias voltage, the multiplication layer is also depleted and acquires a strong electric field. Depending on the placement of the multiplication layer, holes or electrons entering the multiplication layer acquire sufficient kinetic energy to produce secondary electron–hole pairs through *impact ionization* (Dutta et al., 2002, Chap. 9). The secondary carriers can produce more electron–hole pairs though impact ionization, resulting in a terminal current that is increased by a multiplication factor M.

The multiplication factor in APDs depends on the design and materials used, with values ranging between 10 and 500 (Agrawal, 2002). APDs are thus well suited for situations where the optical powers received are very weak and the thermal noise of the receiver electronics dominates other noise sources.

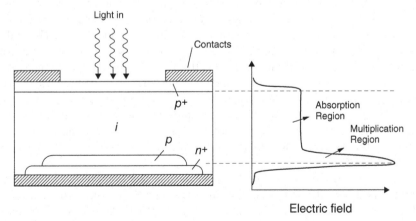

FIGURE 2.8 Basic structure of an avalanche photodiode and the electric field distribution through the absorption and multiplication regions.

The multiplication process is not, however, without penalty. Shot noise, associated with the statistical nature of the photoelectric effect (Sze, 1981), is increased by the statistical nature of the multiplication process. The multiplication also reduces the frequency response of the APD as a result of the time associated with secondary carrier generation.

2.4 OPTICAL MODULATION

Information can be encoded on an optical carrier by varying any of the properties of light: phase, amplitude, polarization, and frequency over time. In practice, only intensity and phase modulation are common in commercial optical communications systems. Of these two, intensity modulation is far more common because of its simplicity and low cost. Recently, there has been extensive research into the use of advanced modulation formats that employ combinations of amplitude, phase, and polarization modulation to increase capacity, reach, and spectral efficiency (Charlet, 2006).

In this section we look at components and methods for modulating light. Modulation in optical systems falls into two categories: *direct modulation*, where the optical source itself is modulated, and *external modulation*, where the source is run in continuous-wave mode and a modulator encodes the information separately. We review the two approaches, and finish with a discussion of modulation formats for digital systems.

2.4.1 Direct Modulation

Direct modulation is often used in optical communication systems because of its ease of implementation and low cost. Under this modulation scheme, a time-varying input current modulates the laser or LED output power directly (Figure 2.9) (Liu, 1996). In digital systems, even if the current were to follow a perfect square wave, the dynamics of the laser produce some distortion. An optical pulse with exaggerated distortions is shown in Figure 2.9. The pulse exhibits a small time delay and then rises with an overshoot of the power followed by a damped oscillation before settling to the on state. Undershoot is not observed when the input current is switched off since the laser can't emit negative power; however, electrical artifacts from photodetectors used to observe the optical pulse can produce an undershoot. A set of rate equations describing the dynamic interactions of photons and carriers inside the active region of the laser is used to predict the transient behavior of the laser (Agrawal, 2004a).

A disadvantage of direct modulation is that the amplitude changes are accompanied by a frequency shift as the drive current changes the refractive index in the active region (Shen and Agrawal, 1986). The refractive index is related to the carrier-dependent gain through the Kramers–Kronig relations. Lasers produce both *adiabatic* and *transient chirp*: the former causing a fre-

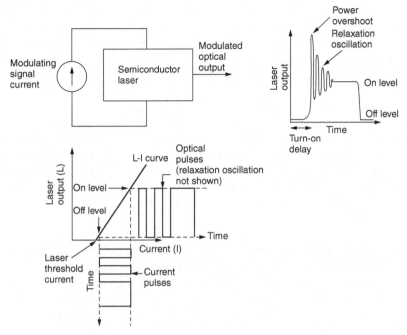

FIGURE 2.9 Direct modulation of a semiconductor laser.

quency shift that is proportional the change in power, the latter having a frequency change proportional to the rate of change of the power (Yariv, 1997, Chap. 15). The time derivative associated with transient chirp makes it more significant at higher frequencies (>1 GHz). In communication systems, laser chirp enhances the optical spectral width beyond the bandwidth associated with amplitude modulation alone. The increased spectral width translates into pulse broadening when combined with the fiber's chromatic dispersion (see Section 2.6.1).

Direct modulation is used extensively for speeds up to 2.5 Gb/s and at distances up to 200 km at 1550 nm (Kaminow and Li, 2002a, Chap. 12). Direct modulation can also be used at 10 Gb/s in short-reach applications. For high-bit-rate transmission over the long distances of conventional single-mode fibers, where chromatic dispersion presents a serious problem, external modulation, with low or negligible frequency chirping, is the preferred method of generating optical signals.

2.4.2 External Modulation

Despite advances made in direct modulation of diode lasers, external modulators are currently preferred for use in longer-distance communication links, particularly for transmitting data at higher bit rates. The main reason for this is that there is much better control of chirp. In most cases, external modulators are designed to produce minimal chirp, although producing a controlled amount of negative chirp can be used to compensate pulse broadening caused by chromatic dispersion at a particular fiber length. In general, external modulators produce significantly less chirp than do directly modulated lasers.

External optical modulators are typically classified into two main categories. Electroabsorption modulators change in the optical transparency of a material to modulate the amplitude of the light, and electrooptic modulators change the refractive index of a material to modulate the phase of light. The phase modulation produced by the electrooptic effect can also be translated into amplitude modulation using an interferometer. Both categories operate by exploiting changes in optical material properties using an electric field.

Electroabsorption Modulators   Electroabsorption modulators (EAMs) for optical communication applications typically employ either the Franz–Keldysh effect or the quantum confined Stark effect (QCSE) and are based on III–V compound semiconductors such as indium phosphide (InP) and gallium arsenide (GaAs). The Franz–Keldysh effect is a mechanism that can be induced in a bulk semiconductor, whereas the QCSE relies on an electric field being applied to quantum confined structures, typically layered quantum wells.

The interband optical absorption mechanism in a semiconductor occurs when a photon excites an electron from the valence band (VB) to the conduction band (CB). For this to occur, a photon must have sufficient energy to displace an electron from the VB to the CB. One of the key characteristics of a semiconductor is that there is an abrupt transition from being transparent to being absorbing at a particular optical wavelength (commonly referred to as the *absorption edge*). The energy of this wavelength corresponds to the bandgap of the semiconductor (E_g). When a slowly modulating electric field (e.g., dc to 100 GHz; slow in comparison to the optical frequency) is applied to a semiconductor, the CB and VB are tilted in response, as illustrated in Figure 2.10. A further consequence of the electric field is that the wave functions of electrons and holes change from plane waves to Airy functions, which have exponential tails that extend outward in space (Pankove, 1971). In particular, the tails *tunnel* into a region of the semiconductor where the hole's energy is above the valence band, while the electron's energy is below the conduction band. If the band tilt is sufficiently large, and hence the tunneling distance d is sufficiently small, absorption can occur for photons that have lower energies than that of the bandgap. If the energy of incident light is close to that of the bandgap, dramatic changes in optical absorption can be obtained by applying a sufficiently strong electric field. The increase in the photo-assisted tunneling rate with applied electric field is known as the *Franz–Keldysh effect*.

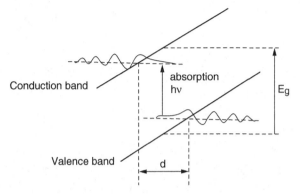

FIGURE 2.10 Electroabsorption mechanism due to the Franz–Keldysh effect. The horizontal axis is the distance through the semiconductor in the direction of the applied electric field. The vertical axis represents energy.

In the second mechanism, the *quantum confined Stark effect* (QCSE), the transparency of a material is altered by controlling the behavior of bound electron–hole pairs known as *excitons*. To create a modulator employing the QCSE, thin layers of ternary and quaternary alloys such as InAlAs, InGaAs, and InAsP are used to form a semiconductor structure with alternating layers of narrow- and wide-bandgap regions. If made sufficiently thin (e.g., 10 nm), the narrow-bandgap regions, referred to as *quantum wells*, confine electron and hole pairs in a small enough region for the electrons and holes to overlap and interact strongly, creating excitons. The excitons produce strong optical absorption at energies near the bandgap of the quantum wells. When an electric field is applied to the quantum wells, the electrons and holes are pulled apart, weakening their interaction. The excitonic absorption peak is weakened, but its spectrum is simultaneously broadened by the separation. The broadening, which shifts the absorption edge to longer wavelengths, is the mechanism by which modulation is achieved using the QCSE (Dagli, 1999).

EAMs that exploit either the Franz–Keldysh or QCSE effects use compound semiconductors with an absorption edge near the system wavelength (e.g., 1550 nm). Design trade-offs typically involve compromises between maximizing the contrast between on and off states, minimizing the optical insertion loss and minimizing the drive voltage required for on-to-off transitions. A drawback of EAMs is that their wavelength sensitivity requires producing multiple designs to accommodate the range of wavelengths used in WDM systems. Furthermore, both the Franz–Keldysh effect and the QCSE result in a change in refractive index as the absorption is modified, and hence produce chirp. For this reason, EAMs are generally used for short-reach, lower-cost applications.

An example of a multi-quantum-well EAM is shown in Figure 2.11. A ridge waveguide is formed by etching into the semiconductor layers, and an

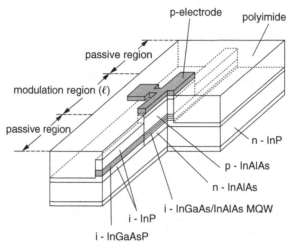

FIGURE 2.11 Structure of a multi-quantum-well electroabsorption modulator for 1550 nm, with a 3-V drive, 50 GHz of bandwidth, and a 8-dB insertion loss. (From Ido et al., 1996, Fig. 1.)

electrode structure is applied to the waveguide to induce a strong electrical field in the quantum wells. The semiconductor compositions are designed with controlled refractive indexes to confine the light in the direction perpendicular to the surface, while the etched ridge confines light in the lateral direction. The dimensions of the etched ridges are similar to those found in lasers and are on the order of 1 to 4 μm in depth and width for modulators designed with operating wavelengths in the region 1.3 to 1.6 μm. The active length of the waveguides are on the order of 100 μm, making EAMs extremely compact (Fukuda, 1999).

The general advantages of EAMs are low driving voltages, good high-frequency performance, and small size. Their drawbacks include limited control of chirp, sensitivity to wavelength and temperature, limited maximum optical input powers, and higher optical insertion loss compared that of electrooptic modulators. Often, several EA modulator designs are required to span the C-band.

Electrooptic Modulators The *linear electrooptic effect*, also known as the *Pockels effect*, produces a change in the refractive index of a crystal in the presence of an applied electric field. The term *electrooptic modulator* (EOM) is often used to refer to modulators exploiting the linear electrooptic effect, although usage varies. The electric field is typically applied with electrodes and has a frequency much below that of optical fields. The linear electrooptic effect exists only in crystals that lack a *center of symmetry* (Yariv and Yeh, 1984, Chap. 7). In these crystals, the applied electric field induces a deformation of the bond charges, which modifies the optical impermeability tensor and hence the refractive index. The most common materials for telecommunications EOMs are lithium niobate (LiNbO$_3$), III–V semiconductors, and organic polymers.

The linear electrooptic effect is a relatively weak effect, resulting in only small changes in the refractive index for a given applied electric field. The change in refractive index Δn is described by the equation

$$\Delta n = \frac{1}{2} n^3 r_{ij} E_j \tag{2.2}$$

where r_{ij} is the appropriate electrooptic coefficient and E_j is the applied electric field. The electrooptic effect in general produces birefringence; Δn here represents a change in refractive index for light polarized along a particular direction and is often accompanied by a simultaneous index change in one or two other perpendicular directions. Designing an EOM requires choosing appropriate crystals, crystal orientations, electric field directions, and optical polarization directions. For a detailed description of the electrooptic effect in various crystals, see Yariv (1997, Chap. 9).

Although the efficiency of electrooptic materials is often judged by the values r_{ij}, particularly with polymers, the factor of n^3 in equation (2.2) must also be taken into account. Materials such as compound semiconductors with high refractive indices can produce strong modulation even with smaller electrooptic coefficients. Regardless of the choice of material, however, the linear electrooptic effect is weak in comparison to the electroabsorption effects. This necessitates using relatively long optical waveguides and electrodes, in the range of tens of millimeters.

The long electrodes used in EOMs presents a design challenge for operation at microwave speeds. The capacitance of these long electrodes is much too large for a simple 50-Ω termination to be placed in parallel with the lumped capacitance. Furthermore, the length of the electrodes exceeds the electrical wavelength at microwave frequencies, so they must be treated as transmission lines, or *traveling-wave electrodes*. The characteristic impedance of the electrodes is designed to be nominally 50Ω, so that they behave as a continuation of the transmission line used to deliver the signal to the modulator.

A design requirement for long electrodes, at least for broadband operation, is that the electrodes and optical waveguide have the same velocity. Velocity matching ensures that light entering the active region of the waveguide will "see" the same electrical pulse as both the light and electrical signal travel down the length of the modulator. The need for precise matching increases at higher bit rates as pulse widths decrease and the accumulated mismatch at the end of the modulator becomes a sizable fraction of the pulse width. A further design requirement related to the long electrodes is the need to minimize both electrical and optical attenuation.

The most common material used for EOMs is LiNbO$_3$. In LiNbO$_3$, the optical refractive index used for high-speed modulators is roughly 2.15.

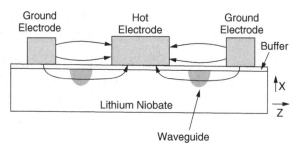

FIGURE 2.12 Lithium niobate Mach–Zehnder modulator cross section in the buff-ered X-cut configuration. (After Wooten et al., 2000, Fig. 3.)

[LiNbO$_3$ is a birefringent crystal that has an *ordinary* index of approximately 2.22 and an *extraordinary* index of 2.15 (Yariv, 1997).] The velocity of the microwave signal is quantified by *a microwave index*, which is the ratio of the speed of light in vacuum to the microwave velocity for a given electrode design. The microwave velocity is a function of the microwave dielectric con-stants of the various materials with which the electric field of the electrodes interacts. For LiNbO$_3$, the microwave dielectric constants are almost an order of magnitude larger than the optical dielectric constants (the square of the refractive index). This fact has led to the development of *fast-wave electrodes* to pull the microwave electric field out of the LiNbO$_3$ and into the air or other lower-dielectric-constant materials. Two common approaches are illustrated in Figure 2.12: Thick electrodes place more of the electric field in the air, while a *buffer layer* such as SiO$_2$ provides a further reduction in dielectric constant (Wooten et al., 2000).

As discussed so far, the linear electrooptic effect provides a mechanism to induce phase modulation by changing the refractive index of a material. The accumulated phase change after a length L is simply

$$\Delta\phi = \frac{2\pi}{\lambda_0} \Delta n L \qquad (2.3)$$

where λ_0 is the free-space wavelength. Although there are a number of methods to convert phase modulation to amplitude modulation, the most common structure used in optical communications is the Mach–Zehnder (MZ) inter-ferometer. The basic structure of an integrated-optic MZ interferometer is illustrated in Figure 2.13. The key parts include a single-mode input waveguide, a Y-branch splitter to divide the optical power in two equal parts, a pair of single-mode waveguides with adjacent electrodes, and a Y-branch coupler to recombine the light into a single-mode output waveguide. The waveguides themselves are high-refractive-index regions, made typically by modifying the LiNbO$_3$ with either a high-temperature indiffusion of titanium or by a proton

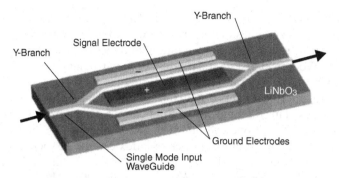

FIGURE 2.13   LiNbO₃ Mach–Zehnder interferometer with coplanar electrodes.

exchange process in which lithium ions are replaced with hydrogen ions (Donaldson, 1991).

When no voltage is applied to the electrodes, the refractive index is the same in both arms of the interferometer, so the light from each arm arrives at the output Y-branch in phase. With this condition, the light is coupled efficiently into the single-mode output with little loss. When a voltage is applied to the center electrode, the interferometer arms see oppositely directed electric fields. This causes the light in one arm to be advanced in phase and the other to be delayed. When the two arms are 180° out of phase, the light from each path interferes destructively at the start of the output waveguide, so no light is transmitted. Instead, the light is radiated out into the bulk crystal at angles corresponding to constructive interference and is ultimately absorbed by any opaque materials used in packaging. Well-designed LiNbO₃ MZ modulators can achieve very high on–off extinction ratios (>20 dB) with low optical insertion loss (3 dB).

The electrooptic transfer function governing the transmission T of the MZ in response to an applied voltage V_a is expressed by

$$T = K\frac{1}{2}\left[1 + \cos\left(\frac{\pi}{V_\pi}V_a + \Delta\phi_0\right)\right] \tag{2.4}$$

where K is the minimum insertion loss (P_{out}^{max}/P_{in}), and ϕ_0 accounts for any residual effective path length difference in the two arms. The *half-wave voltage*, V_π, is given by

$$V_\pi = \frac{\lambda_0}{2n^3 r_{33} L}\frac{g}{\Gamma} \tag{2.5}$$

where r_{33} is 30×10^{-12} m/V for the most commonly used coefficient, g is the gap between the electrodes, and Γ is an overlap factor (<1) that quantifies how effective the electrodes are at creating an electric field in the region of the optical mode. Typical values of V_π range from 3 to 6V and are highly dependent on operating speed, crystal orientation, and electrode configuration. See the article by Wooten et al. (2000) and the book by Chang (2000) for a discussion of the various configurations and their relative merits.

LiNbO$_3$ MZ interferometers are by far the most common of the electrooptic modulators used in communications applications. There are, however, other technologies that exploit one or both of the MZ configuration and the linear electrooptic effect. For example, refractive index changes associated with the QCSE can be exploited to make MZ modulators with lengths of only a few millimeters and V_π values near 2V (Akiyama et al., 2005). The linear electrooptic is also exploited in GaAs to make MZ modulators (Walker, 1991; Dagli, 1999) as well as polarization modulators (Rahmatian et al., 1998; Bull et al., 2004). In GaAs, the optical index is roughly 3.4 at 1550nm, whereas the microwave index on a coplanar strip line is roughly 2.7. *Slow-wave electrodes* are therefore used to slow down the microwave signal to match the optical velocity. Slow-wave electrodes on GaAs typically utilize periodic capacitive T-structures to increase the capacitance of the line without reducing the inductance (illustrated in Figure 2.14).

Table 2.1 summaries some the characteristics of EAMs and EOMs. Both have advantages and disadvantages. For long- and ultralong-haul applications, EOMs are the preferred solution, whereas for short and intermediate reaches, EAMs provide adequate eye quality at a lower cost. Another device, not discussed here, is the electroabsorption modulated laser (EML), which incorpo-

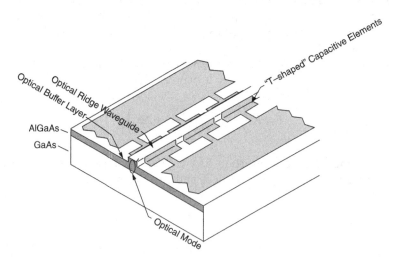

FIGURE 2.14 Structure of a 40-GHz AlGaAs/GaAs polarization modulator with slow-wave electrodes for velocity matching. (After Bull et al., 2004, Fig. 3.)

TABLE 2.1 Typical Parameters for Electroabsorption and Electrooptic Modulation

Parameter	Electroabsorption	Electrooptic Modulation
Bandwidth	MHz to >40 GHz	MHz to >30 GHz
Drive voltage	<3 V	3–6 V
On/off extinction ratio at dc	<18 dB	>20 dB
Optical insertion loss	~8–10 dB	2–6 dB
Chirp parameter	Variable, depends on design and operating conditions	−0.7 or 0
Temperature stability	Requires active temperature control	Stable over at least ± 35 °C
Spectral range of operation	~2–10 nm	>60 nm
Length of active waveguide	<0.5 mm	10 to 40 mm

rates a DFB laser with an EAM on a single chip. Performance of the EML is similar to that of the EAM, but the former is more compact and efficient.

2.5 OPTICAL AMPLIFICATION

Consider the problem of sending an optical signal through fiber across the Atlantic Ocean from New York to London (~5600 km). Even with 0.2 dB/km of optical loss, 1120 dB of accumulated attenuation ensures that essentially no signal arrives at the detector unless amplification or regeneration is provided along the way. Early fiber-optic links used optical-to-electrical and back to optical regeneration; optical signals were converted to the electrical domain, reconditioned, and then retransmitted with local semiconductor lasers. In the mid-1990s, optical amplifiers began to replace electrical-to-optical-to-electrical repeaters in undersea links. Although optical amplifiers still require pump lasers placed along the fiber in 50- to 100-km spacings, keeping the signals in the optical domain increases the bandwidth tremendously, as we will see. Figure 2.15 shows a basic multichannel link using optical amplification.

Optical amplifiers can be built based on various physical principles. The most widely deployed optical amplifier is the erbium-doped fiber amplifier (EDFA). The EDFA's pervasiveness stems from its ability to amplify many channels simultaneously across the C-band, where fiber losses are minimum. Another type of optical amplifier that is of considerable interest for fiber optic communications is the Raman amplifier. Although not nearly as common as the EDFA, the Raman amplifier offers flexibility in the gain spectrum as well

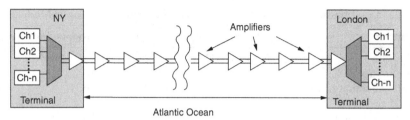

FIGURE 2.15 Multichannel long-haul link between New York and London using optical amplifiers.

as improved noise performance. Another type of optical amplifier is the semiconductor optical amplifier (SOA). The SOA generally provides poorer performance (e.g., noise, crosstalk, and polarization dependence) than the EDFA; however, some of its features, such as strong nonlinearity, make it attractive for optical signal processing (Agrawal 2004a). In this section we restrict our discussion to the more common EDFAs and Raman amplifiers.

2.5.1 Erbium-Doped Fiber Amplifiers

As the name implies, the EDFA uses a length of fiber doped with erbium ions for amplification (Desurvire, 1994; Becker et al., 1999). The basic configuration of the EDFA is shown in Figure 2.16. Structurally, an EDFA consists of an optical coupler, an erbium-doped fiber (~10 to 20 m long), and an optical isolator. The coupler combines the incoming optical data signal with light from a high-powered *pump laser*. The energy of the pump light is transferred to the data signal in the erbium-doped fiber to provide amplification through the multistep process described below. The isolator prevents backward-propagating light (from reflections or other sources) from being amplified and ensures that a resonant cavity cannot occur, which could lead to lasing (analogous to an electrical amplifier oscillating due to positive feedback).

The gain mechanism in EDFAs is similar to that of semiconductor lasers except that electrons are optically pumped into higher-energy states instead of being electrically pumped. For 1550-nm amplification, erbium ions in silica fibers are generally considered to be a three-energy-level system, as shown in Figure 2.16. The bottom level is $^4I_{15/2}$ and the first higher-energy level is $^4I_{13/2}$. When an electron transitions down from the $^4I_{13/2}$ level to the $^4I_{15/2}$ level, a photon with a wavelength roughly between 1530 to 1565 nm is radiated. In the absence of any other radiation in this range, the transition will eventually occur spontaneously with a relatively long lifetime of approximately 10 ms. If an incoming signal is present, the downward transition can produce stimulated emission, providing a coherent amplification of the incoming signal. Since electrons can transition either up or down in energy, gain or absorption will occur, depending on the populations in each energy level. As in the case of

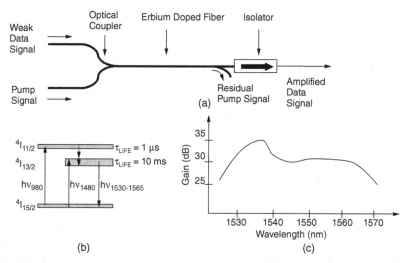

FIGURE 2.16 (a) Structure of an EDFA; (b) energy-level diagram; (c) an example of its optical spectrum.

the laser, population inversion is required so that the stimulated emission exceeds the absorption.

Population inversion in the EDFA can be obtained using either 980- or 1480-nm pump light. In the case of 980-nm pumping, electrons are excited from the ground state to the $^4I_{11/2}$ state. The electrons then quickly drop (non-radiative) to the $^4I_{13/2}$ energy level with a characteristic lifetime of about 1 μs. The $^4I_{13/2}$ level will then persist because of its longer lifetime, and provide stimulated emission with the presence of an input signal. The EDFA can also be pumped at 1480 nm to excite the $^4I_{13/2}$ level directly. The energy levels have some spectral width, in part because of the influence of the amorphous silica on the erbium. Thus, 1480-nm pump light will excite the upper energy range of the $^4I_{13/2}$ level. The width of the energy levels gives rise to the broadband characteristics of the EDFA. Figure 2.16 shows the typical gain spectrum of the C-band EDFA, which covers wavelengths from roughly 1530 and 1565 nm. A single EDFA can thus amplify many channels at different wavelengths simultaneously, making it a key enabling technology for WDM systems. L-band EFDAs are also available and employ long fibers (>100 m) combined with low inversion levels to amplify in the region 1570 to 1610 nm (Agrawal, 2004a).

In practice, EDFAs use either 980- or 1480-nm pumps, and in multistage EDFAs, they can be used in conjunction with each other. The 980-nm pump can be used for a preamplifier with a low noise figure, and the 1480-nm pump provides the power amplifier stage. Because 1480 nm is closer in energy to 1550 nm than to 980 nm, the 1480-nm pump can be optically more energy efficient in terms of signal output power to pump power. Output powers for

EDFAs can range from 10 mW to several watts, with gains typically in the range 20 to 35 dB.

Unlike electrical amplifiers, the rise time of the EDFA is practically instantaneous in comparison to the rise time of data pulses. The EDFA will amplify a high-speed bit stream with a time-invariant gain, meaning that pulse shapes are preserved and consecutive pulses receive the same gain. The EDFA is thus transparent to bit rate and modulation format. At low frequencies, however, nearing the 10-ms spontaneous emission lifetime of the $^4I_{13/2}$ erbium energy level, the gain varies depending on the optical power of the input signal, causing significant distortion. It is therefore important to keep the average input optical power constant over millisecond and longer time spans.

One of the challenges with EDFAs is that they naturally produce a nonuniform gain spectrum. As illustrated in Figure 2.16, gain around 1535 nm is significantly higher than that at 1560 nm. In a long-haul WDM system with many amplifiers, the gain differences can accumulate, leading to excessive power imbalances among the channels. Gain flattening can be obtained by manipulating the glass composition, using hybrid techniques with multiple parallel fibers, or using passive filters such as the *long-period fiber grating* with complementary absorption spectrums (Becker et al., 1999). EDFAs can also be combined with Raman amplifiers to flatten gain.

Although the EDFA is a very useful device and is used abundantly in long-haul links, the gain comes at the expense of added noise. The noise is the result of amplified spontaneous emission (ASE). Although most of the photons emitted by the erbium ions amplify the input signal, some are emitted spontaneously in all directions and have no correlation to the input signal. A few of these spontaneously emitted photons will have their direction aligned with the fiber axis and will be guided in the forward and reverse directions along the fiber. These spontaneously emitted photons will then be amplified through stimulated emission as they travel out of the EDFA, giving rise to ASE. The ASE has an average power that shows up as a direct current when incident on a photodetector, but also has an associated noise spectrum. The photodetector, being a square-law device, mixes the ASE noise spectrum with itself, giving rise to *spontaneous–spontaneous current noise*. The ASE will also mix with signal's spectrum, giving rise to *signal–spontaneous current noise*. A detailed analysis of the origins and implications of these two noise sources can be found in a book by Becker et al. (1999).

ASE noise is typically specified for EDFAs as a noise figure. The noise figure is defined as

$$NF = \frac{SNR_{in}}{SNR_{out}}$$

where SNR_{in} is taken as the SNR that you would measure by detecting the light at the input to the EDFA in a shot-noise-limited detector (i.e., ignoring the thermal noise of the detector as well as the RIN of the light source). SNR_{out} is assumed to be dominated by signal–spontaneous noise. These assumptions lead a simple expression for the noise figure (Becker et al., 1999):

$$NF = 2n_{sp}\frac{G-1}{G} \approx 2n_{sp}$$

where n_{sp} is the spontaneous emission factor, which quantifies population inversion of the Er ions, and G is the amplifier gain. The theoretical limit is $n_{sp} = 1$, which gives rise to a best possible noise factor of 2, equivalent to 3 dB. For typical EDFAs, NF is in the range 3.5 to 6.5 dB.

2.5.2 Raman Amplifiers

Raman amplifiers utilize stimulated Raman scattering (SRS), one of the non-linear effects found in optical fiber (see Section 2.6.4), to provide gain for the data signal (Islam, 2003; Headley and Agrawal, 2005). Raman scattering is an inelastic scattering process: The photon energy changes during the scattering, converting one wavelength to another. The energy diagram in Figure 2.17 illustrates the process. A higher-energy pump wave is launched into the fiber with photons of energy $h\nu_p$. A portion of these photons undergo scattering from the silica molecules, giving up their energy to a virtual energy level at $h\nu_p$. The virtual energy level then radiates a new photon with a lower-energy $h\nu_s$ value, giving the balance of the initial energy to the silica molecules as a lattice vibration (E_{phonon}). This process is called *Raman scattering*, and the resulting lower-energy wave is called a *Stokes wave* (a reverse process also

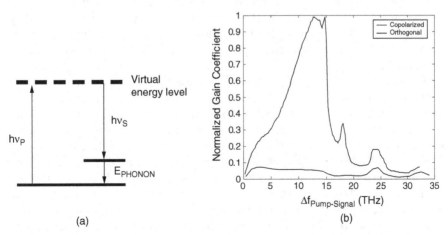

(a)   (b)

FIGURE 2.17 (a) Energy-level diagram for SRS; (b) gain spectra for the Raman process. (After Stolen et al., 1989.)

exists that produces an anti-Stokes wave). When an input signal is present with photons at the lower-energy $h\nu_s$, combined with strong pump powers, the Raman scattering events can become stimulated, resulting in coherent amplification of the signal (SRS).

Energy conversion with SRS is unfortunately inefficient at low optical power, requiring a long optical fiber (tens of kilometers for SMF) and a high-power pump source. Pump powers on the order of 1 W are used to achieve gains comparable to those of EDFAs. Raman amplifiers nonetheless have a number of advantages over EDFAs. One of the distinguishing features of Raman amplifiers is that simply shifting the pump wavelength can shift the gain spectrum, allowing amplification across the broad range from 1280 to 1530 nm, which is inaccessible with EFDAs (Islam, 2002). The Raman scattering from silica molecules produces a characteristic downshift of about 13 THz from the pump frequency, or about 100 nm in a 1550-nm band. The spectrum is roughly triangular in shape, the gain increasing linearly with wavelength offset, peaking at about 100 nm, then dropping rapidly with increasing offset (see Figure 2.17). The width of the peak results from the spread of vibrational energies of the amorphous silica. The gain spectrum can be further broadened and flattened by using multiple pump sources with appropriate wavelengths and powers (Bromage, 2004). For 1550-nm amplification, the pump wavelengths will be near 1450 nm.

One of the advantages of the Raman amplifier is that the same fiber as that used for transmission can be used as the gain medium. Thus, existing SMF links can become an integral part of the amplifier. This type of amplification is *distributed*, in contrast to *discrete amplifiers* such as EDFAs, where the gain-producing fiber is simply coiled up in a box. Raman amplifiers can also be configured as discrete amplifiers using shorter lengths of highly nonlinear fiber. Raman amplification can, for example, be combined with a dispersion-compensating fiber, which as a result of its increased nonlinearity produces a 10-fold increase in Raman gain efficiency over SMF (Headley and Agrawal, 2005, Chap. 4). However, distributed amplification is advantageous from a noise perspective and so is of great interest for communication systems.

The advantage of distributed gain from a noise perspective comes from the general result that it is better to place gain before loss to minimize noise in an amplifier chain. With gain before loss, the loss attenuates both the signal and the amplifier noise equally, whereas loss before gain attenuates the signal but leaves the added noise of the amplifier at full strength. Because of nonlinearity in the fiber and catastrophic damage limits, with discrete amplification, there are practical limits to how much power can be launched into the fiber. With distributed gain it is possible to increase the fiber's output power without increasing signal power at the input. Even when the pump is launched into the input of the fiber, the wavelength difference of the pump and signal limit many of the deleterious nonlinear effects that would otherwise occur if the signal(s) contained all the power at the input.

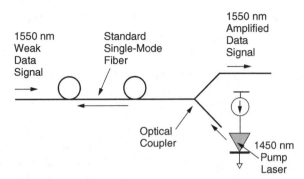

FIGURE 2.18   Distributed Raman amplifier with a counterpropagating pump.

Fortunately, SRS does not depend on the relative directions of the pump and signal, so it is possible to have the pump traveling forward or in reverse, or to pump from both directions. Figure 2.18 shows a simple case of a reverse-pumped Raman amplifier. The counterpropagating pump produces maximum local gain near the output of the fiber, which decays toward the input as the pump is depleted or absorbed. Distributing the gain in this manner means that the signal sees some of the SRS gain ahead of some of the fiber loss, thereby improving the noise figure of the link. It is common to define an effective noise figure for the distributed Raman amplifier that represents the gain of a fictional discrete amplifier placed at the end of the fiber that provides an equivalent gain. The effective noise figure of this fictional amplifier is generally negative, representing the advantage of the distributed gain. The combined noise figure of the amplifier and fiber loss is nonetheless still positive, so no quantum limits are violated. For more on noise figure calculations with Raman amplifiers, see the article by Bromage (2004).

2.5.3   EDFA vs. Raman Amplifier

SRS does have a few characteristics that make amplifier implementation more complex than with EFDAs. The long lifetime of the $^4I_{13/2}$ level in the EFDA has the effect of averaging the pump power, thereby preventing coupling of noise from the pump to the signal. SRS gain responds practically instantaneously (compared to data pulse widths) to pump power fluctuations, however, allowing noise from the pump to couple to the signal (Islam, 2002). This noise coupling can be suppressed using a counterpropagating pump to limit the interaction time that a pump fluctuation has to influence a data pulse. Another difference between Raman amplifiers and EDFAs is that SRS is highly polarization dependent. The gain is an order of magnitude stronger when the pump is copolarized with the signal. For this reason, a depolarized pump source is used for Raman amplification to minimize

TABLE 2.2   Comparison Between EDFA and Raman Amplifiers

Feature	Erbium-Doped Fiber Amplification	Raman Amplification
Amplification band	1530 to 1610 nm	1280 to 1620 nm and beyond
Pump power	10 mW to 5+ W	500 mW to 5+ W
Pump wavelength	980 or 1480 nm	100 nm shorter than the signal wavelength
Gain[a]	20 to 30+ dB	10 to 30+ dB
Topology	Discrete	Discrete or distributed
Noise figure	3.5 to 6.5 dB	4 to 7+ dB (discrete)
		−3 to 1 dB (distributed)

[a]Because the gain of the distributed Raman amplifier is located inside an otherwise lossy fiber span, the gain is specified as on–off gain, representing the change in signal level at the output with the pump turned on as compared with the pump turned off. The actual gain from the input to the output of the Raman pumped fiber depends on the fiber loss and will be 0 dB if the amplifier cancels the loss perfectly.

gain drift as the polarization state of the signal varies along the fiber and drifts over time.

Both EDFA and Raman amplifiers have several advantages and some disadvantages, but fortunately for optical networking, the devices are largely complementary. They can be used in conjunction with each other for controlling gain flatness, increasing the gain bandwidth, or combining discrete and distributed amplification approaches. The basic properties of EDFA and Raman amplifiers are summarized in Table 2.2.

2.6   FIBER TRANSMISSION IMPAIRMENTS

Distortion in optical data transmission is caused primarily by dispersive and nonlinear effects. We have already seen one type of dispersive effect, multimode dispersion in multimode fibers. In single-mode fibers, multimode dispersion is eliminated; however, two other types of dispersion ultimately limit bandwidth and distance: chromatic dispersion and polarization mode dispersion. As with multimode dispersion, these two types of dispersion result in parts of a transmitted signal arriving at the receiver at different times, due to the frequency and polarization dependence of the velocity of light though the fiber. The result is distortion and increased bit error rate in digital links. Another mechanism responsible for fiber transmission impairments is the nonlinear refractive index in silica. The nonlinear refractive index can induce distortion dependent on the optical power. As we will see, nonlinear effects and dispersive effects can interact, giving rise to system designs that incorporate careful consideration of each effect to optimize link reach and speed.

2.6.1 Chromatic Dispersion

As used in the field of fiber-optic communications, the term *chromatic dispersion* (CD) is an effect more accurately described as group velocity dispersion.[†] To understand the implications of CD, its helpful to look at how waves travel in a dispersive medium such as glass, where the refractive index is a function of the frequency ω. The phase velocity is the velocity of the peaks and troughs of a single-frequency wave and is given by

$$v_p = \frac{c}{n(\omega)} \tag{2.6}$$

where c is the speed of light in vacuum and $n(\omega)$ is the refractive index; in the case of an SMF, $n(\omega)$ can be replaced with the effective index $n_{eff}(\omega)$. Given that we are interested in sending information using modulated light waves, not just single frequencies, the question arises: At what speed will a modulated wave such as a pulse travel? If the pulse has a spectral bandwidth that is narrow compared to the optical carrier frequency, as is generally the case, the envelope of the pulse will travel at the group velocity, which is given by (Yariv and Yeh, 1984)

$$v_g = \frac{c}{n + \omega(dn/d\omega)} \tag{2.7}$$

We see that if $dn/d\omega$ is positive (called *normal dispersion*), the group velocity will be slower than the phase velocity. The difference between phase velocities and group velocities means that the peaks and troughs of the carrier across the pulse are shifting in time with respect to the pulse envelope. The pulse's envelope will slow down but will not suffer distortion provided that v_g is constant for all frequencies. If v_g changes with frequency, we have group velocity dispersion (CD), and the pulse not only travels at a different velocity from the phase velocity, but also becomes stretched in time. Figure 2.19 illustrates this process.

Pulse broadening due to CD is typically explained as originating from the lower and higher frequencies contained in the pulse arriving at different times at the end of the fiber. Although this is an intuitively appealing explanation, note that the group velocity can be constant with frequency even though n is a function of frequency, as long as the denominator in equation (2.7) remains constant. The particular behavior of n that leads to distortion is actually a nonzero second derivative of n with respect to λ. The *dispersion parameter* quantifies this effect and is given by

---

[†]In other areas in optics, *chromatic dispersion* is applied to the variation in refractive index and hence phase velocity with wavelength rather than the variation of group velocity. Phase velocity dependence on wavelength leads to the splitting of colors by a prism.

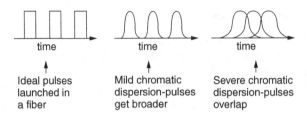

FIGURE 2.19 Chromatic dispersion phenomena leading to intersymbol interference if left uncompensated.

$$D = \frac{d}{d\lambda}\left(\frac{1}{v_g}\right) = -\frac{d^2 n}{d\lambda^2}\frac{\lambda}{c} \tag{2.8}$$

where λ is the wavelength in vacuum. The units of D are typically expressed in ps/(nm·km) to indicate how many picoseconds will separate different parts of the spectrum separated in nanometers over a distance in kilometers.

The impact of chromatic dispersion on an optical pulse train depends on the fiber length, dispersion parameter, and bit rate of the pulse train. With DFB lasers and external modulation, the spectral width is determined by the modulation of the pulse train and we can ignore the spectral width of the DFB in the absence of modulation. Unfortunately, for high-bit-rate systems, the degradation due to CD increases as the square of the bit rate. The square relationship results from a linear increase in spectral width with bit rate, combined with a linear increase in the proximity of adjacent pulses with bit rate. The increased spectral width gives rise to more pulse stretching, while the closer proximity of the pulses means that the wider pulses will quickly overlap and become distorted. Fortunately, dispersion compensation techniques are available for high-speed long-haul links. We discuss these techniques below.

Figure 2.20 shows the CD characteristics of a typical SMF. The CD is made up of two components: waveguide dispersion and material dispersion. The *waveguide dispersion* results from the slight variation in optical path length with wavelength as the optical fields adjust to satisfy Maxwell's equations at the core–cladding interface. The *material dispersion* represents the contribution from refractive index variation of fused silica. As seen in Figure 2.20, the waveguide dispersion is negative and cancels some of the material dispersion. The total CD is zero near 1310 nm and is an attractive feature for otherwise CD-limited systems that do not suffer from nonlinear effects. As we will see in Section 2.6.4, a zero-CD fiber strengthens the deleterious effects of nonlinear interactions. In Section 2.6.2 we look at dispersion compensation techniques that allow the use of dispersive fiber combined with CD cancellation in discrete points along the link. For long-haul links, this approach exploits CD to reduce nonlinear impairments.

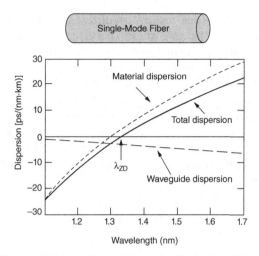

FIGURE 2.20 Composition of fiber chromatic dispersion. (From Agrawal, 2002, Fig. 2.10.)

Before moving on to other dispersion topics, let's look at the amount of pulse broadening that is expected from typical values of CD. We can roughly estimate the extent of the pulse broadening directly from the dispersion parameter D using (Agrawal, 2002)

$$\Delta T = \frac{d}{d\lambda}\left(\frac{L}{v_g}\right)\Delta\lambda = DL\,\Delta\lambda = DL\frac{\lambda^2}{c}\Delta f \qquad (2.9)$$

where $\Delta\lambda$, or Δf, represents the pulse's spectral width. The value of $\Delta\lambda$ depends on the optical source's intrinsic spectral width as well as the modulation spectrum. If the source's intrinsic spectral width is much larger than the bit rate, we can ignore the impact of the modulation on $\Delta\lambda$. The amount of pulse broadening that we can tolerate is still related directly to the bit rate. We can roughly estimate the limit of broadening by imposing the condition that $\Delta T < 1/B$ (i.e., if a pulse has spread by the duration of 1 bit), we can expect problems recovering the signal. We can then rearrange equation (2.9) to give (Agrawal, 2002)

$$BL < \frac{1}{D\,\Delta\lambda} \qquad (2.10)$$

Equation (2.10) shows the importance of low dispersion and narrow spectral width for bandwidth and link length. For multimode lasers with spectral widths between 2 and 5 nm, if we operate near 1550 nm with $D = 17\,\text{ps/(nm·km)}$, we anticipate BL to be limited to between 12 and 30 Gb·km/s. Considerably different values will, of course, result with different types of optical sources at different wavelengths. In the case where $D = 0$, for example at 1310 nm, equation (2.10) is no longer applicable, and we must consider higher-order dispersion effects. [See the book by Agrawal (2004b) for a thorough analysis.]

Let's look at the case where the optical spectrum width is limited by the modulation, such as the case of a DFB laser with an external 40 Gb/s modulator. The narrow spectral width of a DFB laser (in continuous-wave mode) is typically quoted in megahertz instead of nanometers, with values between 10 and 30 MHz. We can crudely estimate the spectral width of the modulation to be equivalent to the bit rate. Rearranging equation (2.9) with $\Delta T < 1/B$ and $\Delta f = B$, we obtain

$$B^2 L < \frac{c}{D\lambda^2} \qquad (2.11)$$

We see that the bit rate is now squared while the length is still first order. The squaring results from the spectral width increasing with bit rate, while simultaneously, the allowable pulse spreading decreases with bit rate. In this regime, a fourfold increase in bit rate requires a 16-fold reduction in dispersion to maintain the same link length. This fact makes dispersion compensation a key issue for long-haul systems at 10 Gb/s and much more so at 40 Gb/s and beyond. Evaluation of equation (2.11) with $B = 40$ Gb/s at 1550 nm with $D = 17$ ps/(nm·km) produces a rather limited length of 4.6 km if left uncompensated.

Another parameter that influences the effect of CD is chirp. As we have seen in Section 2.4, the optical frequency can change across the duration of a pulse, depending on the particular technology used for modulation. In the case of directly modulated lasers, a 2.5-Gb/s pulse may be accompanied by a chirp of more than 10 GHz (Kaminow and Li, 2002b, Chap. 14), clearly violating the assumption above that $\Delta f = B$. Chirp can, however, be beneficial in certain instances. If, for example, the chirp is such that the leading edge of a pulse experiences an opposite change in frequency from the trailing edge, the chirped pulse can either broaden or narrow in time, depending on the sign of the chirp relative to the sign of the dispersion. If the pulse's leading edge is *red-shifted* (longer wavelength) while the trailing edge is *blue-shifted*, and if the CD is such that longer wavelengths travel more slowly, the pulse will compress as it travels along the fiber for some distance. Eventually, the pulse will broaden again as the red-shifted energy slows further, to become the trailing edge of the pulse, causing broadening at a faster rate than an unchirped pulse with a narrow spectrum.

2.6.2 Dispersion Management Techniques

The square dependence of dispersion impairment with bit rate has resulted in a rapid change in the relevance of dispersion management as bit rates have

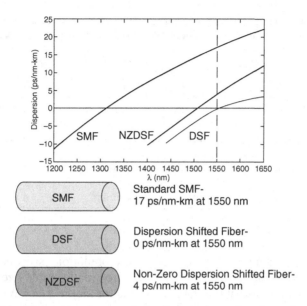

FIGURE 2.21 Dispersion characteristics for single-mode, dispersion-shifted, and nonzero dispersion-shifted fibers.

increased. Prior to the development of multichannel WDM systems, dispersion was managed by operating only a single channel placed at the zero-dispersion point of SMF at 1310 nm. Without optical amplifiers, fiber loss required electrical signal regenerators to be placed closer than the dispersion-limited distance. With the change to the lower-loss C-band (1550 nm), dispersion increased from zero to ~17 ps/nm·km (Figure 2.21). Narrow-spectrum DFB lasers replaced the multifrequency-mode Fabry–Perot lasers, making the dispersion at 1550 nm tolerable for low bit rates. The dispersion limited distance for 2.5 Gb/s at 1550 nm is about 1000 km in SMF. At higher rates, this distance falls rapidly to 60 km at 10 Gb/s, and 3.5 km at 40 Gb/s (Kaminow and Li, 2002b, Chap. 14). With the ability to overcome fiber loss with EDFAs, dispersion becomes the dominant obstacle unless compensation is used. In this section we explore some of the various dispersion management and compensation techniques, including dispersion-shifted fiber, dispersion-compensating fiber, and fiber Bragg gratings.

Dispersion-Shifted Fiber   During the 1980s, fiber manufacturers developed a seemingly ideal product that combined zero dispersion with the minimum-loss region at 1550 nm (Kaminow and Li, 2002b, Chap. 14). Dispersion-shifted fiber (DSF) was achieved by modifying the refractive index profile to make the waveguide dispersion cancel the material dispersion at 1550 nm (Figure 2.21). Although DSF is quite effective for a single channel, nonlinear effects

TABLE 2.3 Chromatic Dispersion Characteristics in Various Types of Fiber with International Telecommunication Union (ITU) Designation and Historical Deployments

Fiber Type	Chromatic Dispersion over the C-Band	Historical Deployment
Standard single-mode fiber (G.652)	Large, 16 to 17 ps/nm·km	Early 1980s
Dispersion shifted fiber (G.653)	Very small, <1 ps/nm·km	Late 1980s
Nonzero dispersion shifted fiber (G.655)	Small, 2 to 6 ps/nm·km	Early 1990s
Large effective area NZDSF	Small, 2 to 6 ps/nm·km	Late 1990s

such as four-wave mixing and cross-phase modulation (see Section 2.6.4), which cause crosstalk between the channels, are greatly emphasized with negligible dispersion. DSF was thus not well suited for the emerging WDM systems in the early 1990s.

To combat nonlinearities, nonzero dispersion-shifted fiber (NZDSF) was developed with a small but nonzero dispersion value near 1550 nm to detune the phase-matching conditions of the nonlinear effects (Pal, 2005). The dispersion curve for NZDSF is also shown in Figure 2.21. The first NZDSF fibers obtained their dispersion shift at the expense of a reduced effective area of the core. Although nonzero dispersion more than compensated for increased power density and associated nonlinear susceptibility from the reduced effective area, a large effective area NZDSF was soon developed. Large effective area NZDSFs such as Corning's LEAF were deployed in the late 1990s.

As a result of historical deployments, worldwide fiber installations consist of a mixture of standard single-mode, dispersion-shifted, and nonzero dispersion-shifted fibers. Methods are therefore needed to compensate for dispersion in a wide variety of installed fibers to avoid costly infrastructure upgrades. Furthermore, even with NZDSF, the accumulated dispersion will severely limit distances at high bit rates in the absence of compensation. The historical deployment trends are summarized in Table 2.3.

Dispersion-Compensating Fiber   Dispersion-compensating fiber (DCF) is the standard approach to canceling dispersion in fibers such as SMF or NZDSF (Pal, 2005). The dispersion of DCF is designed to be very large and negative. The total chromatic dispersion is a sum of the material and waveguide dispersion. *Material dispersion* arises from the dependence of the refractive index on wavelength, whereas *waveguide dispersion* results from the changes in light distribution within the core–cladding structure of the fiber when the wavelength changes. For DCF, the core–cladding refractive index difference is near 2% (for SMF, the difference is <0.4%) and a smaller core diameter is used compared to that of SMF. More recent designs also use multiple claddings around the core to tune the fiber mode profile for maximum negative dispersion and minimum insertion loss. DCF designs can also be tailored to provide *dispersion-slope* compensation so that multiple wavelengths in a WDM system are all compensated simultaneously (Kaminow and Li, 2002b, Chap. 14).

Commercial DCFs have dispersion values around −100 ps/(nm·km) and insertion loss values in the range 0.4 to 0.7 dB/km. Although the dispersion is large compared to that of standard fiber, it is still much smaller than is desirable. A 50-km span of SMF, for example, requires about 8 km of DCF to cancel the dispersion. Since existing fiber installations already cover the necessary distance, the length of the DCF adds loss to the link without increasing the actual distance of the span. The DCF is therefore simply coiled up and placed at the input (precompensation), the output (postcompensation), or placed at periodic intervals along the fiber (Figure 2.22). If the fiber were perfectly linear, the dispersion could be effectively canceled independently of the location of the DCF. Because of the interplay by dispersion and nonlinear effects (see Section 2.6.4), the location of the dispersion compensation plays a key role in long-haul systems. A practical choice is to place DCF periodically at

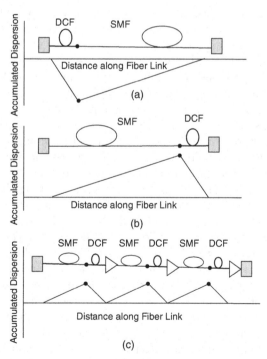

FIGURE 2.22 Accumulation of dispersion along a link using DCF for (a) precompensation, (b) postcompensation, and (c) periodic compensation.

the amplifier sites, which have a spacing of about 80 km for terrestrial systems and about 50 km for submarine systems (Agrawal, 2002).

In addition to causing excess insertion loss, DCF also suffers from increased nonlinearity as a result of its small core area ($\sim$20 μm²). If the DCF is placed after an EDFA, the high launch power exacerbates the nonlinearity problems. If the DCF is placed before the EDFA, the "loss before gain" problem hurts the noise figure. One compromise is to place the DCF in the middle of a dual-stage EDFA (Kaminow and Li, 2002a, Chap. 7). The shortcomings of conventional DCFs have promoted research into alternative dispersion compensation approaches. One of the more promising approaches is the fiber Bragg grating dispersion compensator. Bragg grating compensators are now commercially available for 10- and 40-Gb/s systems.

Dispersion Compensation with Fiber Bragg Gratings   Fiber Bragg grating (FBG) is a wavelength-sensitive reflector created by introducing a modulation of the refractive index (the grating) along the core of an optical fiber (Hill and Melty, 1997). The grating reflects light at the wavelength λ_B that satisfies the Bragg condition:

$$\lambda_B = 2n_{\text{eff}}\Lambda$$

where n_{eff} is the guided mode effective index and Λ is the period of the grating. Wavelengths not satisfying the Bragg condition pass through the grating with negligible loss. The grating itself is formed by exposing the fiber to an intense ultraviolet image of the grating, which in the case of germanium-doped fiber produces a refractive index change between 10^{-5} and 10^{-3}. FBGs are used in numerous applications, including add–drop multiplexers, wavelength filters, wavelength lockers, sensors, and dispersion compensators.

Chirped FBGs, where the grating period changes with position along the fiber length, are used to compensate chromatic dispersion in optical fibers. A basic chirped FBG dispersion compensator is shown in Figure 2.23. Light enters the compensator through a circulator that directs the input light into the FBG. Because the grating spacing Λ decreases along the FBG, longer wavelengths are reflected near the start of the grating while shorter wavelengths are reflected nearer the end of the grating. For conventional single-mode fibers operating in the anomalous dispersion regime ($D > 0$), where the group velocity of shorter wavelengths is faster than that of longer wavelengths, the extra time taken for the shorter wavelengths to return from the grating counteracts the fiber dispersion. After a round trip through the chirped FBG, pulses emerge from the third port of the circulator with reduced width as their spectral components are realigned. More complex grating patterns make multichannel dispersion slope compensation possible (Figure 2.24).

Dispersion compensators based on FBGs are commercially available as single- or multi-channel (>40) units. Modules can compensate various fiber lengths up to and beyond 100 km (–1700 ps/nm) with insertion loss below 3 dB.

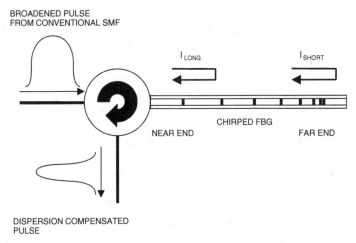

FIGURE 2.23 Chirped fiber Bragg grating used for dispersion compensation.

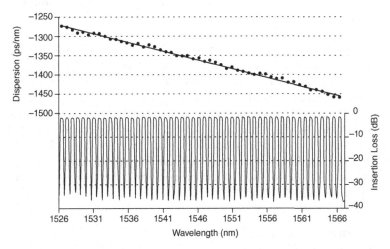

FIGURE 2.24 Dispersion and insertion loss of a multichannel FBG dispersion compensator module for the C-band. (Courtesy of *TeraXion*.)

Tunable dispersion compensation is also possible by stretching the grating with stress or by thermally tuning the grating (Lachance et al., 2003). Tunable dispersion is becoming more important as fiber networks move toward multichannel reconfigurable systems, where the amount of compensation required can change rapidly.

Electronic Dispersion Compensation Another very promising method for combating dispersion is electric dispersion compensation (EDC) (Winters and

Gitlin, 1990; Haunstein et al., 2004; Buchali, 2006). EDC can be performed by predistorting the transmitted signal to minimize distortion at the receiver or by processing the received signal to undo the distortion caused by dispersion. One of the intrinsic challenges with EDC is that the phase information of the optical signal is lost with low-cost direct amplitude detection. The result is that chromatic dispersion, a linear distortion, produces nonlinear distortion after the square-law detector that is more difficult to correct. Despite the challenges with high-speed electronic signal processing, commercial products are now available at 10 Gb/s that use MLSE (maximum likelihood sequence estimation)–based detectors.

One of the advantages of EDC is that its compensation mechanism is not specific to chromatic dispersion. Other impairments, such as polarization mode dispersion (PMD) and self-phase modulation (SPM) can also be improved. EDC is well suited for extending the reach of 10-Gb/s metropolitan networks to a few hundred kilometers without any optical dispersion compensation (Elbers et al., 2005). For long-haul links, EDC can be combined with DCF to improve performance; however, the benefit of EDC is limited when strong SPM and cross-phase modulation are present (Chandrasekhar and Gnauck, 2006).

Summary of Dispersion Compensation The benefits and shortcomings of the dispersion techniques discussed above are summarized in Table 2.4. Also listed is optical–electrical–optical (O-E-O) regeneration, which prior to EDFAs, was required to overcome fiber loss. O-E-O regenerators detect the optical signal, regenerate the pulses in the electrical domain, and retransmit with a laser. With the introduction of EDFAs, O-E-O regeneration became an expensive way to overcome dispersion and DCF became the preferred approach. FBGs and EDCs are more recent compensation approaches. For a more complete discussion of dispersion-compensation approaches, see the book by Agrawal (2002, Chap. 7).

2.6.3 Polarization Mode Dispersion

Before we discuss polarization mode dispersion (PMD), we need to discuss briefly the fundamentals of electromagnetic polarization. A wide beam of light that travels in a single direction without converging or diverging (i.e., a plane wave) is made up of electric and magnetic field oscillations that are perpendicular to the direction of propagation. Out of convention, polarization is defined as the orientation of the electric field instead of the magnetic field.[†] If

---

[†]Most materials interact strongly with electric fields, whereas few interact with magnetic fields, making the electric field the more practical choice for the definition of polarization. The magnetic field of the plane wave is perpendicular to both the electric field and the propagation direction, and only the single quantity "polarization" is needed to specify the orientation of both fields.

TABLE 2.4   Comparison of Various Dispersion Compensation Techniques

Dispersion Compensation Technique	Benefits	Shortcomings
Dispersion-compensating fiber	Multichannel (dispersion–slope compensation), smooth spectral response	High insertion loss, high nonlinear coefficient, bulky, static compensation
Fiber Bragg gratings	Low insertion loss, multichannel, tunable compensation, compact	Phase ripple can cause distortion
Electronic dispersion compensation	Low cost, compact, dynamic compensation; can compensate for other impairments (PMD, SPM)	High-speed electronics challenging at high bit rates, insufficient compensation for long haul
O-E-O regeneration	Robust technique: prevents accumulation of dispersion and nonlinear impairments	Expensive for multichannel systems; limited bit rate (2.5 Gb/s) with practical repeater distances

the electric field direction changes randomly at a rate faster than we care to detect, we say that the light is *depolarized*. If the electric field oscillates consistently in one direction, we say that the light is *linearly polarized*. For a complete description of polarization, we can resolve the electric field into two perpendicular components with some orientation that is convenient for the problem we wish to analyze; horizontal and vertical are frequently used. If we have a consistent oscillation in both horizontal and vertical directions, the resultant is linear polarization at 45°. If, however, the horizontal and vertical polarizations have a phase delay between them, we end up with elliptical polarization or with circular polarization if the phase delay is 90° and the horizontal and vertical polarizations have equal amplitudes.

In optical communications, some of the components we use are polarization sensitive, whereas others are not. Lasers produce polarized light, while most photo detectors are insensitive to polarization. The velocity in optical fibers is very weakly polarization sensitive; however, at bit rates of 10 or 40 Gb/s, fluctuations on the order of tens of picoseconds in the arrival time of a bit can mean link failure. The origin of polarization-induced timing fluctuations is *birefringence*, which broadly means having two refractive indices for orthogonal polarizations (Nye, 1985). In fibers, birefringence refers to having two different effective indices and is given by

$$b = n_s - n_f \qquad (2.12)$$

where n_s is the "slow" axis effective index and n_f is the "fast" axis effective index. The terms *fast* and *slow* are relative; a perturbation could speed up or slow down both axes, but our main interest here is just the difference between the slower and faster axes. Although fiber manufacturers go to great efforts to make circularly symmetric fibers with $b = 0$, small unintentional deviations in the circular shape of the core give rise to *form birefringence* or *geometric birefringence* (Kaminow and Li, 2002b). A slightly elliptical core will produce different speeds for light polarized along the major and minor axes of the ellipse. Furthermore, stresses and bends applied to the fiber will also cause birefringence. Unfortunately, these external sources of birefringence change over time, typically on the order of hours for installed fiber links (Kaminow and Li, 2002b), giving rise to unpredictable behavior.

To analyze the propagation of light in a short fiber with birefringence, it is convenient to resolve the electric field of the incident light into two components that are aligned with the fast and slow axes of the fiber. The relative power in the axes remains constant, but the relative phase changes as a result of the birefringence. This means that the polarization state emerging from the fiber is generally different from that which entered the fiber unless the phase is a multiple of 2π. Since conventional photo receivers (noncoherent) are insensitive to polarization, the change in polarization along the fiber is not in itself problematic. More problematic is the difference in travel times (from different group velocities) for a pulse that carries energy in both the fast and slow axes. This effect, illustrated in Figure 2.25, causes a pulse to broaden by an amount equal to the differential group delay (DGD), which is given by (Kaminow and Li, 2002b)

$$\Delta\tau = L\frac{d}{d\omega}\left(\frac{b\omega}{c}\right) = L\left(\frac{b}{c} + \frac{\omega}{c}\frac{db}{d\omega}\right) \qquad (2.13)$$

The DGD given by equation (2.13) applies to the simple situation where a fiber has consistent axes along its length. In practice, the orientation of the axes as well as the birefringence will vary in an unpredictable manner along the fiber length, mixing up the powers among the differing successive fast and slow axes along the fiber. As a result, a portion of the DGD gets canceled along the fiber. In long fibers, this cancellation reduces the DGD dependence on length from a linear to a square root relation according to

$$\Delta\tau_{rms} = D_{PMD}\sqrt{L} \qquad (2.14)$$

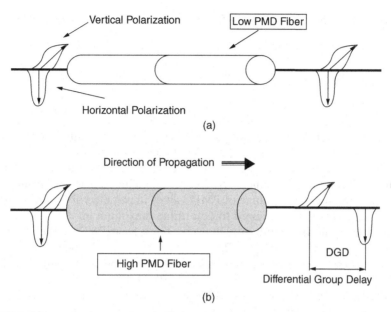

FIGURE 2.25 Principle of polarization mode dispersion showing the creation of a differential group delay as a result of different group velocities for orthogonal polarizations. The perfectly circular fiber core (a) has no DGD, whereas the oval-shaped core (b) induces DGD.

where D_{PMD} is the *first-order PMD coefficient*, which is usually expressed in $ps/km^{1/2}$. Fibers installed in the 1980s have PMD coefficient values on the order of $1\,ps/km^{1/2}$, whereas current fibers produced today have values less than $0.1\,ps/km^{1/2}$ (Kaminow and Koch, 1997).

Despite the randomness of the birefringence along the fiber, there are still two orthogonal polarization states that can be launched into the fiber that will emerge from the far end at distinct times as if the fiber had uniquely defined fast and slow axes with a DGD give by $\Delta\tau_{rms}$ (Kaminow and Koch, 1997, Chap. 6). This PMD model is referred to as the *principal polarization states model* (PSP). PSPs have the property that when launching into one of them at the input of the fiber, the output polarization state is invariant to first order with changes in wavelength. As a result, the PSP model describes what is known at *first-order PMD*. Second-order PMD can also come into play: when the signal bandwidth is large and each spectral component must be considered to have different PSPs.

The "rms" label in equation (2.14) reflects the statistical nature of PMD. Thermal and mechanical variations over time will affect the extent to which PMD accumulates or cancels along the length of the fiber, producing DGD values that are much larger than $\Delta\tau_{rms}$ for some time lapse. To ensure that the probability of link failure is low (e.g., 5 minutes per year), $\Delta\tau_{rms}$ must be specified to be between 10 and 15% of a bit duration (Agrawal, 2002). This condition gives

$$B\sqrt{L} < \frac{1}{10D_{PMD}}$$
(2.15)

At $B = 40\,Gb/s$, the fiber length is limited to roughly $600\,km$ with $D_{PMD} = 0.1\,ps/km^{1/2}$, or only $6\,km$ with older fibers having $D_{PMD} = 1\,ps/km^{1/2}$. This example shows the importance of the fiber PMD coefficient for high-bit-rate systems.

PMD was recognized to be a critical parameter for fiber manufacturers around 1995, and PMD coefficients began to be included in fiber specifications. There is, however, a wide range of PMD values in installed fiber, which means that characterization is required to determine maximum bit rates on any particular fiber span. In a study by Deutsche Telekom in 2003 of their installed fibers, 98.5% were suitable for $2.5\,Gb/s$ to $1000\,km$, 86% were suitable for $10\,Gb/s$ to $1000\,km$, and 60% were suitable for $40\,Gb/s$ to $600\,km$ (Breuer et al., 2003). In the same study, the average PMD was $0.32\,ps/km^{1/2}$ for fibers installed between 1985 and 1991, $0.13\,ps/km^{1/2}$ for fibers installed between 1992 and 1998, and $0.05\,ps/km^{1/2}$ for fibers installed between 1999 and 2001. Thus, with careful selection of existing installed fibers for 10- and 40-Gb/s links, PMD need not pose limitations in many instances.

Finally, in this section we compare the characteristics of chromatic dispersion and polarization mode dispersion. Table 2.5 summarizes the mechanisms and scaling factors for these two types of dispersion. In the next section we discuss how dispersion effects can be compensated.

TABLE 2.5   Comparison Between Chromatic and Polarization Mode Dispersion

Feature	Chromatic Dispersion	Polarization Mode Dispersion
Origin	Wavelength dependence of refractive index and relative waveguide size	Imperfect circular symmetry of waveguide core and stresses
Dependence on transmission speed	Square: 16 times worse at 10 Gb/s compared to 2.5 Gb/s	Linear: 4 times worse at 10 Gb/s than at 2.5 Gb/s
Dependence on distance	Linear: doubles when distance doubles	Square root: increases by 40% when distance doubles
Other influencing factors	Spectral width of light source, chirp, modulation format	Fiber age, statistical behavior with fluctuations in temperature, bends, and stresses

Polarization Mode Dispersion Compensation There are several approaches
to PMD compensation in the electrical and optical domains. As mentioned
above, electrical compensators are attractive for low-cost simultaneous compensation of various impairments, including both CD and PMD. One aspect
of PMD that lends itself to electrical compensation is that because the two
PSPs are orthogonal in polarization, their relative optical powers are linearly
reproduced in the electrical domain despite using a square-law detector
(Kaminow and Li, 2002b, Chap. 7).

Optical PMD compensators can be classified into three main types:
principal-state-of-polarization (PSP) transmission, PMD nulling, and fixed
DGD (Kaminow and Li, 2002b, Chap. 7). In the PSP transmission method, the
polarization state at the input of the fiber is adjusted to match one of the fiber's
PSPs. In this way, a pulse emerging from the fiber is composed of only one of
the PSPs and so is not spread out in time. One of the challenges with this
approach is that feedback is required from the receiver to the transmitter to
adjust the polarization state transmitted as the orientation of the PSPs change
over time. The PMD nulling and fixed DGD methods do not require limiting
the pulses to traveling in only one of the PSPs. Instead, they try to correct for
the PSPs' time delay at the output by introducing elements with opposite
DGD. Although feasible for single-channel operation, these approaches are
very complicated when trying to accommodate multiple channels since each
channel has different PSPs.

The area of PMD compensation has received considerable interest, with
system upgrades from 2.5 Gb/s to 10 and 40 Gb/s (Kaminow and Li, 2002a,
Chap. 7). Although PMD is less limiting than chromatic dispersion, its effects
cannot be ignored when deploying long-haul links, particularly at 40 Gb/s.
Ideally, a low-cost PMD compensator could be added to each link that could
accommodate the wide variability in PMD of installed fiber, as well as to deal
with the statistical variability of PMD over time. Technical and economic challenges have so far limited the deployment of PMD compensators in commercial systems (Rosenfeldt, 2005), leaving the use of low-PMD fiber as the
preferred mitigation approach. In one study on 10- and 40-Gb/s systems, PMD
compensators were found to be cost-effective only at 40 Gb/s over fibers with
medium levels of PMD (deployed between 1992 and 1998) (Eiselt, 2004). It is
also anticipated that PMD compensators may be required for ultralong-haul
systems at 40 Gb/s (Lanne and Corbel, 2004).

2.6.4 Nonlinear Effects

Although optical power levels used in fiber communication systems are small
in terms of absolute power, typically milliwatts, the power density can be
extremely large as a result of the small core size. A power of 10 mW in a
typical single-mode fiber, with an 80-μm^2 effective area, corresponds to an
intensity of 12.5 kW/cm^2! At such high intensity levels, it is perhaps not surprising that the fiber properties are influenced by the light itself. Furthermore,
fiber links are amazingly long compared to the optical wavelength, ~10^9 wave-

lengths/km, which means that small perturbations to the fiber can accumulate into large changes in optical phase. In this section we look at two types of nonlinear behavior: nonlinear phase modulation and stimulated scattering processes.

Nonlinear Phase Modulation   Although silica is a highly linear material at optical frequencies, it exhibits an optical Kerr effect that cannot be neglected at high powers or over long distances. The optical Kerr effect, originating from third-order susceptibility (Toulouse, 2005), gives rise to a refractive index that is dependent on the optical intensity according to

$$n = n_0 + n_2 P/A_{\text{eff}} \tag{2.16}$$

where n_0 is the refractive index at low powers, n_2 the nonlinear index coefficient, P the optical power, and A_{eff} the effective area of the fiber. A_{eff} is similar in value to the actual area of the core, however, because the optical fields are not distributed uniformly inside the core, the two quantities differ. A_{eff} is quoted by fiber manufacturers, but it can also be calculated directly from the mode profile (Agrawal, 2002). Even though n_2 is extremely small, $\sim2.6\times10^{-20}\,\text{m}^2/$ W for silica fibers (Agrawal, 2002), its effect can become appreciable over long fiber spans, particularly when cascaded optical amplifiers are used. Because the nonlinear refractive index is a function of intensity (P/A_{eff}), the resulting effects can be reduced by using large effective area fibers.

Self-Phase Modulation   Since the optical path length is a function of the refractive index, the effect of n_2 on a single optical carrier is to modify the phase in proportion to the power. This process is called *self-phase modulation* (SPM). The amount of phase change after a distance L can be calculated from the nonlinear Schrödinger equation and is given by (Agrawal, 2004b)

$$\phi_{\text{SPM}} = \gamma P_0 L_{\text{eff}} \tag{2.17}$$

where P_0 is the optical power at the input to the fiber, L_{eff} is the *effective length* of a lossy fiber, given by

$$L_{\text{eff}} = \frac{1 - \exp(-\alpha L)}{\alpha} \tag{2.18}$$

and γ is the is the fiber nonlinear coefficient, given by

$$\gamma = \frac{n_2 \omega}{c A_{\text{eff}}} \tag{2.19}$$

where ω is the optical carrier frequency.

In amplitude-modulated systems, the optical power varies with time, and hence the amount of SPM becomes a function of the envelope of the carrier, or the pulse shapes for digital modulation. At the middle of the pulse, where the power is largest, the maximum amount of SPM phase change occurs, with progressively less phase change toward the edges of the pulse. Since the pulse is typically not square, especially at higher bit rates, the SPM phase change varies across the pulse, causing a frequency chirp:

$$\Delta \omega = \frac{d\phi_{SPM}}{dt} = -\gamma \frac{dP_0}{dt} L_{eff}.$$ (2.20)

The sign of the chirp is such that the rising edge of the pulse (positive dP_0/dt) acquires a decrease in optical frequency, or is red-shifted, whereas the trailing edge becomes blue-shifted. Although the spectrum of a pulse train is modified by SPM, in the absence of dispersion the changes appear in the time domain only as changes to the phase of the carrier, leaving the pulse envelope unaltered. When SPM occurs in a dispersive fiber, the interaction of the two effects is quite complicated. If the chromatic dispersion is in the *normal regime* ($D < 0$), the red-shifted leading edge of a pulse travels faster than the center frequency, and the blue-shifted trailing edge travels slower than the center frequency, stretching the pulse out in time. In the *anomalous regime* ($D > 0$) typical of SMF at 1550 nm, the two effects counteract each other and the pulse can be compressed for some distance along the fiber. As a result, in a dispersion-compensated link, the optimum transmission performance occurs when the dispersion is not completely compensated (Kaminow and Koch, 1997, Chap. 8). A special case of dispersion–SPM interaction is *soliton propagation*, where under certain conditions, a particular RZ pulse shape can travel without distortion. A review of solitons and their application to fiber optic communications may be found in a book by Agrawal (2002, Chap. 9).

Cross-Phase Modulation   An implication of nonlinear propagation is that we can no longer rely on linear superposition to analyze individual channels in isolation. Two effects that become important in multichannel systems are cross-phase modulation and four-wave mixing. *Cross-phase modulation* (XPM) originates from the same mechanism that gives rise to SPM, third-order non-linear electric susceptibility. It is possible to modify equation (2.17) to include XPM for the jth channel, to give (Agrawal, 2004b, Chap. 4)

$$\phi_j = \gamma L_{eff}\left(P_j + 2\sum_{k \neq j} P_k \right)$$ (2.21)

where the first term accounts for SPM in the jth channel and the second term includes the XPM contributions from all other channels, k. Equation (2.21) is valid in the absence of dispersion but can nonetheless be used to determine if XPM will be significant at given powers and effective fiber lengths. Equation (2.21) assumes that all channels are copolarized; in comparison, cross-polarized channels experience a three-fold reduction in XPM (Agrawal, 2004, Chap. 4). Since XPM accumulates with propagation distance, the interaction of two pulses depends on the overlap of each pulse as they propagate. For closely spaced channels, with similar group velocities, two pulses in adjacent channels can travel together for long distances, allowing the XPM to accumulate, while for more separated channels, the pulses will not overlap for sufficient distance to accumulate a significant phase change. Nonzero group velocity dispersion (CD) is therefore a key factor in reducing XPM effects.

Four-Wave Mixing   Finally, we discuss *four-wave mixing* (FWM), which again is a consequence of the third-order susceptibility of the fiber. When three optical frequencies are present in a fiber, the nonlinearity results in a beating of the optical fields to produce new spectral components at frequencies satisfying $\omega_4 = \omega_1 \pm \omega_2 \pm \omega_3$ (Yariv and Yeh, 1984). Dispersion plays a key role in determining how efficient the FWM process is since phase matching of the contributing fields is required for power to build up at ω_4 along the length of the fiber. Since the frequencies closest together will travel at the most similar group velocities, FWM is particularly prominent when $\omega_1 = \omega_2$. In a WDM system where $\omega_3 = \omega_1 + \Delta\omega$ is the neighboring channel to ω_1, the new frequency given by $\omega_4 = 2\omega_1 - \omega_3 = \omega_1 - \Delta\omega$ appears in the neighboring channel below ω_1. The result is crosstalk between channels, which can produce ghost pulses and transmission errors. Fortunately, FWM can be suppressed by employing fibers that have high local dispersion to disrupt phase matching, combined with periodic dispersion compensation to limit pulse broadening.

Stimulated Scattering   Nonlinear scattering processes have important consequences for fiber-optic transmission. In this section we discuss stimulated Brillouin scattering and stimulated Raman scattering, both of which result from inelastic scattering processes, in contrast to the elastic process involved in Rayleigh scattering.

Stimulated Brillouin Scattering   As light propagates along a fiber, random acoustic noise in the fiber produces density and hence refractive index variations that scatter some of the light in random directions. This process is known as *Brillouin scattering*. In an optical fiber, some of the scattered light will propagate backward along the fiber toward the light source. This backward wave will interfere with the forward wave to produce a beating pattern of high- and low-electric-field strength regions along the fiber. The beating pattern causes a mechanical deformation as a result of electrostriction (Agrawal, 2002), which in turn causes a periodic refractive index variation. The refractive index variation in turn behaves as a Bragg grating and causes more light to be reflected backward, further strengthening the refractive index variation. This positive feedback process is known as *stimulated Brillouin scattering* (SBS).

The frequency of the reflected wave (called the *Stokes wave*) from SBS is shifted with respect to the frequency as the incident light. The grating caused by the electrostriction, being mechanical in origin, propagates at the velocity of sound in the fiber (~6 km/s) in the direction of the forward light. The motion of the grating produces a Doppler shift in the reflected light that is down-shifted by approximately 11 GHz for 1550-nm light (Agrawal, 2002). This change in frequency satisfies the phase-matching criteria such that the beating pattern of the forward and backward waves travels at exactly the acoustic velocity, allowing the strength of the grating to grow. Another defining characteristic of SBS is the width of its gain spectrum, which is a relatively narrow value of ~20 MHz. The finite spectral width is the result of acoustic damping, which limits the length of the grating (Kaminow and Koch, 1997, Chap. 8).

SBS imposes some practical limits on how much optical power can be transmitted through a fiber. Solution of the coupled differential equations governing the growth of the Stokes wave and the corresponding decay of the forward wave produces an equation for the threshold power for SBS as (Agrawal, 2002)

$$P_{th} \approx \frac{21 A_{eff}}{g_b L_{eff}} \tag{2.22}$$

where g_b is the Brillouin gain and is approximately 5×10^{-11} m/W. For a typical effective fiber length of 20 km (determined by the fiber loss for long fibers) and an effective fiber area of 80 μm², the threshold power is ~2 mW. Above the SBS threshold, the power reflected grows rapidly while transmitted power saturates. In practice, the SBS threshold increases by roughly a factor of 2 when data are transmitted (Agrawal, 2004b) due to the distribution of the signal's power in the frequency domain. Nonetheless, for single-channel systems, the SBS imposes severe limit on launch power unless line width broadening techniques are used (Kaminow and Koch, 1997, Chap. 8). Fortunately for WDM systems where individual channel powers are relatively low, the channel spacings are sufficiently far apart that the SBS threshold applies to each channel individually, making SBS a relatively insignificant problem.

Stimulated Raman Scattering We have already seen that stimulated Raman scattering (SRS) provides a mechanism for optical amplification. Unfortunately, SRS can also be detrimental in WDM systems. The shorter-wavelength channels behave as pump signals that give up their power to amplify the longer-wavelength channels. SRS will create crosstalk in channels that are separated up to 120 nm, with the crosstalk increasing almost linearly with channel separation as a result of the SRS gain spectrum. As with other crosstalk effects, chromatic dispersion can mitigate the effects of SRS crosstalk by reducing the amount of time that two bits in different channels will spend traveling together. Power equalization filters can also be used to counteract the loss of power of the shorter-wavelength channels. See the book by Kaminow and Koch (1997, Chap. 8) for an analysis of SNR degradation caused by SRS.

KEY POINTS

Fiber-optic transmission:

- Optical fiber is a much better medium than electrical wires for transmitting data over long distances, due to its higher bandwidth and lower attenuation levels.
- Long-haul networks use the 1550-nm wavelength window (called the C-band), due to the availability of EDFAs and corresponding low attenuation value of 0.2 dB/km.
- Multimode fiber has a larger core than single-mode fiber, which allows the propagation of multiple modes.
- MMF is used in very short-reach applications, such as local area networks (LANs) and storage area networks, because of its low cost and easy fiber coupling.
- Single-mode fiber has a very small core that allows for propagation of a single mode. SMF does not suffer from multimode dispersion, and as such is used for high-bit-rate and long-distance applications.

Light sources and detectors:

- Light-emitting diodes are the simplest and lowest-cost light source used for fiber optic communications. They have the broadest spectrum of the sources used, which limits the distance due to fiber dispersion, and have relatively slow transient behavior.
- Laser sources include edge-emitting structures such as Fabry–Perot lasers and distributed feedback lasers, as well as vertical-cavity surface-emitting lasers (VCSELs).
- Fabry–Perot lasers are used for low bit rates and short- to medium-reach applications because of their broad optical spectrum.
- Distributed feedback lasers have narrow spectrums and as such are used for high bit rates in wave-division-multiplexing systems and long-haul networks.
- VCSELs are used in LANs as they operate primarily in the 850-nm wavelength window, which has higher fiber transmission loss.
- Light receivers use photodiodes to convert the envelope of the optical signal into an electrical baseband signal. PIN (p-type/intrinsic/n-type) diodes are used when low sensitivity is sufficient; APDs (avalanche photodiodes) are used when received powers are low.

Optical modulation:

- Lasers are directly modulated low-bit-rate or short-reach applications. Laser chirp combined with fiber dispersion is the limiting impairment.

- External modulators are used for high-bit-rate and long-reach applications.
- Electroabsorption modulators change the absorption in a semiconductor to modulate light. They are very compact, with low drive voltage, but have high insertion loss and wavelength sensitivity.
- Electrooptic modulators (EOMs) change the refractive index in a crystal, typically lithium niobate, in response to an applied voltage. A Mach–Zehnder interferometer is typically used to convert phase modulation into amplitude modulation. EOMs have low insertion loss and are designed with either zero or low negative chirp, making them the choice for long-haul applications.

Optical amplification:

- There are two main types of optical amplifiers that are used: erbium-doped fiber amplifiers (EDFAs) and Raman amplifiers.
  - EDFAs are used widely, require very little pump power, and provide high gains across the C- and L-bands.
  - Raman amplifiers require large pump powers but offer an adjustable gain spectrum and distributed amplification over installed fibers, leading to low noise figures.

Fiber transmission impairments:

- There are three fiber dispersion phenomena that limit transmission bit rate and distance: modal dispersion, chromatic dispersion, and polarization mode dispersion.
  - Modal dispersion limits only multimode fibers.
  - Chromatic dispersion is caused by the wavelength-dependent group velocity, which has a material component and a waveguide component.
  - Polarization mode dispersion is caused by the various group velocities for orthogonal polarizations of light when traveling in noncircular symmetric fibers.
- Nonlinear effects (SPM, XPM, FWM) can cause distortion in single channels as well as crosstalk in multichannel systems.
- Chromatic dispersion is important for suppressing nonlinear impairments.

Acknowledgments

We would like to thank Jeff Bull, Hiroshi Kato, and Benjamin Tsou, all of Versawave Technologies Inc., for writing most of this chapter.

REFERENCES

Agrawal, G. P., *Fiber-Optic Communication Systems*, 3rd ed., Wiley, Hoboken, NJ, 2002.

Agrawal, G. P., *Lightwave Technology: Components and Devices*, Wiley, Hoboken, NJ, 2004a.

Agrawal, G. P., *Lightwave Technology: Telecommunication Systems*, Wiley, Hoboken, NJ, 2004b.

Agrawal, G. P., and N. K. Datta, *Semiconductor Lasers*, 2nd ed., Van Nostrand Reinhold, New York, 1993.

Akiyama, S., H. Itoh, T. Takeuchi, A. Kuramata, and T. Yamamoto, Wide-wavelength-band (30 nm) 10-Gb/s operation of InP-based Mach–Zehnder modulator with constant driving voltage of $2V_{pp}$, *IEEE Photonics Technology Letters*, vol. 17, no. 7, pp. 1408–1410, 2005.

Becker, P. C., N. A. Olsson, and J. R. Simpson, *Erbium-Doped Fiber Amplifiers*, Academic Press, San Diego, CA, 1999.

Breuer, D., H.-J. Tessmann, A. Gladisch, H. M. Foisel, G. Neumann, H. Reiner, and H. Cremer, Measurements of PMD in the installed fiber plant of Deutsche Telekom, *Digest of the LEOS Summer Topical Meetings*, July 14–16, 2003.

Bromage, J., Raman amplifier for fiber communication systems, *Journal of Lightwave Technology*, vol. 22, no. 1, pp. 70–93, 2004.

Buchali, F., Electronic dispersion compensation for enhanced optical transmission, paper OWR5, presented at OFC 2006, March 2006.

Bull, J. D., N. A. F. Jaeger, H. Kato, M. Fairburn, A. Reid, and P. Ghanipour, 40 GHz electro-optic polarization modulator for fiber optic communications systems, *Proceedings of the SPIE*, vol. 5577, pp. 133–143, 2004.

Chandrasekhar, S., and A. H. Gnauck, Performance of MLSE receiver in a dispersion-managed multispan experiment at 10.7 Gb/s under nonlinear transmission, *Photonics Technology Letters*, vol. 18, no. 23, pp. 2448–2450, 2006.

Chang, W. S. C., *RF Photonic Technology in Optical Fiber Links*, Cambridge University Press, New York, 2002.

Charlet, G., Progress in optical modulation formats for high-bit-rate WDM transmissions, *IEEE Journal of Selected Topics in Quantum Electronics*, vol. 12, no. 4, pp. 469–483, 2006.

Coldren, L. A., and S. W. Corzine, *Diode Lasers and Photonic Integrated Circuits*, Wiley, Hoboken, NJ, 1995.

Dagli, N., Wide-bandwidth lasers and modulators for RF photonics, *IEEE Transactions on Microwave Theory and Techniques*, vol. 47, no. 7, 1999.

Desurvire, E., *Erbium-Doped Fiber Amplifiers: Principles and Applications*, Wiley, Hoboken, NJ, 1994.

Donaldson, A., Candidate materials and technologies for integrated optics: fast and efficient electro-optic modulation, *Journal of Physics D: Applied Physics*, vol. 24, no. 6, pp. 785–802, 1991.

Dutta, A. K., N. K. Dutta, and M. Fujiwara, *WDM Technologies: Active Optical Components*, Academic Press, San Diego, CA, 2002.

Eiselt, M. H., PMD compensation: a system perspective, presented at OFC 2004, February 23–27, 2004.

Elbers, J.-P., H. Wernz, H. Griesser, C. Glingener, A. Faerbert, S. Langenbach, N. Stojanovic, C. Dorschky, T. Kupfer, and C. Schuliea, Measurement of the dispersion tolerance of optical duobinary with an MLSE-receiver at 10.7 Gb/s, presented at OFC 2005, March 2005.

Fukuda, M., *Semiconductor Optical Devices*, Wiley, Hoboken, NJ, 1999.

Haunstein, H. F., W. Sauer-Greff, A. Dittrich, K. Sticht, and R. Urbansky, Principles for electronic equalization of polarization-mode dispersion, *Journal of Lightwave Technology*, vol. 22, no. 4, pp. 1169–1182, 2004.

Headley, C., and G. P. Agrawal, Eds., *Raman Amplification in Fiber Optical Communication Systems*, Academic Press, San Diego, CA, 2005.

Hill, K. O., and G. Meltz, Fiber Bragg grating technology fundamentals and overview, *Journal of Lightwave Technology*, vol. 15, no. 8, pp. 1263–1276, 1997.

Ido, T., S. Tanaka, M. Suzuki, M. Koizumi, H. Sano, and H. Inoue, Ultra-high-speed multiple-quantum-well electro-absorption optical modulators with integrated waveguides, *Journal of Lightwave Technology*, vol. 14, no. 9, pp. 2026–2034, 1996.

IEEE, *IEEE Standard for Information Technology*, Part 3, Carrier sense multiple access with collision detection (CSMA/CD) access method and physical layer specification, Std. 802.3-2005, IEEE, Press, Piseataway, NJ, 2005.

Islam, M. N., Raman amplifiers for telecommunications, *IEEE Journal of Selected Topics in Quantum Electronics*, vol. 8, no. 3, pp. 548–559, 2002.

Islam, M. N., *Raman Amplifiers for Telecommunications*, Vols. 1 and 2, Springer-Verlag, New York, 2003.

Kaminow, I. P., and T. L. Koch, Eds., *Optical Fiber Telecommunications*, Vol. IIIA, Academic Press, San Diego, CA, 1997.

Kaminow, I. P., and T. Li, Eds., *Optical Fiber Telecommunications*, Vol. IVA, *Components*, Academic Press, San Diego, CA, 2002a.

Kaminow, I. P., and T. Li, Eds., *Optical Fiber Telecommunications*, Vol. IVB, *Systems and Impairments*, Academic Press, San Diego, CA, 2002b.

Kazovsky, L., S. Benedetto, and A. Willner, *Optical Fiber Communication Systems*, Artech House, Norwood, MA, 1996.

Lachance, R. L., S. Lelièvre, and Y. Painchaud, 50 and 100 GHz multi-channel tunable chromatic dispersion slope compensator, presented at OFC 2003, March 2003.

Lanne, S., and E. Corbel, Practical considerations for optical polarization-mode dispersion compensators, *Journal of Lightwave Technology*, vol. 22, no. 4, pp. 1033–1040, 2004.

Liu, M. K., *Principles and Applications of Optical Communications*, McGraw-Hill, New York, 1996.

Miya, T., Y. Terunuma, T. Hosaka, and T. Miyashita, Ultimate low-loss single-mode fibre at 1.55 μm, *Electronics Letters*, vol. 15, pp. 106–108, 1979.

Nye, J. F., *Physical Properties of Crystals*, Oxford University Press, New York, 1985.

Pal, B. P., Ed., *Guided Wave Optical Components and Devices*, Academic Press, San Diego, CA, 2005.

Pankove, J. I., *Optical Processes in Semiconductors*, Dover, New York, 1971.

Paul, D. K., K. H. Greene, and G. A. Koepf, Undersea fiber optic cable communications system of the future: operational reliability, and systems considerations, *Journal of Lightwave Technology*, vol. 2, no. 4, pp. 414–425, 1984.

Peters, M. G., B. J. Thibeault, D. B. Young, J. W. Scott, F. H. Peters, A. C. Gossard, and L. A. Coldren, Band-gap engineered digital alloy interfaces for lower resistance vertical-cavity surface-emitting lasers, *Applied Physics Letters*, vol. 63, no. 25, pp. 3411–3413, 1993.

Pulfrey, D. L., and N. G. Tarr, *Introduction to Microelectronic Devices*, Prentice Hall, Upper Saddle River, NJ, 1989.

Rahmantian, F., N. A. F. Jaeger, R. James, and E. Berolo, An ultrahigh-speed AlGaAs–GaAs polarization converter using slow-wave coplanar electrodes, *IEEE Photonics Technology Letters*, vol. 10, no. 5, pp. 675–677, 1998.

Ramo, S., J. R. Whinnery, and T. Van Duzer, *Fields and Waves In Communication Electronics*, 3rd ed., Wiley, Hoboken, NJ, 1994.

Rosenfeldt, H., Deploying optical PMD compensators, presented at OFC 2005, March 2005.

Sackinger, E., *Broadband Circuits for Optical Fiber Communications*, Wiley, Hoboken, NJ, 2005.

Shen, T. M., and G. P. Agrawal, Pulse-shape effects on frequency chirping in single-frequency semiconductor lasers under current modulation, *Journal of Lightwave Technology*, vol. 4, no. 5, pp. 497–503, 1986.

Snyder, A. W., and J. D. Love, *Optical Waveguide Theory*, Kluwer Academic Publishers, Boston, MA, 1983.

Stolen, R., J. P. Gordon, W. J. Tomlinson, and H. A. Haus, Raman response function of silica-core fibers, *Journal of the Optical Society of America* B, vol. 6, no. 6, pp. 1159–1166, 1989.

Sze, S. M., *Physics of Semiconductor Devices*, Wiley, New York, 1981.

Toulouse, J., Optical nonlinearities in fibers: review, recent examples, and systems applications, *Journal of Lightwave Technology*, vol. 23, no. 11, pp. 3625–3641, 2005.

Walker, R. G., High-speed III–V semiconductor intensity modulators, *IEEE Journal of Quantum Electronics*, vol. 27, no. 3, pp. 654–667, 1991.

Winters, J. H., and R. Gitlin, Electrical signal processing techniques in long-haul fiber-optic systems, *IEEE Transactions on Communications*, vol. 38, pp. 1439–1453, 1990.

Wooten, E. L., K. M., Kissa, A., Yi-Yan, E. J., Murphy, D. A., Lafaw, P. F., Hallemeier, D., Maack, D. V., Attanasio, D. J., Fritz, G. J., McBrien, and D. E., Bossi, A review of lithium niobate modulators for fiber-optic communications systems, *IEEE Journal of Selected Topics in Quantum Electronics*, vol. 6, no. 1, pp. 69–82, 2000.

Yariv, A., *Optical Electronics in Modern Communications*, 5th ed., Oxford University Press, New York, 1997.

Yariv, A., and P. Yeh, *Optical Waves in Crystals: Propagation and Control of Laser Radiation*, Wiley, New York, 1984.

Yu, S. F., *Analysis and Design of Vertical Cavity Surface Emitting Lasers*. Wiley, Hoboken, NJ, 2003.

Zory, P. S., Jr., *Quantum Well Lasers*, Academic Press, San Diego, CA, 1993.

3

WAVELENGTH-DIVISION MULTIPLEXING

3.1 INTRODUCTION

The goal of this chapter is to introduce a powerful optical networking technology, wavelength-division multiplexing (WDM). The basic principle of WDM is simple and was explained briefly in Chapter 1. WDM is a technology that maker it possible to send multiple optical signals, each on a different wavelength, along a single piece of optical fiber. This property caused WDM

Network Infrastructure and Architecture: Designing High-Availability Networks,
By Krzysztof Iniewski, Carl McCrosky, and Daniel Minoli
Copyright © 2008 John Wiley & Sons, Inc.

technology to revolutionize optical networking when it was introduced in the 1990s.

This chapter is structured as follows. First, we discuss the WDM principle of operation, followed by calculations of optical fiber capacity using WDM. Two classes of WDM technology are introduced: dense WDM (DWDM) and coarse WDM (CWDM). DWDM uses several optical wavelengths, requires tight control of optical properties but results in powerful bandwidth capacity. CWDM uses only a few channels. This means that it is much easier to build but has less bandwidth capacity than DWDM.

In the second part of the chapter we discuss various pieces of optical networking hardware. WDM regenerators, optical cross-connects, and switches are described.

A key element of any network is its ability to switch data traffic quickly. There are, in principle, two different types of data switching: those in the optical and electrical domains. As the signals travel in optical fibers, optical switching seems more straightforward since electrical switching requires additional conversion of the optical signals into the electrical domain. In this section we contrast optical switching with electrical switching.

The hardware equipment section is followed by a discussion of WDM network issues. Dynamic provisions using tunable lasers, optical amplifications, wavelength blocking, and optical–electrical–optical conversions are illustrated. Included is a case study of a hypothetical long-haul link between New York and San Francisco constructed using WDM technology.

3.2   WDM TECHNOLOGY

3.2.1   WDM Basics

WDM is a technology that combines multiple wavelengths on a single-mode fiber. The ability to multiplex wavelengths is of critical importance for creating high-bandwidth optical pipes to transport data. Imagine that each wavelength is worth 10 Gb/s of data and that 100 wavelengths can be multiplexed on a single fiber. The total bandwidth of the data traveling in this fiber is 1 Tb/s.[†] At the present moment, this is more than sufficient to carry all Internet traffic from New York to London under the Atlantic Ocean.

Besides multiplication of the data rate in a single fiber, WDM systems have an advantage in that the number of channels can be adapted according to actual demand. Assume that in our example, only 50 Gb/s of capacity is needed today. The provider of the transmission system does not have to invest today in transmission capacity that may not be required until some years later, so today he can buy the equipment to serve five wavelengths at 10 Gb/s only. As

---

[†]During the 2007 Optical Fiber Conference, an Alcatel-Lucent research team demonstrated technologies to achieve 25-Tb/s fiber transmission over a three-span network of 240 km. The system uses Raman amplifiers and dispersion-compensation modules.

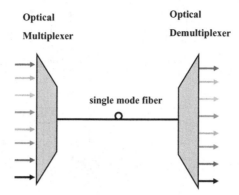

FIGURE 3.1 WDM principle of operation.

needs arise in the future, the enterprise can purchase additional hardware to serve more wavelengths. As this example indicates, telecommunications service providers that want to have the ability to expand their bandwidth quickly in the future are often interested in deploying WDM.

How is wave multiplexing done? In a WDM system the optical signals at different frequencies are launched into the inputs of a wavelength multiplexer (mux), as illustrated in Figure 3.1. At the output of the wavelength multiplexer all wavelengths are effectively combined and coupled into a single-mode fiber. At the end of the transmission link the optical channels are separated again by means of a wavelength demultiplexer (demux) and thus arrive at the various outputs. We should point out that optical mux and demux devices are really the same devices: the mux with its input and output ports reversed works as the demux, and vice versa.

Each WDM wavelength requires its own laser source and light detector. The particular choice of wavelength used is governed by standards established by the ITU. As a result, the WDM equipment from different vendors has a chance to interoperate. The krypton line at 193.10 THz was selected to be the reference line. The 100-GHz number was selected as a *channel spacing*, which is the term for the distance between two neighboring WDM lines. An example of a WDM signal is shown in Figure 3.2 for the case of 16 channels in a 1530.33- to 1553.86-nm optical window. Note that the wavelengths are 0.8 nm apart, which, converted to frequency domain, gives the 100-GHz number.

3.2.2 WDM Bandwidth Capacity

How much bandwidth can be harvested using WDM technology? Using standard telecommunications single-mode fiber, a wavelength range of 1280 to 1650 nm can be utilized. The reasons behind selection of these particular wavelength numbers was explained in Chapter 2. Simply put, below 1280 nm and

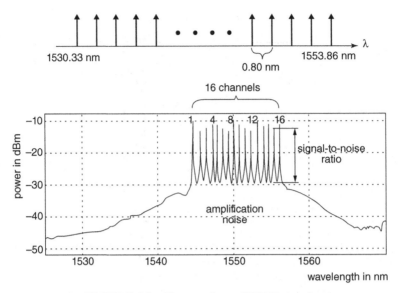

FIGURE 3.2   Sixteen-channel WDH signal.

above 1650 nm, attenuation levels in optical fibers are too high for long-haul transmission.

Let us convert 1280- and 1650-nm wavelength numbers to frequencies using the well-known relationship between frequency and wavelength:

$$\text{frequency} = \frac{\text{speed of light}}{\text{wavelength}}$$

Using this formula we obtain 235 and 182 THz, respectively, for the highest and lowest bounds of that range. Recall that a terahertz is 10^{12} Hz, 1000-fold greater than a gigahertz. To determine the amount of information that can be carried, we need to make an assumption about how many bits of data correspond to 1 hertz. The answer depends on the particular modulation scheme used, but for simplicity's sake we can assume that 1 hertz corresponds to 1 bit. Subtracting 182 from 235, we get 53 THz. This result is quite amazing, implying that potentially 53 Tb/s of bandwidth can be harvested from one strand of fiber. Contrast that numbers with 40 Gb/s, which at present is the amount of data that a single state-of-the-art electrical signal can carry. That is a three-order-of-magnitude difference!

Without WDM capability, optical fiber would be extremely unutilized. Assume in our example that only one 10-Gb/s signal can be send in a single fiber. Because 10 Gb/s out of 53 Tb/s is less than 0.02%, it is clear that there is highly unutilized potential. However, if 100 wavelengths are sent at

10 Gb/s using WDM, which is feasible today, the aggregate bandwidth becomes 1 Tb/s, which is a very impressive result. In fact, the largest transmission capacity has recently been demonstrated to be about 10 Tb/s, so we are not that far from our roughly calculated theoretical maximum of 53 Tb/s. In fact, complex quantum physics calculations indicate that the theoretical maximum is about 100 Tb/s. This indicates that our first-order approximation is pretty good.

By now it should be quite obvious that the transmission capacity of a single-mode fiber can be much better exploited via the simultaneous transmission of several wavelengths. However, there are two reasons that it is currently not possible to make unlimited use of the total wavelength bandwidth from 1280 to 1650 nm. The first problem is signal attenuation. At certain frequencies optical attenuation is much higher, due to the presence of certain impurities in the fiber. This is the water peak problem discussed in Chapter 2. The second problem is that of optical amplification. Widely used erbium-doped fiber amplifiers (EDFAs) operate efficiently in only a portion of the 1280- to 1650-nm window.

To overcome the problem of the water peaks, optical fiber manufacturers have developed a specialized fiber that does not have the levels of undesired impurities that caused the problem in the first place. This specialized fiber can be used across wide optical bands but is obviously more expensive to manufacture. The problem of limited wavelength range for EDFAs is currently being addressed by physicists. Their work has resulted in promising research on optical amplifiers working in different wavelength ranges than that of EDFAs. We can safely assume that with time both problems will eventually be solved and the entire optical wavelength range will be available for use for optical transmission.

3.2.3 Coarse vs. Dense WDM Systems

Depending on the number of wavelengths desired to be sent on a single fiber, WDM can be classified either as *coarse WDM*, for eight or fewer wavelengths, or *dense WDM*, for systems with eight or more wavelengths. Coarse WDM (CWDM) equipment is cheaper, as the spacing between wavelengths is large but has limited scalability potential. Dense WDM (DWDM) equipment is much more expensive to build, as precise and stable separation between wavelengths is required, but its capacity is scalable.

For DWDM systems, a universal ITU standard was developed so that each manufacturer would use the same wavelength grid. The reference line at 193.10 THz was considered to be a starting point, and a 100-GHz channel spacing was selected as typical; 100 GHz corresponds to an 0.8-nm wavelength difference and allows for transmission of at least 100 wavelengths. To get even larger numbers of wavelengths in ultradense systems, spacings of 50 and 25 GHz have been used in some systems. Although 50 or 25 GHz may sound

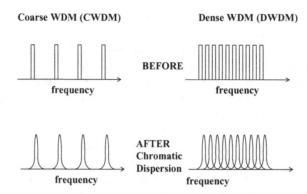

FIGURE 3.3 Chromatic dispersion effect in DWDM and CWDM systems.

large for those of us familiar with electronics, in the optical world these are very small numbers. Remember that a 25-GHz deviation for a 235-THz wavelength is about 0.01%, very high precision indeed.

DWDM transmission puts high demands on the components of the stabilized wavelength system and its parameters. These demands concern primarily the output wavelengths of the laser sources, as they must emit exactly at the center wavelength. Therefore, the laser diode for the respective channel needs to be selected individually. The preselected laser can then be fine-tuned to the exact center wavelength (e.g., by changing the chip temperature). To avoid interference from adjacent transmission channels, significant deviations from the center frequencies are not allowed. At a frequency grid of 100 GHz, the deviation allowed is about 20 GHz or 0.16 nm. That is why DWDM lasers have to be extremely stable in wavelength and must provide a small linewidth. In addition, the laser temperature must be stabilized, as the wavelength emission process is strongly temperature dependent, as discussed in Chapter 2.

CWDM lasers, on the other hand, are not subjected to the same stringent requirements. Wavelength spacing in CWDM systems is much larger, so laser stability and its small linewidth are not huge concerns. As a result, much cheaper components can be used to build cost-effective CWDM equipment. Compared to DWDM, CWDM is also much less sensitive to chromatic dispersion, as illustrated in Figure 3.3. Chromatic dispersion and other forms of optical signal impairments have been covered in Chapter 2.

3.2.4 Future Extensions of DWDM Capacity

Laying new fiber can be expensive. As a result, current trends aim to better utilize the intrinsic bandwidth of the single-mode fibers that have already been deployed. This can be achieved in three ways. One way is to decrease the channel spacing. Although 100-GHz spacing is recommended as the standard,

50- and 25-GHz systems have already become available. Recent experiments tried to reduce the channel spacing to as little as 12.5 GHz. However, very low wavelength spacing drastically increases the performance and stability demands for all components of the system.

Another way to increase WDM channel capacity is to open up to a larger wavelength range for transmission. This can be accomplished by using a special fiber that has eliminated the water peak problem and can operate in a wider optical window. Yet another alternative way to increase WDM channel capacity is to increase a signal transmission rate. Although 10 Gb/s is the standard maximum transmission rate today, a lot of work has been done for 40-Gb/s systems. Expensive electronics and large dispersion problems ensure that 40-Gb/s systems remain somewhat unattractive at this moment, but the situation might change in the next few years.

To satisfy increasing needs for higher bandwidth, the network has to evolve to increase its capacity. A comparison between the approaches to increasing network capacity that were discussed above can be summarized as follows:

1. *More Fiber.* In this scenario a telecommunication company keeps the same transmission rate and keeps using the same equipment but increases the number of fibers. If there is sufficient space in existing conduits to pull more fiber, this approach is manageable, although it might be expensive. However, if new conduits must be laid, it becomes very expensive.

2. *More Bandwidth.* In this scenario the telecommunications company keeps the same transmission rate but uses new optical equipment to utilize a wider optical bandwidth window. This approach works best if a new fiber without the water peak problem is already deployed; otherwise, high attenuation at certain frequencies have to be dealt with.

3. *More Wavelengths.* Currently, DWDM equipment uses 100-GHz channel spacing, resulting typically in 40 channels for the given optical band. Future equipment will use 25 GHz or smaller spacing, resulting in at least 160 channels. As a result, a fourfold increase in channel density should be possible, due to performance advances in optical components. Ultimately, systems will move outside the traditional C-band around 1550 nm and extend to L-band, or possibly operate at the combined C + L-band, which would increase the number of channels even further (beyond 200). In this scenario the carrier would need to upgrade its DWDM equipment.

4. *More Bits.* In this scenario the carrier increases the bit transmission rate through its fiber links. However, there are practical limitations on the distance over which the optical signal is transmitted, and as a result the maximum transmission distance gets shortened significantly in exchange for the higher transmission rate. For example, the chromatic dispersion limit is about 1000 km for 2.5 Gb/s but becomes just 60 km at 10 Gb/s. A similar limitation, although not as severe, exists due to the polarization-mode dispersion effect. Both

chromatic and polarization mode dispersion limits depend on the type of fiber being used, so the attractiveness of this scenario depends on the type of fiber infrastructure used in the particular network.

In practice, some or all of the solutions noted above might be combined to increase bandwidth. For example, if present DWDM equipment can operate at a rate of 10 Gb/s, it would offer 400-Gb/s capacity when 40 wavelengths are utilized. If in the future each channel operates at a rate of 40 Gb/s, the new solution might offer a 8-Tb/s bandwidth for 200 wavelengths.

Two possible scenarios are illustrated in Figure 3.4. Let us select 100 channels at 50-GHz spacing and a 10-Gb/s transmission rate as being state of the art. Using these assumptions, today's WDM capacity is about 1 Tb/s. Let us follow a horizontal path first to see how 10 Tb/s can be achieved. Assuming that the channel spacing can still be maintained at 50 GHz, by increasing a transmission rate from 10 Gb/s to 40 Gb/s, a fourfold increase in WDM capacity can be accomplished. A further increase to 10 Tb/s can happen if a wider optical transmission window is used and sophisticated modulation schemes are exploited that provide more than 1 bit per clock cycle. In practice, it might be easier to maintain a 10-Gb/s transmission rate and instead, to increase the number of channel wavelengths by decreasing the channel spacing to 25 GHz and eventually to 12.5 GHz. This scenario is shown on the diagonal path in Figure 3.4. Coupled with the use of a wider transmission window, the WDM capacity number would probably increase to 10 Tb/s in this case as well.

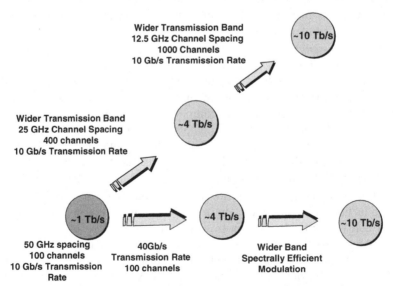

FIGURE 3.4 Hypothetical paths to increase WDM capacity.

3.3 NETWORKING EQUIPMENT FOR WDM

3.3.1 WDM Regenerators

While traveling along fiber, an optical signal undergoes undesired changes. First, its power, or magnitude, decreases, an effect called *signal attenuation.* Second, its shape changes and it becomes "fuzzy," an effect called *dispersion.* Both effects were discussed in Chapter 2. Depending on the parameters of the fiber, laser characteristics, and photodiode performance, either attenuation or dispersion becomes a limiting factor in optical transmission. If attenuation is a limiting factor, the signal needs to be amplified by means of optical amplification. If dispersion is a limiting factor, the signal needs to be dispersion compensated. To restore the quality of the signal, one can "clean up" the signal using optical or electrical components.

Optical amplification can be handled well by EDFAs and Raman amplifiers, both of which work in a reasonably broad optical spectrum, thus generally necessitating use of only one amplifier per fiber. This is very economical for WDM systems. If more than one optical amplifier is needed, optical-to-electrical and return electrical-to-optical (O-E-O) regeneration might be a better option.

A WDM O-E-O regenerator is shown in Figure 3.5. First, the optical signal is demultiplexed into a series of individual wavelengths. Each optical wavelength is subsequently converted from the optical to the electrical domain using a cascade of optical transceivers. We discuss optical transceivers in detail in Chapter 8, but for now it is sufficient to know that they convert optical signals into electrical ones, and vice versa. Once the signal is available in the electrical domain, all sort of electronic processing becomes available. However, in a simple WDM regenerator an electrical signal is immediately converted back to an individual wavelength. Finally, at the output, all wavelengths are multiplexed back into a combined WDM signal that can be sent on an individual fiber.

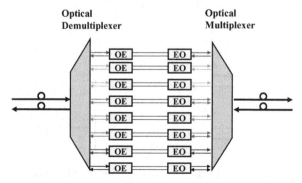

FIGURE 3.5 WDM regenerator with O-E-O conversion.

There are a few interesting observations that we can make at this point about the O-E-O WDM regenerator. First, the outgoing wavelength does not have to be of the same frequency as the incoming one. Wavelength conversion is free in this O-E-O conversion process, and as we will see later, that characteristic can prove beneficial in solving network problems. Second, once an electrical signal is available in the "middle" of the regeneration process, its contents can easily be modified by adding or subtracting certain bits. Third, the optical-to-electrical conversion process is required for each optical signal at both its input and output ports. This requirement can make WDM regenerators very expensive if a large number of optical channels are used, which is the case in DWDM systems. One might therefore conclude that DWDM regenerators with O-E-O conversion are used only if absolutely necessary: that is, where optical amplification alone would not be sufficient, due to dispersion problems.

3.3.2 Optical Cross-Connects and Switches

Simple Optical Cross-Connect   In WDM optical networks there might be a need to cross-connect optical signals at various points in the network. We use the term *cross-connect* here instead of *switch* to reserve the terms *switch* and *switching* for layer 2 (OSI model) operations. However, once the context has been made clear, we can use the terms *cross-connect* and *switch* interchangeably. The simplest optical cross-connect one can think of is shown in Figure 3.6. The device can accept *n* signals at its inputs and provide *n* outputs at the required port positions. There are many optical devices that are capable of *n* × *n switching*, the most common ones relying on optical mirrors, as discussed later. The simple conceptual device sketched in Figure 3.6 has no wavelength-conversion capabilities, and the lack of this feature can become a problem in real networks.

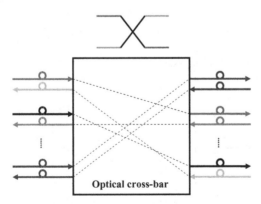

FIGURE 3.6   Optical *n* × *n* cross-connect with optical core.

Optical Switching Technologies　　Multiple technologies for optical switching in a crossbar fashion have been developed in research labs: liquid crystals, ink bubbles, holograms, or devices based on thermooptical or acoustooptical effects. In liquid crystals the electric current alters its properties in a manner that changes the polarization of the light passing through them. The technology has been "borrowed" from laptop screen technology but has not entered the commercial scene for telecommunication applications. Ink bubble technology has been borrowed from the ink jet printer. The surfaces of tiny bubbles formed by printer ink pens act like tiny mirrors, glancing light onto alternative paths. Again, no commercial deployment followed this research announcement.

Hologram technology uses ribbonlike holograms within crystals that are arranged in rows and columns, like a trellis. When charged electrically, the holograms can deflect specific colors selectively onto new paths, while the rest of the light wavelengths pass through unaffected. Finally, in thermooptical devices heating is being used to change the refractive index, which in turns affects the direction of the light beam.

Despite all these exciting developments, the most promising approach to manufacturing is based on a simple mechanical technology called *microelectronic mechanical systems* (MEMSs). Optical cross-connects in MEMSs are arrays of tiny mirrors which can be moved using either electrostatic or magnetic forces. Similar technologies have been used for years to manufacture airbag sensors and have been explored in labs for numerous other applications, but only fairly recently have they been applied to optical switching. The major advantage of MEMSs is their silicon-friendly processing, promising a highly automated and high-yielding MEMS manufacturing process in the future.

MEMS Technology for Optical Switching　　As stated above, among many optical switching technologies, MEMS-based mirrors seem to be the most promising technology for optical cross-connects (OXCs). The subcomponents of the OXC module are configured using MEMS mirrors, arrays of collimated optical fibers, and control electronics to position MEMS mirrors. Using simple mirrors that have the ability to move between two positions, as shown in Figure 3.7, or to rotate about an axis, can lead to unmanageably large arrays of $n \times n$ mirrors, limiting the switch size.

A more powerful three-dimensional architecture would require $2n$ mirrors, where each mirror accommodates n states as shown in Figure 3.8. Mirrors achieving n states require the ability to rotate on two orthogonal axes and usually are implemented using gimbal-type suspension and complex control electrodes (Greywall et al., 2003), as shown in Figure 3.9. To achieve proper positional accuracy of mirrors, electrostatic (Greywall et al., 2003; Yamamoto et al., 2003) and electromagnetic actuation (Taylor et al., 2003) have been proposed.

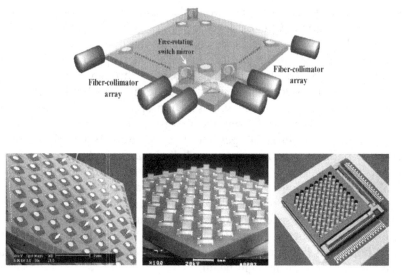

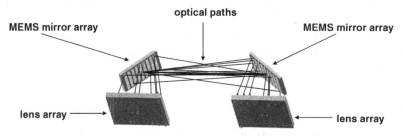

FIGURE 3.7 Two-dimensional switching architecture using an $n \times n$ mirror array. (From Chu et al., 2001.)

FIGURE 3.8 Three-dimensional switching architecture.

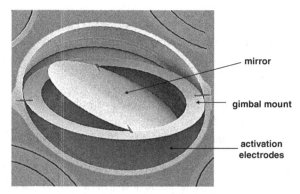

FIGURE 3.9 Gimbal-type suspension mirror for three-dimensional switching. (From Greywall et al., 2003, with permission © IEEE 2003.)

Optical Switches with Wavelength-Conversion Capability   To convert a simple cross-connect from Figure 3.6 into a switch with wavelength-conversion capability, an optical-to-electrical and return electrical-to-optical (O-E-O) conversion can be added as shown in Figure 3.10. Although the device shown has O-E-O conversion added at both inputs and outputs, in some cases conversion only at the output ports might be sufficient. What is interesting in this device is the fact that although the cross-connect is purely optical, all signals are available in electrical form at some point inside the device.

Although adding wavelength conversion improves the functionality of the device greatly, it comes at the price of more dissipated power, higher costs, and loss of transparency. Somewhat more efficient devices can usually be built using electrical cross-bar functions, as shown in Figure 3.11. This optical

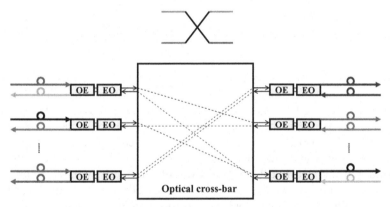

FIGURE 3.10 Optical $n \times n$ cross-connect with optical core and wavelength-conversion capability.

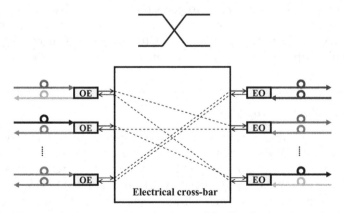

FIGURE 3.11 Optical $n \times n$ cross-connect with electrical core and wavelength-conversion capability.

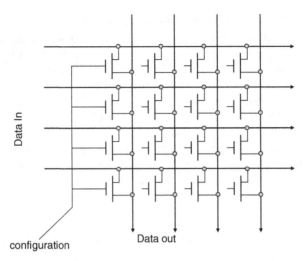

FIGURE 3.12   Electronic $n \times n$ cross-connect.

switch device has the capability to cross-connect wavelengths; change colors, if needed; restore optical signals completely; and even provide quick monitoring of passed data. For these reasons it is the most popular device deployed in real WDM networks today, although in many cases hybrid devices, which have both optical and electrical switching cores, are also deployed. One has to wonder at the fact that the optical switch from Figure 3.11 is actually using electronic means to switch, not optical! This tells you that one has to be careful reading marketing brochures, as product names can be quite misleading. It is always worthwhile to determine the underlying switching technology.

An example of a simple electronic cross-bar is shown in Figure 3.12, where control signals are applied to gates of transistors that connect rows to columns as desired. We describe VLSI transistor technology in more detail in Chapters 7 and 8. Let us now move to optical add–drop multiplexers that combine optical switching, electronic grooming, and WDM transport capabilities.

3.3.3   Optical Add–Drop Multiplexers

With small modifications at the conceptual level, but fairly major growth in a practical sense, the WDM regenerator from Figure 3.5 can be transformed into an optical add–drop multiplexer (OADM), shown in Figure 3.13. An OADM can also serve as a WDM terminal at network entry points. OADMs and WDM terminals are always needed in WDM systems in order to launch and terminate WDM signals. They have to provide the ability to add and drop electrical

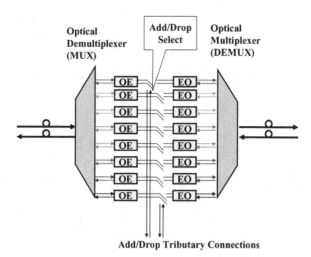

FIGURE 3.13 Optical add–drop multiplexer or WDM terminal with add–drop capability.

signals as required. Some of these network elements also have additional grooming capability to convert some low-bandwidth signals into one larger bandwidth signal in order not to waste the capacity of the WDM system. Note that from the WDM system capacity point of view, one channel is used completely whether it carries a 10-Gb/s electrical signal or only 1 Mb/s, so it is advantageous to groom electrical signals to the largest denominator prior to WDM transmission.

In the past, telecommunication companies have used OADMs with fixed functionality just to add or drop wavelengths at a node. Because these devices have generally used fixed-wavelength add–drop filters, they turn wavelength reconfiguration into a very extensive manual task, requiring a large effort to implement network changes if requested. Recently introduced *remotely reconfigurable OADMs* (ROADMs) solve that problem.

In ROADMs add–drop wavelength functionality can be programmed from a remote location. To change network configuration a service provider no longer has to send its service workers to perform the change manually. As a result, the great added value that ROADMs bring is their introduction of increased flexibility and speed by providing remote reconfigurability, simpler service provisioning, and faster fault detection. The benefits for telecommunication companies are obvious and include simplified, remote network management, fast service delivery, and significant savings in network operating costs. An example of a commercial product that can be classified as an ROADM is shown in Figure 3.14. Wavelengths in a main optical ring can be carried as 2.5-, 10-, or 40-Gb/s signals. The photonic switch can be used to cross-connect these

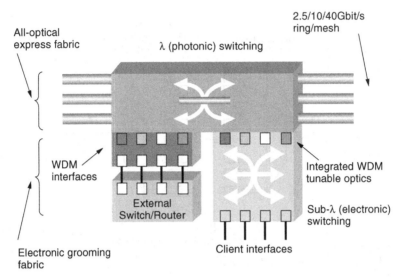

FIGURE 3.14 Reconfigurable optical add–drop multiplexer. (Courtesy of Ciena Inc.)

wavelengths as needed. Two smaller switches, one electrical and one optical, are used to add–drop tributary signals. To better understand how the cross-connect function can benefit optical networks, we need to have a closer look at WDM networks.

3.4 WDM NETWORKS

When building long-haul WDM networks, telecommunication companies have to answer many questions. What is the most cost-effective way to transport traffic across hundreds to thousands of kilometers? What is the most effective way to maintain signal strength over long distances and multiple nodes in the typical fiber network? What network elements should be employed to provide efficiency, reliability, and flexibility for their services? In this section we discuss briefly some these network aspects.

3.4.1 WDM Network Provisioning

Current DWDM networks are typically static since a unique laser is required for each wavelength. As a result, network provisioning is a slow, manual process that can easily take days or even weeks to set up. To make networks more efficient in provisioning, a better, dynamic way of setting up the service is required. Fortunately, devices that enable dynamic WDM networks do exist and are known as *tunable lasers*. Unlike their static cousins, tunable lasers have the ability to change the frequency at which they emit optical signals. Tunable lasers use a variety of structures and tuning methods. There are thermally tuned lasers, vertical-cavity surface-emitting lasers (VCSELs) with MEMS tuning, distributed Bragg reflector chips that are tuned by electrical current, and external cavity lasers that are tuned mechanically.

We have no intention of discussing the details of operation of tunable laser devices here; this is a very active R&D area that is bound to produce new devices and techniques in the very near future. If you would like to read more about tunable lasers, consult the references provided at the end of this chapter or Chapter 2 for basic information on lasers. At this point it is important to realize that tunable lasers provide the network with flexible, remote service provisioning which can be conveniently placed under software control. This flexibility leads inherently to cost savings by reducing laser inventory and producing new revenue-generating opportunities for rapidly provisioned services.

Tunable lasers can also be used in optical networks in various other applications beyond tuning WDM transmission wavelengths. They can be employed in transparent photonic switches that operate as all-optical bypasses without O-E-O conversion. Tunable lasers can be utilized in photonic layer control planes for automatic discovery of network topologies and to provide wavelength assignment on demand. Tunable lasers have become the "Holy Grail" for the next generation of WDM networks.

3.4.2 Wavelength Blocking

All optical networks have one common problem: wavelength blocking. Let us explain it quickly using an example. Consider network ABCDEFG, shown in Figure 3.15. Assume that node A is sending a signal to node C, while node G wants to send a signal to node D. Assume that both originating points A and F are unaware of each other and that they both decide to use exactly the same wavelength. Since A is already using that wavelenght at point B, that particular signal is blocked. Point F has no way of sending information to D in a given available connectivity.

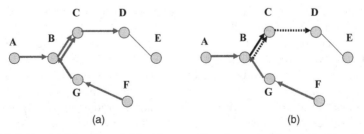

(a)   (b)

FIGURE 3.15 (a) Wavelength blocking problem between B and C; (b) solution by converting wavelength at point B (dashed lines).

If the equipment at node B has a wavelength-conversion capability, it can convert a signal to a different wavelength and still deliver both signals to node B as required, as indicated in Figure 3.15(b). However, if the equipment at node B does not have wavelength-conversion capability, network scheduling becomes more difficult. Obviously, in the example given, the node F could had started with a different wavelength in the first place, and that would have solved the problem. In practice, this wavelength selection might not be straightforward if hundreds of wavelengths and thousands of networking points are involved. Sometimes an alternative path around can also be found, but this cannot be guaranteed and is highly dependent on the level of network congestion. Wavelength conversion seems like the most robust solution.

The problem described above is called a *wavelength blocking problem*. Clearly, the wavelength-conversion ability is an important feature that helps in solving network engineering problems. For these reasons, optical switches with wavelength-conversion capability are deployed in much larger numbers than are purely optical cross-connects. The most robust solution for wavelength conversion is O-E-O conversion. Although O-E-O is discussed in detail in Chapter 8, we review briefly here the use of O-E-O in WDM networks.

3.4.3   O-E-O Conversion in WDM Networks

One of the important issues in WDM networks is signal degradation. As discussed in Chapter 2, optical signals suffer from attenuation and dispersion problems. Signal attenuation can be counteracted by optical signal amplification. There are two types of optical amplification: that used in EDFAs and that used in Raman amplifiers. Raman amplifiers show some advantages over EDFAs in high-performance WDM systems, as they enable ultralong reach, dynamic gain flattening, and automatic gain control.

However, no amount of signal amplification can resolve dispersion problems. Although there are some optical techniques for the reduction of signal dispersion, they are not yet cost-efficient. The most robust technique that solves attenuation and dispersion problems at the same time is O-E-O regeneration, which converts an incoming optical signal into an electrical signal, regenerates the electrical signal, converts it back to optical, and sends it on its way. Behind many vendors' claims of providing "all-optical" networks, you'll find O-E-O conversion at many points in their network architectures—with related latency, complexity, and costs.

Despite these shortcomings, O-E-O regeneration cannot be replaced by any other technology. Its provides complete regeneration of an optical signal, eliminates attenuation and dispersion signal degradation, and provides a means of wavelength conversion. As long as O-E-O transceivers continue to reduce their scale to higher frequencies and reduce the power dissipated, it will be difficult for any technology to replace them in WDM networks.

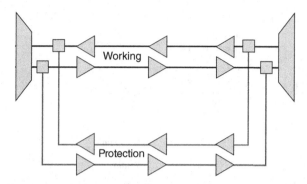

FIGURE 3.16   Working and protection two-fiber-pair configuration.

3.4.4   WDM Network Protection

One of the greatest WDM advantages, its huge bandwidth capacity, is also one of its greatest problems. Recall that you can send as much as 1 Tb of data per second over a single strand of single-mode fiber. What if that fiber is cut and becomes unavailable for 1 hour? Well, we have just lost 1 Tb of data, or 1 billion 1 k-long bit packets. Potentially, millions of customers would be affected, causing a huge networking problem. WDM data traffic needs to be protected, but this might be more difficult than you think, as WDM technology is mostly deployed as point-to-point links. Running the traffic in two fibers in the same physical location is not going to help much since during the fiber cut, it is likely that both strands will be cut simultaneously.

A possible solution involving the use of two pairs of fibers, located in different physical space, is shown in Figure 3.16. The first pair is called a *working pair* and is used unless the fiber is cut, in which case the second, *protecting pair* takes over. Needless to say, one needs a control plane mechanism to tell the network to switch from the working to the protecting pair. This mechanism is not easily implementable in WDM networks, as they do not "see" control signals as naturally as do SONET networks, discussed in Chapter 4. As a result, WDM network service protection and restoration is a difficult challenge. Smart control plane schemes and more efficient network configuration, such as mesh, do help. This subject is discussed in more detail in Chapter 12.

3.5   CASE STUDY: WDM LINK DESIGN

A typical configuration of a point-to-point WDM system is comprised of the following:

- Multiple optical transmitters
- An optical multiplexer

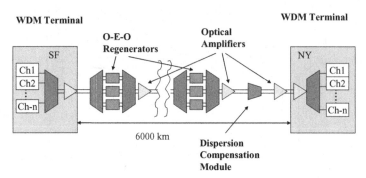

FIGURE 3.17 San Francisco-to-New York WDM point-to-point link.

- Spans of optical single-mode fiber
- Optical amplifiers
- Dispersion-compensating devices
- An optical demultiplexer
- Multiple receivers

Such a system is illustrated in Figure 3.17 for a hypothetical San Francisco to New York long-haul link. The real network link will use a combination of O-E-O regenerators, EDFA and Raman optical amplifiers, and dispersion-compensation elements to solve attenuation and dispersion problems for light transmission from San Francisco to New York. To simplify the analysis but illustrate to some degree the network engineering problems involved, the following example uses a number of simplifying assumptions.

Problem Formulation Design a long-haul link between New York and San Francisco (approximate distance of 6000 km) for 40-Gb/s bandwidth capacity. Assume that you have the following 10-Gb/s components available:

- Laser source
  - Output power 1 mW and spectral width 0.5 nm
  - Laser driver to drive the laser
- Photoreceivers
  - APD (avalanche photodiode), optical input sensitivity of 1 μW
  - PIN (positive–intrinsic–negative photodiode), optical input sensitivity of 10 μW
  - TIA amplifier to convert output from either APDs or PIN photodiodes
- Optical fiber
  - 1550-nm wavelength band with an optical attenuation loss of 0.24 dB/km

- Single-mode fiber dispersion of 1 ps/(nm·km) and O-E-O regenerator that can recover half of the data eye

Assume that optical mux and demux devices are available to convert four 10-Gb/s optical signals into one 40-Gb/s signal. Keep in mind that the link can be attenuation or dispersion limited. Calculate the distance between neighboring O-E-O repeater stages if (1) an APD is used, and (2) a PIN diode is used.

Problem Solution   As the output laser's optical power is 1 mW and its APD can resolve 1 μW, the optical budget loss is 30 dB. Keep in mind that the attenuation in decibels is calculated as $10\log(P_{out}/P_{in})$. For a PIN diode with optical sensitivity of 10 μW, the optical budget is only 20 dB. Assuming that the only losses are due to 1550 nm fiber, we can calculate the distance between the repeaters to be 30 dB divided by 0.24 dB/km, which is 125 km for the APD. The corresponding distance becomes only 83.3 km for the PIN diode.

Let us check the dispersion limits. Keep in mind that we are sending 10-Gb/s signals for which the data eye is 100 ps wide (1 over 10 GHz). Consider first the PIN diode. Since the fiber dispersion is 1 ps/(nm·km) and the laser spectral width is 0.5 nm, the total dispersion value for 83.3 km is 41.6 ps. As a result, 41.6% of the data eye is smeared due to dispersion. However, as the O-E-O regenerator can recover up to 50%, we should be okay in this case.

Things get more complicated with the APD. The total dispersion for the distance of 125 km is 62.5 ps. That is 62.5% of the data eye, which is clearly too much for the O-E-O to handle. As a result, we have to shorten the distance to such a value that the total dispersion is not larger than 50 ps. The distance can be calculated to be 100 km.

Final Answer

- For the APD the long-haul link is dispersion limited and the distance between O-E-O regenerators should be shorter than 100 km. At least 60 regeneration stages are required.
- For the PIN diode the long-haul link is attenuation limited and the distance between O-E-O regenerators should be shorter than 83.3 km. At least 73 regeneration stages are required.

Final note: In practice the calculations must involve some second-order effects, which have been neglected here. In particular, a system margin will have to be included in a real-life situation. Also, it is worthwhile to note that instead of using O-E-O regenerators, the long-haul link will probably be more economical using EFDAs rather than O-E-O regenerators, or using a combination of regenerators and amplifiers, as optical amplifiers do not solve dispersion problems.

KEY POINTS

WDM technology:

- A suitable optical transmission window is in the range 1280 to 1650 nm, making the bandwidth of the single-mode fiber around 50 THz. The only way to take advantage of this large bandwidth is by using transmitters of different wavelengths.
- WDM is a technology of sending multiple wavelengths over one fiber.
- Lasers in WDM systems usually operate in the general region of 1550 nm, because that is the range in which optical fiber has the lowest attenuation.
- When multiple wavelengths are carried on one fiber, the wavelengths are typically separated by a multiple of 100 GHz with 0.8-nm spacing, sometimes referred to as the *ITU grid*, after the standards body that set this figure (ITU-T G.692 Recommendation). The 100-GHz grid is further subdivided into 50- and 25-GHz spacings, which allow tighter packing but increase interference.
- WDM increases the maximum capacity of one fiber connection up to about 1 Tb/s, which will probably increase to about 10 Tb/s in the very near future.
- WDM technology can be implemented using both large (coarse WDM) and small wavelength spacing (dense WDM). Typically, coarse WDM carries fewer than eight wavelengths, and dense WDM carries more than eight wavelengths.

Networking equipment for WDM:

- Signal degradation in optical networks is caused primarily by fiber attenuation and dispersion. Optical EDFAs and Raman amplifiers can be used effectively to amplify signals, but they do not help with dispersion problems.
- WDM regenerators are used to regenerate WDM signals completely; both attenuation and dispersion problems are solved concurrently. WDM regenerators with O-E-O conversion require a large number of electrical-to-optical converters.
- Optical cross-connects without O-E-O capability have limited use due to wavelength-blocking problems. Optical cross-connects with O-E-O capability might have either optical or electrical crossbars.
- An optical add–drop multiplexer is similar to a WDM regenerator, but with the added ability to add–drop electrical signals.
- Microoptical electromechanical system devices are enabling technologies to build optical cross-connects. Two technologies can be used. In the two-

dimensional solution, mirrors with two positions (on and off) are used to achieve nonblocking optical paths in two-dimensional space. In three-dimensional solutions three-dimensional, case mirrors that tilt freely about two axes are used to achieve cross-connects in three-dimensional space.

WDM networks:

- Traditional WDM networks use fixed-wavelength lasers. As a result, new service provisioning requests can take a very long time to set up. Tunable lasers enable dynamic WDM networks with fast service provisions.
- WDM networks encounter wavelength-blocking problems where the networking node is blocked for transmission at the particular wavelength. Wavelength-conversion capability deployed at certain network nodes is very helpful to eliminate wavelength-blocking problems.
- O-E-O conversion must be deployed frequently in WDM networks. O-E-O converters regenerate optical signals, a process that eliminates signal distortions caused by attenuation and dispersion effects. They also provide wavelength-correction capability.
- WDM networks are difficult to protect against fiber cuts.

REFERENCES

Agarwal, G. P., *Fiber-Optic Communication Systems*, Wiley, Hoboken, NJ, 1997.

Chu, P. B., S. Lee, S. Park, et al., *MOEMS-enabling technologies for large optical cross-connects, Proceedings of SPIE*, vol. 4561, pp. 55–65, 2001.

Freeman, R., *Fiber-Optic Systems for Telecommunications*, Wiley, Hoboken, NJ, 2002.

Goff, R., *Fiber Optics Reference Guide*, Focal Press, distributed by Siluer Pixel Press, Hauppauge, NY, 2002.

Greywall, D., P. Bush, and F. Pardo, Crystalline silicon tilting mirrors for optical cross-connect switches, *Journal of Microelectromechanical Systems*, vol. 12, no. 5, pp. 708–712, 2003.

ITU-T-Recommendation G.692; *Optical Interfaces for Multichannel Systems with Optical Amplifiers*.

Mukherjee, B., *Optical Communication Networks*, McGraw-Hill, New York, 1997.

Ramaswami R., and K. Sivarajan, *Optical Networks: A Practical Perspective*, Academic Press, San Diego, CA, 1998.

Rogers, A., *Understanding Optical Fiber Communications*, Artech House, Norwood, MA, 2001.

Taylor, W., J. Bernstein, J. Brazzle, and C. Corcoran, Magnet arrays for use in a 3-D MEMS mirror array for optical switching, *IEEE Transactions on Magnetics*, vol. **39**, no. 5, pp. 3286–3288, 2003.

Yamamoto, T., J. Yamaguchi, N. Takeuchi, A. Shimizu, E. Higurashi, R. Sawada, and Y. Uenishi, A three-dimensional MEMS optical switching module having 100 input and 100 output ports, *IEEE Photonics Technology Letters*, vol. **15**, no. 10, pp. 1360–1362, 2003.

4

SONET

4.1 INTRODUCTION

The *synchronous optical network* (SONET) is a standard for optical telecommunications transport established by ANSI in the 1980s. Its European and Japanese counterpart, the *synchronous digital hierarchy* (SDH), was formulated by the ITU standards body. The differences between the two standards are reasonably small for our purposes, so in this chapter, we will, for the most part, use only the term SONET. However, almost anything written about

Network Infrastructure and Architecture: Designing High-Availability Networks,
By Krzysztof Iniewski, Carl McCrosky, and Daniel Minoli
Copyright © 2008 John Wiley & Sons, Inc.

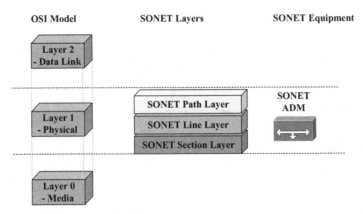

FIGURE 4.1   OSI model, SONET Layers, and examples of networking equipment.

SONET will apply to SDH, sometimes with small modifications, which occasionally are mentioned.

SONET uses time-domain multiplexing (TDM) as a way to multiplex lower-bandwidth signals into one high-capacity signal that can be sent over optical fiber. Being a networking standard, it provides a set of specifications to determined frame formats, bit rates, and optical conditions to ensure multivendor interoperability. It also provides fast restoration schemes, meaning that in the case of a fiber cut or some other network failure it has the capability to restore the service, typically within 50 ms. Finally, SONET has extensive built-in features for operations, administration, maintenance, and provisioning (OAM&P). These features are used to provision connections in point-to-point and ring topologies, to detect defects, and to isolate network failures. It is important to realize that the comprehensive SONET/SDH standard has provided transport infrastructure for worldwide telecommunications for the last 20 years. For example, it is estimated that about 200,000 SONET rings are deployed in North America.

The goal of this chapter is to introduce SONET as an example of layer 1 networking protocol. We discuss SONET layering structure (Figure 4.1), explain how a SONET frame is built, and describe the functions of frame overhead bytes. We review the basic classes of SONET networking equipment and show how SONET equipment is connected to form a transport network. We pay particular attention to the SONET add–drop multiplexer as an important example of SONET networking hardware.

4.2   SONET NETWORKS

4.2.1   SONET Transmission Rates

Today, SONET uses mostly gigabit transmission rates of 2.5, 10, and 40 Gb/s. However, its digital hierarchy starts at 51 Mb/s and covers transmission rates

TABLE 4.1   SONET and SDH Bandwidth Rates

Bandwidth	SONET	SDH	Optical Carrier	Number of Voice Channels
51.84 Mb/s	STS-1	—	OC-1	672
155.52 Mb/s	STS-3	STM-1	OC-3	2,016
622.08 Mb/s	STS-12	STM-4	OC-12	8,064
2.488 Gb/s	STS-48	STM-16	OC-48	32,256
9.953 Gb/s	STS-192	STM-64	OC-192	129,024
39.813 Gb/s	STS-768	STM-256	OC-768	516,096

below 1 Gb as well: 155 and 622 Mb/s. As you can see, the SONET rate multiplier is 4; every higher bandwidth connection is four times faster than the one before. SDH is built in a very similar way, and the comparison between the two is shown in Table 4.1.

SONET line signals can be optical or electrical. Electrical signals are used for short distances: for example, as interconnections of networking equipment on the same site or between racks of the networking gear. Electrical signals are sent on copper wires, whereas optical signals used for longer distances require optical fibers. We will use the terms *synchronous transport signal* (STS) for electrical signals and *optical carrier* (OC) for optical signals.

As its name implies, SONET is synchronous. That means that a highly stable reference point or clock is provided for every piece of SONET equipment through an elaborate clock synchronization scheme. Since a very precise clock is available, there is no need to align the data streams or synchronize clocks. SONET provides extensive monitoring and error recovery functions. Substantial overhead information is provided in SONET to allow for quicker troubleshooting and the detection of failures before they degrade to serious levels that can bring the entire network down.

4.2.2   SONET Network Architectures

SONET links can be thought of as highways, as SONET is used in metropolitan and wide area networks. Using this highway analogy, we can say that a SONET transport network is hierarchical, as a collection of small roads, medium-sized roads, and highways for long distance. Unlike highway systems, the typical architecture for SONET is a collection of rings, although point-to-point connections are used as well. Figure 4.2 illustrates the following three classes of SONET networking equipment: O-E-O regenerators, add–drop multiplexers, and terminal multiplexers.

O-E-O regenerators are used to regenerate optical signals that travel long distances. The most straightforward, although not necessarily the most effective way to regenerate is to convert to the electrical domain and then back to the optical domain, as discussed elsewhere in the book. SONET O-E-O regenerators might differ from protocol-independent O-E-O regenerators by

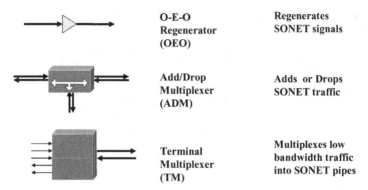

FIGURE 4.2   SONET networking equipment.

implementing additional performance-monitoring functions. To understand performance monitoring, we need to know how a SONET frame is built, which comes later in the chapter.

Add–drop multiplexers (ADMs) are the most versatile pieces of SONET networking gear, as they can add or drop any amount of SONET traffic, as desired by the network operations. SONET ADMs are used to create SONET transport networks consisting of SONET rings and point-to-point connections.

Terminal multiplexers (TMs) are a specialized class of ADMs used at the edges of SONET networks. They have the capability to multiplex lower-bandwidth signals coming from SONET or non-SONET access networks. Terminal muxes are used to aggregate lower-bandwidth traffic into higher-bandwidth SONET pipes for transmission over optical fibers.

An example of SONET transport architecture is shown in Figure 4.3. This particular SONET ring consists of five ADMs, which are collecting traffic from other ADMs and TMs residing outside the ring. In addition to ADMs, the rings contain O-E-O regenerators which are inserted in the network when the distance between two ADMs becomes too large. SONET regenerators are much simpler and cheaper than ADM regenerators, but unlike ADMs, they have no ability to branch out traffic. In addition to O-E-O regenerators, optical amplifiers can be used in the ring as well.

Another example of a simple SONET network is shown in Figure 4.4. A device called a *digital cross-connect* (DCC) cross-connects SONET data streams between two rings of ADMs. ADMs and DCCs are both very popular pieces of SONET networking equipment. A more realistic example of SONET networking is shown in Figure 4.5. SONET rings and point-to-point links are connected together to form SONET transport networks. SONET transport is our main interest in this chapter. However, in addition to transporting data, the SONET network has to be able to perform two additional functions. First, the network needs to be monitored and provisioned. Monitoring and provisioning functions are provided by a separate OAM&P network, shown in

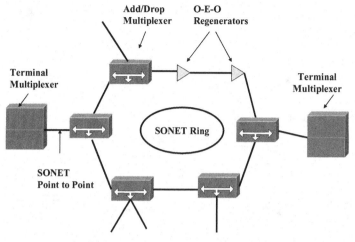

FIGURE 4.3   SONET transport network.

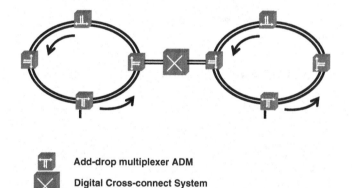

FIGURE 4.4   SONET digital cross-connect connecting two ADM rings.

Figure 4.6. Second, the SONET network needs to be synchronized, as each piece of SONET gear needs to understand a common reference timing. The network clocking information required is provided by a separate synchronization network, as shown in Figure 4.5.

To describe various network segments properly, the following terminology is used by SONET:

- *Section:* the distance between two O-E-O regenerators or between the regenerator and another piece of SONET equipment
- *Line:* the distance between two ADMs
- *Path:* the distance from a starting terminal location to the ending terminal location

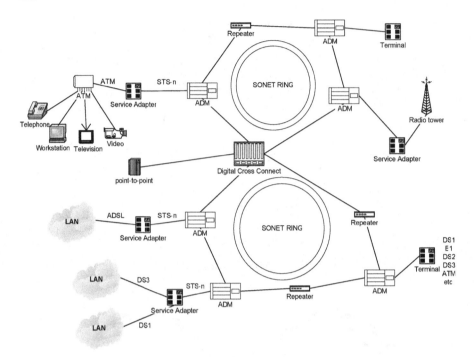

FIGURE 4.5   Example of a SONET network.

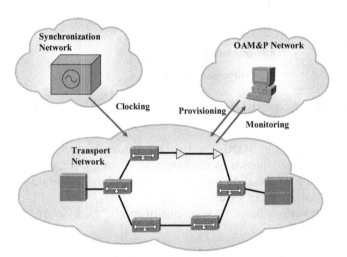

FIGURE 4.6   SONET network overview.

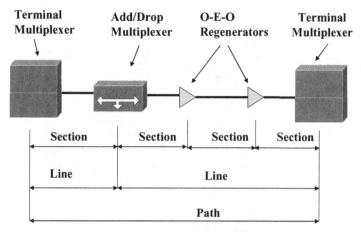

FIGURE 4.7   SONET network segment terminology.

As a result, the SONET path consists of multiple lines, and the lines, in return, can consist of multiple sections. This terminology is illustrated in Figure 4.7.

4.3   SONET FRAMING

SONET defines a technology for carrying many signals of different capacities through a synchronous, flexible optical hierarchy. This is accomplished by means of a byte-interleaved multiplexing scheme. Byte-interleaving simplifies multiplexing and offers end-to-end network management. To explain how SONET framing is performed, we will illustrate first what the smallest building block, called STS-1, looks like.

4.3.1   STS-1 Building Block

The first step in the SONET multiplexing process involves generation of the lowest-level, or base, signal. In SONET, the base signal is referred to as synchronous transport signal–level 1 (STS-1), which operates at 51.84 Mb/s. Higher-level signals are integer multiples of STS-1, creating the family of STS-N signals. An STS-N signal is composed of n byte-interleaved STS-1 signals. The STS-n hierarchy was shown in Table 4.1, where the optical counterpart for each STS-N signal, designated optical carrier level N (OC-N), was included as well.

The frame format of the STS-1 signal is shown in Figure 4.8. In general, the frame can be divided into two main areas: transport overhead and synchronous payload. Transport overhead is divided, in turn, into section overhead (SOH), line overhead (LOH), and path overhead (POH). STS-1 is a

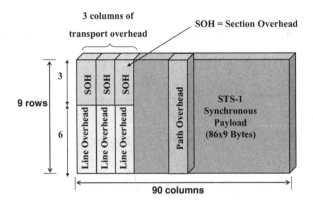

STS-1 = 90 Bytes x 9 = 810 Bytes = 6,480 bits
6,480 bits x 8 kHz = 51.84 Mb/s

FIGURE 4.8   STS-1 frame format.

specific sequence of 810 bytes (6480 bits), which includes various overhead bytes and an envelope capacity for transporting payloads. It can be depicted as a 90-column by 9-row structure (you could make a connection here to the DS0 sampling period). With a frame length of 125 μs (8000 frames per second), STS-1 has a bit rate of 51.840 Mb/s. The order of transmission of bytes is row by row from top to bottom and from left to right (most significant bit first).

The synchronous payload envelope can also be divided into two parts: the STS POH and the payload. The payload is the revenue-producing data traffic being transported and routed over the SONET network. Once the payload is multiplexed into the synchronous payload envelope, it can be transported and switched through SONET without having to be examined and demultiplexed at intermediate nodes. Thus, SONET is said to be service independent or transparent.

Transport overhead is composed of section overhead and line overhead. The STS-1 POH is part of the synchronous payload envelope. As shown in Figure 4.9, the first three columns of the STS-1 frame are for the transport overhead. The three columns contain 9 bytes. The first 9 bytes create overhead for the SOH and the next 18 bytes create overhead for the LOH. The remaining 87 columns constitute the STS-1 envelope capacity (payload and POH).

The interesting challenge in every protocol is how to find the beginning of the frame. In SONET this process is quite straightforward, as illustrated in Figure 4.8. The first two bytes in the section overhead are called the A1 and A2 framing bytes. A1 is 11110110, and A2 is 00101000. A SONET station simply looks for these bytes, and once it finds them, declares the beginning of

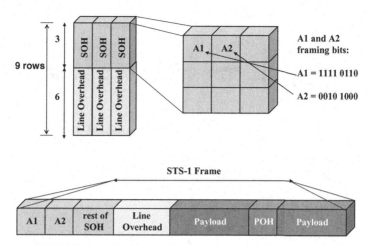

FIGURE 4.9   Finding the beginning of the SONET frame.

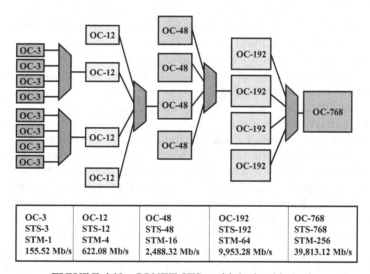

OC-3	OC-12	OC-48	OC-192	OC-768
STS-3	STS-12	STS-48	STS-192	STS-768
STM-1	STM-4	STM-16	STM-64	STM-256
155.52 Mb/s	622.08 Mb/s	2,488.32 Mb/s	9,953.28 Mb/s	39,813.12 Mb/s

FIGURE 4.10   SONET STS multiplexing hierarchy.

the frame. Since an STS-1 frame always has 810 bytes, A1 and A2 bits have to continue to appear in regular intervals. If they do not, a framing error is declared.

The basic structure of STS-1 is repeated for the higher rates. Three STS-1 cells are multiplexed to create STS-3, these in turn are multiplexed to create STS-12, and so on. The process of multiplexing from STS-3 up to STS-768 is illustrated schematically in Figure 4.10. When doing the multiplexing, a SONET

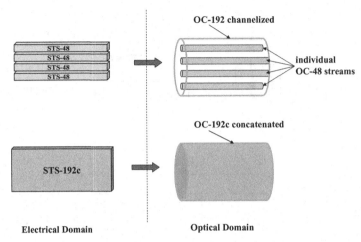

FIGURE 4.11   Channelization vs. concatenation.

station has a choice as to whether to merge overhead information or leave it untouched in separate SONET streams. Both options, referred to as channelization and concatenation, are illustrated in Figure 4.11. In a *channelized* case, all lower-rate streams retain their overhead structure. In a *concatenated* case, the overhead structure is merged.

What are the advantages of channelized vs. concatenated, or vice versa? The concatenated format is clearly more efficient, as some redundant overhead information is removed; therefore, the concatenated OC-192c stream has a higher effective data rate than the channelized OC-192 stream. However, concatenation, or as some would say, merging into a bigger SONET pipe, takes some processing to do when the concatenated data reaches the common destination. Also, a channelized structure is more accessible for individual data streams on the receiving end if subsequent demultiplexing is required. To use the analogy of trains, we can say that an express train carries a concatenated payload as all of the payload has the same destination. A regular train stopping at every station, on the other hand, carries a channelized payload. This is because some of the payload may be unloaded at each station. Each station unloading is associated with a required demultiplexing. Clearly, in situations where traffic efficiency is the key, concatenation is preferred, while in other cases channelization might be preferred.

4.3.2   Synchronous Payload Envelope

The synchronous payload envelope (SPE) is an interesting concept in SONET. Let us first explain what it is and later discuss why a "strange" mechanism of floating SPE was introduced in SONET. The STS-1 synchronous payload envelope (SPE) may begin anywhere in the STS-1 envelope capacity, as

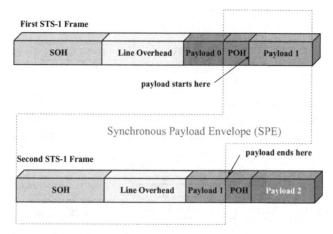

FIGURE 4.12 Synchronous payload envelope position in the STS-1 frame.

shown in Figure 4.12. Typically, it begins in one STS-1 frame and ends in the next. The STS payload pointers contained in the transport overhead designates the location of the byte where the STS-1 SPE begins. STS POH is associated with each payload and is used to communicate various information from the point where a payload is mapped in the STS-1 SPE to where it is delivered.

Why does SPE need to float its position within an STS-1 frame? As a SONET system is synchronous, it would seem that the position of the data payload can be fixed with a SONET frame. Yes, a SONET system is synchronous, but the synchronization is not perfect. Even in a very tightly specified clock distribution system there will be minor differences between clocks assigned to different stations.

Assume that two 10-GHz clocks have a 1-ppm (part per million) difference, as illustrated in Figure 4.13. One clocks the input data at the rate of precisely 10 Gb/s, while the other is faster by 1 ppm, so it clocks the data at the rate of 10.00001 Gb/s. As a result of the clock difference, 1 kb of data would be accumulated between output and input in 1 second of continuous data transfer. Up to this point, the difference of 1000 bits can probably be accommodated by a FIFO (first-in first-out) buffer. However, in 1000 seconds the difference would increase to 1 Mb, more than the entire STS-1 frame. Clearly, at some point in time the FIFO buffer would be completely empty. In the opposite case of the clock running too slowly (Figure 4.13), the FIFO buffer would be completely full. As a result, a mechanism is needed to accommodate slight clock differences in synchronous systems.

Such a mechanism was introduced to SONET and is called a floating SPE using H pointers. The H1 and H2 pointers reside in the transport overhead at the top of the STS-1 frame and indicate the position of the beginning of the SPE. If the output rate is higher than the input rate, the H3 pointer is used to

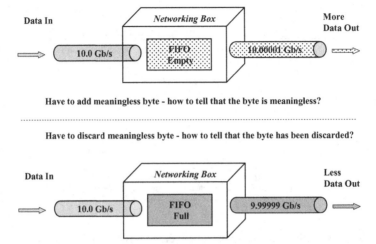

FIGURE 4.13 Pointer functions to accommodate clock difference between input and output.

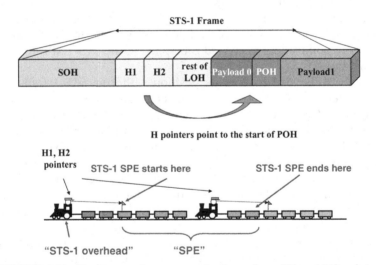

FIGURE 4.14 Finding synchronous payload envelope: H1 and H2 pointers.

add a meaningless byte to accommodate the clock difference. If the output rate is lower than the input rate, the H3 pointer is used to discard the meaningless byte.

A schematic illustration of the actions of H pointers is shown in Figure 4.14. A SONET data stream can be envisioned as a convoy consisting of many trains (frames). Each train has its own locomotive (overhead) and cars (data payload).

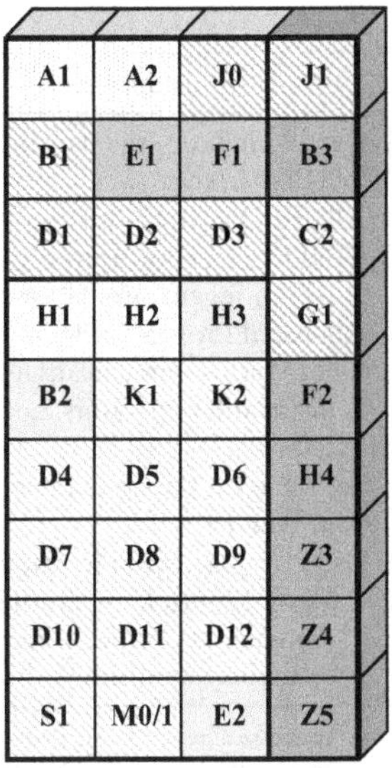

A1	A2	J0	J1
B1	E1	F1	B3
D1	D2	D3	C2
H1	H2	H3	G1
B2	K1	K2	F2
D4	D5	D6	H4
D7	D8	D9	Z3
D10	D11	D12	Z4
S1	M0/1	E2	Z5

• A1, A2 are used to recognize frame boundary

• J0 is section trace to verify continuity

• B1, B2 represent bit interleaved parity (BIP-8)

• D1 to D12 are used for network management

• H1, H2 are used point to the SPE beginning

• K1, K2 are used for failure messaging (APS)

• S1, M0/1 are used for synchronization

• J1 is responsible for path tracing

•C2 and G1 are indicating path status

FIGURE 4.15   SONET overhead information.

The train conductors (H pointers) sit in the locomotives and indicate using colored flags where a "floating" car sequence starts.

SONET provides substantial overhead information, allowing simpler multiplexing and greatly expanded OAM&P capabilities. The overhead information has several layers and is summarized in Figure 4.15. Path-level overhead is carried from end to end; it is added to DS-1 signals when they are mapped into VTs and for STS-1 payloads that travel end to end. Line overhead (Table 4.2) is used for the STS-N signal between STS-N multiplexers. Section overhead (Table 4.3) is used for communications between adjacent network elements, such as regenerators. Enough information is contained in the overhead to allow the network to operate and allow OAM&P communications between an intelligent network controller and the individual nodes.

One of the important features of SONET is automatic protection switching (APS). APS functionality is typically accomplished by using two pairs of optical fiber on a given SONET link between two ADM stations, as shown in Figure 4.16. One pair is the working pair, the other is a protection pair, used when the working pair fails. The K1 and K2 bytes are responsible for maintaining communication between both ADMs. An example of a typical network failure is shown in Figure 4.17. The network failure starts with an accidental fiber cut by a construction crew. One of the ADMs detects network failure thanks to B1/B2 byte information. Service is then restored with the help of K1 and K2 bytes (e.g., the traffic can be circulated in the opposite direction on

TABLE 4.2   Section Overhead[a]

Byte	Name	Function
A1/A2	Framing bytes	Used to indicate the beginning of an STS-1 frame
J0/Z0	Section trace (J0) and section growth (Z0)	Allocated to trace origins of a frame
B1	Section bit-interleaved parity code (BIP-8) byte	Used to check for transmission errors over a regenerator section
E1	Section orderwire byte	Allocated for local orderwire channel for voice communication for installation operators; not used today
F1	Section user channel byte	Set aside for purposes of network provider
D1/D2/D3	Section data communications channel bytes	Used from a central location for alarms, control, monitoring, administration, and other communication needs

[a]Contains 9 bytes of the transport overhead accessed, generated, and processed by section-terminating equipment. This overhead supports functions such as performance monitoring, framing, and data communication for operation, administration, management, and provisioning.

TABLE 4.3   Line Overhead[a]

Byte	Name	Function
H1/H2	STS payload pointers	Allocated to a pointer that indicates an offset between the pointer and the first byte of the SPE
H3	Pointer action byte	Allocated for SPE frequency justification purposes
B2	Line bit-interleaved parity code (BIP-8) byte	Used to determine if a transmission error has occurred over a line; calculated over all bits of the line overhead
K1/K2	Automatic protection switching bytes	Used for protection signaling, detecting alarm indications, and remote defect indication signals
D4–D12	Line data communications channel bytes	Used for OAM&P information (alarms, control, maintenance, remote provisioning, monitoring, and administration)
S1	Synchronization status	Allocated to convey the synchronization status of the network element
Z1	Growth byte	Allocated for future growth
M0	M0 byte	Allocated for a line remote error indication
Z2	Growth byte	Allocated for future growth
E2	Orderwire byte	Allocated for local orderwire channel, voice communication, and installation operators; not used today

[a]Contains 18 bytes of overhead accessed, generated, and processed by line-terminating equipment. This overhead supports functions such as locating SPE in the frame, multiplexing signals, line maintenance, automatic protection switching, and performance monitoring.

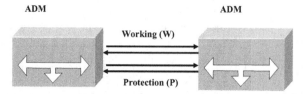

FIGURE 4.16   Work and protection links for automatic protection switching.

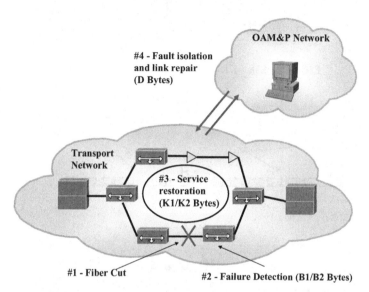

FIGURE 4.17   Example of SONET network failure.

the ring). Finally, the fault is isolated and the link can be repaired thanks to D-byte communication with an OAM&P SONET subnetwork.

4.3.3   SONET Virtual Tributaries

One of the benefits of SONET is that it can carry large payloads. However, the existing lower-speed T/E digital hierarchy can be accommodated as well, thus protecting investments in equipment that has already been installed. To achieve this capacity, the STS SPE is subdivided into smaller components, known as *virtual tributaries* (VTs), for the purpose of transporting and switching payloads smaller than the STS-1 rate. The lower-speed traditional T1/E1 hierarchy was introduced briefly in Chapter 1. The subdivision into VT structure is shown schematically in Figure 4.18.

To accommodate mixes of different VT types within an STS-1 SPE, the VTs are grouped together. An STS-1 SPE that is carrying VTs is divided into seven

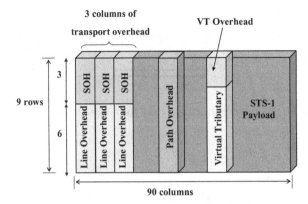

FIGURE 4.18   STS-1 frame and virtual tributaries.

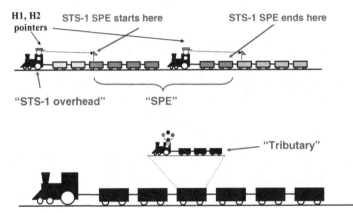

FIGURE 4.19   Relationship between SONET SPE and virtual tributaries.

VT groups, with each VT group using 12 columns of the STS-1 SPE. Each VT group can contain only one size (type) of VT, but within an STS-1 SPE, there can be a mix of different VT groups. For example, an STS-1 SPE may contain four VT1.5 groups and three VT6 groups, for a total of seven VT groups. Thus, an SPE can carry a mix of any of the seven groups. The groups have no overhead or pointers; they are just a means of organizing the various VTs within a single STS-1 SPE.

As illustrated in Figure 4.18, each VT has its own overhead. How are STS-1 and VT overheads distinguished? To understand this point properly, let us utilize another train analogy. Consider a SONET data stream to be a train as shown in Figure 4.19. The "locomotive" represents STS-1 transport overhead and the "cars" represent columns of STS-1 data payload. The locomotive contains H1 and H2 pointers which point out where the new SPE starts. This represents the mechanism of finding the position of the SPE, as discussed earlier. Each car in the train represents VT tributaries. As illustrated in Figure 4.19, each car is really like a small train. It contains its own small locomotive that represents VT overhead, as well as its own small cars that represent VT data payload.

4.3.4 SDH vs. SONET

Synchronous digital hierarchy (SDH) embraces most of SONET and is an international standard defined by the ITU. It is used everywhere outside North America, but it is often regarded as a European standard because of its origins. As discussed previously, SONET is based on an STS-1 that has 51.84 Mb/s of bandwidth. SDH, on the other hand, is based on a threefold larger unit called a synchronous transport module, STM-1, that has a 155.52-Mb/s bandwidth (3 × 51.84 = 155.84). STM-1 corresponds to STS-3 in SONET. Table 4.1 shows the correspondence between SONET and SDH rates.

As a result of this 3:1 relationship between SONET and SDH and other protocol details, SDH is compatible with a subset of SONET, and traffic interworking between the two is possible. However, interworking for alarms and performance management is generally not possible between SDH and SONET systems. Broadly speaking, SONET networking equipment is deployed in North America and SDH equipment elsewhere. Like STS-1, the STM-1 frame has a repetitive structure with a period of 125 μs and consists of nine equal-length segments, as shown in Figure 4.20. The STM-1 frame has 270 columns, three times more than STS-1. The data payload can be divided into smaller, autonomous units with their own overhead. In SONET, these units are called

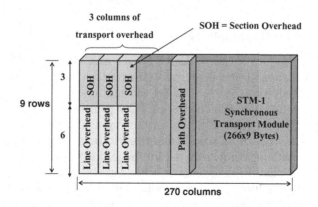

STM-1 = 270 Bytes x 9 = 2430 Bytes = 19,440 bits
19,440 bits x 8 kHz = 155.52 Mb/s

FIGURE 4.20 STM-1 frame for SDH protocol.

virtual tributaries (VTs), whereas in SDH they are called *virtual containers* (VCs). In both cases the principle for division is the same but the details of implementation between VTs and VCs differ greatly, as they carry over from differences between T- and E-carrier hierarchies.

4.4 SONET EQUIPMENT

4.4.1 SONET O-E-O Regenerator

SONET signals have to be converted from the electrical domain to the optical domain, and vice versa, using optical modules. The basic principle of operation and the implementation details of optical modules are discussed elsewhere in the book and are not repeated here except to note that optical modules used in SONET networks have to comply with jitter requirements that are specific to SONET standards.

The simplest piece of SONET equipment is the O-E-O regenerator. These regenerators restore optical signals by converting them to the electrical domain and retransmitting. How far can SONET optical signals go along optical fibers? The issue is the same as in WDM systems; in fact, once in the optical domain the signal is carried by photons that do not understand frame formatting. Depending on what types of optical elements (lasers and photodiodes) and optical fibers (single or mulitimode) are used, a different reach of the optical signal along a fiber can be obtained. Typical reach ranges in SONET include the following selection: SR: short reach, up to 2 km; IR: intermediate reach, up to 40 km; LR: long reach, up to 80 km; and ULR: extra long reach, over 80 km. In metropolitan area networks, IR and LR modules are used as typical spacings in SONET rings ranging from 40 to 80 km. Some long-haul links can send signals over 100 km before regeneration is applied.

4.4.2 SONET ADM Multiplexer

An add–drop multiplexer can be described as a multiplexing hub. Like a hub, an ADM has the capability to aggregate traffic; in fact, it can add or drop traffic as required. In addition, SONET ADM can multiplex the lower-speed data into a higher-bandwidth connection using TDM principles. A single-stage multiplexer/demultiplexer can multiplex various inputs into an OC-N signal. At an add–drop site, only those signals that need to be accessed are dropped or inserted. The remaining traffic continues through the network element without requiring special pass-through units or other signal processing.

The add/drop multiplexer also provides interfaces between the various network signals and SONET signals. In addition, the ADM can provide low-cost access to a portion of the traffic that is passing though. Most designs of ADM are suitable for incorporation in rings to provide increased service flexibility in both urban and rural areas. ADM ring design also employs alternative routing for maximum availability to overcome fiber cuts and equipment failures. A group of ADMs, such as in a ring, can be managed as an entity for distributed bandwidth management. Although ADMs manufactured by various companies are compatible at the OC-N level, they may differ in features from vendor to vendor. The SONET standard does not restrict manufacturers to providing a single type of product, nor does it require them to provide all types. For example, one vendor might offer an ADM with access at DS-1 only, whereas another might offer simultaneous access at DS-1 and DS-3 rates.

4.4.3 SONET Terminal Multiplexer

The path terminating element (PTE), an entry-level path-terminating terminal multiplexer, acts as a concentrator for T-carrier (or E-carrier) as well as other tributary signals. Its simplest deployment would involve two terminal multiplexers linked by fiber with or without a regenerator in the link. This implementation represents the simplest SONET link, consisting of a section, line, and path all in one link. Hub multiplexers provide an entry point to the SONET network. They are built in a fashion similar to ADMs. A single unit can act as an ADM on a ring while serving as a hub multiplex for a number of fiber spurs off the ring.

4.5 SONET IMPLEMENTATION FEATURES

4.5.1 SONET Scrambling

The digital transition of data signals is based on the timed change of the signal state at a time coinciding with the change of a control signal, generically called a *clock*. The clock frequency determines the data rate of a transmitted signal and controls the receiving node or, more specifically, the digital circuit that is responsible for the recovery of the data received. Occasionally, the clock is a separate signal (especially when the transmitter and the receiver are within the same networking node), but more often it is embedded within the data signal transmitted. In optical fibers the clock is always embedded in the data stream, as a separate transmission of the clock would be too expensive. However, it is possible to extract the clock information from the data stream by examining the data state changes. For this procedure to work properly, data transmitted have to follow certain coding standards to maximize the number of changes in the data stream. In this section we explain a mechanism called data scrambling that is used by SONET to increase the number of data transitions.

A typical modulation scheme used in SONET systems is called *non-return-to-zero* (NRZ). NRZ transmits an optical pulse for the duration of the bit interval for a data 1 and transmits no energy for a data 0. The NRZ line code is the simplest to implement and the most bandwidth-efficient binary

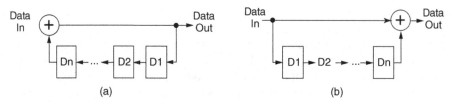

FIGURE 4.21 SONET scrambling process: (a) $x^n + 1$ scrambler; (b) $x^n + 1$ descrambler.

line code. The primary drawback to NRZ, however, is that long strings of 1's or 0's in the data cause long periods in which there is no transition in the signal level. During that period the circuitry at the receiver[†] can drift out of synchronization with the data transmitted so that the receiver might begin sampling data away from the optimum point. Data error can eventually occur.

For this reason it is important to have the signal level make frequent transitions from 1 to 0 and to maintain a roughly balanced number of 0's and 1's. Each networking protocol accomplishes that task in a different way. Ethernet uses 8b/10b coding, where 2 extra bits of data are added to every 8-bit-long sequence, with the resulting 10-bit sequence always having a property of dc balance. Although foolproof, this coding technique carries a 25% overhead penalty.

SONET architects chose a different path. They decide to rearrange transmitting bits in a process called *scrambling* in such a way that the resulting data sequence has the property of dc balance. As no extra bits are added, there is no overhead penalty in this case, although the system is not completely error proof, as we will discuss in a moment. The scrambling process is illustrated in Figure 4.21. In a typical implementation, the scrambler output is the XOR of the current data with the data that preceded it by n bits. Mathematically, this operation amounts to dividing the data stream by an $x^n + 1$ scrambler polynomial. This operation creates a pseudorandom sequence that is reached in data transitions. For SONET, n has been chosen equal to 43 to minimize the probability of data killer patterns. As a result, an $x^{43} + 1$ scrambler is applied to the data payload. The receiver effectively multiplies the payload data by the $x^{43} + 1$ scrambler polynomial to reverse the process and recover the original data.

4.5.2 SONET Clock Distribution

In synchronous digital networks, the clock is originated from the primary reference source clock (master clock), which is then cascaded down the synchronous clock hierarchy to achieve the highest level of timing accuracy. Of course,

---

[†]The specialized analog circuitry used for this purpose, called *clock and data recovery*, is discussed in Chapter 8.

as the clock is passed down through the clock hierarchy, some increase in clock frequency variation is inevitable. These variations can be classified as timing jitter and wander. *Timing jitter* is a short-term variation (within a period of less than 0.1 s) of the clock frequency. *Wander* is a long-time frequency variation (within a period of more than 0.1 s). It can also be described as persistent jitter in one direction. Wander may be caused by variations in clock frequency delay through the transmission path or by bit-stuffing operations.

To prevent problems with data timing variations, various protocols have provisions to adjust the timing as data are passing through the node. An example of such a provision is evident in the SONET/SDH standard, where the H1, H2, and H3 bytes of the line overhead indicate the pointer to the "floating" payload (specifically, H3 indicates the pointer adjustment action byte), allowing the payload data rate to be adjusted slightly with respect to a line clock. Timing variations depend largely on the accuracy of the SONET clock in a given node of the network. SONET classifies a variety of clocks, depending on how accurate they are. Stratum 1 is the highest-quality clock and has an accuracy specified at $\pm 1 \times 10^{-11}$. Stratum 2 is next at $\pm 1.6 \times 10^{-8}$, then stratum 3 at $\pm 4.6 \times 10^{-6}$. A SONET minimum clock (SMC) has an accuracy of $\pm 20 \times 10^{-6}$. Strata 1 and 2 require an atomic clock source, and stratum 3 and SMC can be achieved by conventional oscillators.

The S1 byte of the SONET TOH is used to communicate the quality of the reference clock being used by the transmitting node. The receiving node can then determine whether the signal received is acceptable for use in deriving its own reference clock. When a node derives its timing from a SONET signal, it is referred to as *line timing*, or in some cases, *loop timing*. To minimize the effects of timing variations, different methods of clock exchange are employed. Depending on the location of the source of the reference clock as well as the type of network configuration, the network timing can be arranged in various ways. We briefly mention the following general techniques for clock distribution.

1. *Free-Running Clock Exchange.* In this approach each of the nodes uses its own very accurate free-running crystal oscillator to generate clocks. The SONET chip implementation requires large buffers, or FIFOs (sometimes referred to as *elastic stores*), that can hold enough data to be able to accommodate frame slips or frame repeats. In this case the slips are reported to the controlling microprocessor so that the system can initiate retransmission of the "slipped" frame or, in the case of a repeated frame, can drop that frame in the downstream node. This method may be used in point-to-point networks with asynchronous interfaces.

2. *External-Timing Clock Exchange.* This method is similar to free-running clock exchange except that it uses an externally generated clock reference source at each node instead of crystal oscillators.

3. *Through-Timing Clock Exchange.* In this approach the timing source generated in one node is passed through to the next node without changing

the timing. It requires clock and data recovery on each receive line interface and the recovered clock is used to outgoing time signals. It can be used in signal regeneration units.

4. *Line-Timed Clock Exchange.* In this scheme clocks are derived from a receive line interface and are used to time all other interfaces. This approach cannot be used in a cascade of nodes, due to the cumulative effect of the timing variation when passing through each node.

5. *Loop-Timed Clock Exchange.* Loop-timed clock exchange is also called the *master–slave timing mode*. It can be understood as a special case of line-timed configuration where the master node originates the timing and receive line interface of the slave node, recovers the clock, and loops it back through the upstream transmit interface to the master node.

A network example employing various timing configurations is shown in Figure 4.22. Each of the timing methods described above requires special attention when designing a clock distribution unit for a networking chip. This is because the nodes can be reconfigured and need to be flexible enough to support all clock distribution methods. Implementation of various timing methods requires parts of the circuitry to be clocked by different clocks, depending on the network-timing configuration. VLSI chip design of SONET devices is discussed in Chapter 7.

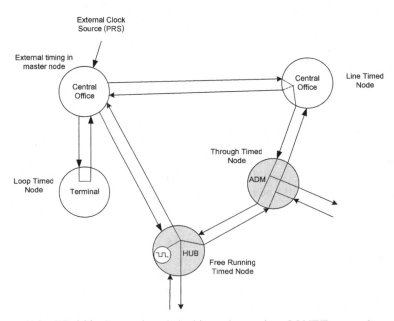

FIGURE 4.22 Examples of clocking schemes in a SONET network.

4.5.3 SONET Byte Stuffing

As discussed earlier, SONET uses a concept called *pointers* to compensate for frequency and phase variations. Pointers allow the transparent transport of synchronous payload envelopes (either STS or VT) between nodes with separate network clocks using almost the same timing. The use of pointers avoids the delays and loss of data associated with the use of large (125-µs frame) slip buffers for synchronization. Pointers provide a simple means of dynamically and flexibly phase-aligning STS and VT payloads, thereby permitting ease of dropping, inserting, and cross-connecting these payloads in the network. Transmission signal wander and jitter can also be minimized readily with pointers.

H1 and H2 pointers in the STS-1 allow the SPE to be separated from the transport overhead. The pointer is simply an offset value that points to the byte where the SPE begins. If there are any frequency or phase variations between the STS-1 frame and its SPE, the pointer value will be increased or decreased accordingly to maintain synchronization. In addition, either a meaningless byte is added (stuffed) or removed (discarded). SONET uses the terms *positive byte stuffing* and *negative byte stuffing*; both processes are described in detail below.

Positive Stuffing When the frame rate of the SPE is too slow in relation to the rate of the STS-1, bits 7, 9, 11, 13, and 15 of the pointer word are inverted in one frame, thus allowing 5-bit majority voting at the receiver. These bits are known as the *increment bits*. Periodically, when the SPE is about 1 byte off, these bits are inverted, indicating that positive stuffing must occur. An additional byte is stuffed in, allowing the alignment of the container to slip back in time. In positive stuffing the stuff byte is made up of noninformation bits. The actual positive stuff byte directly follows the H3 byte (i.e., the stuff byte is within the SPE portion). The pointer is incremented by one in the next frame, and subsequent pointers contain the new value. Simply put, if the SPE frame is traveling more slowly than the STS-1 frame, every now and then stuffing an extra byte in the flow gives the SPE a 1-byte delay.

Negative Stuffing Conversely, when the frame rate of the SPE frame is too fast in relation to the rate of the STS-1 frame, bits 8, 10, 12, 14, and 16 of the pointer word are inverted, thus allowing 5-bit majority voting at the receiver. These bits are known as the *decrement bits*. Periodically, when the SPE frame is about 1 byte off, these bits are inverted, indicating that negative stuffing must occur. Because the alignment of the container advances in time, the envelope capacity must be moved forward. Thus, in negative stuffing, actual data are written in the H3 byte, the negative stuff opportunity (within the overhead).

The pointer is decremented by 1 in the next frame, and subsequent pointers contain the new value. Simply put, if the SPE frame is traveling more quickly than the STS-1 frame, every now and then pulling an extra byte from the flow and stuffing it into the overhead capacity (the H3 byte) gives the SPE a 1-byte advance. In either case, there must be at least three frames in which the pointer remains constant before another stuffing operation (and therefore a pointer value change) can occur.

KEY POINTS

SONET main characteristics:

- TDM multiplexing onto high-capacity fiber systems
- Standard bit rate and frame format
- Fast-restoration (50-ms) schemes; various network topologies (mostly ring)
- Operations, administration, maintenance, and provisioning (OAM&P)

SONET/SDH origins:

- Voice is digitized by 8-kHz sampling of analog signal; 8 bits is used for pulse code modulation. The DS0 bank contains one voice call and as a result has a bandwidth of 8 bits × 8 kHz = 64 kb/s.
- Twenty-four DS0 frames are combined, and by adding one extra bit for control purposes, a DS1 frame is created. The DS1 bandwidth is (24 × 8 bits + 1 bit) × 8 kHz = 1.544 Mb/s. The DS1 frame used in the North American PDH (plesiosynchronous digital hierarchy) system is called the T1 carrier.
- Europeans created a more elegant system, without bit-robbing complications, by combining 30 DS0 frames and adding two control frames. The resulting E1 signal has a bandwidth of 32 × 8 bits × 8 kHz = 2.048 Mb/s.
- A SONET/SDH frame is always 125 μs long (a result of 8-kHz sampling) regardless of the bit transmission rate.
- SONET uses a synchronous transport signal, STS-1, of 51 Mb/s as the basis of its hierarchy. SDH uses an STM-1 of 155 Mb/s, which is three times larger than STS-1, as its basic unit.

SONET characteristics:

- SONET/SDH data are scrambled, with the exception of the A1, A2, and J0 bytes in the section overhead. After scrambling, the signal is called OC (optical carrier): for example, OC-192.
- SONET/SDH uses time-division multiplexing, in which parallel signals A, B, and C can be multiplexed into a higher-speed serial signal by allocating appropriate time slots.

- A SONET/SDH frame has 810 bytes. Its overhead is divided among section, line, and path overheads.
- Section overhead is used in point-to-point connections (regenerator). Line overhead is used by add–drop muxes. Path overhead is used by the terminating equipment (terminal mux).

REFERENCES

Ballart, R., and Yau-Chau Ching, SONET: now it's the standard optical network, *IEEE Communications Magazine*, vol. 27, 1989.

Freeman, R., *Fiber-Optic Systems for Telecommunications*, Wiley, Hoboken, NJ, 2002.

Goralski, W., *SONET, A Guide to Synchronous Optical Networks*, McGraw-Hill, New York, 1997.

ITU-T Recommendation G.707, *Synchronous Digital Hierarchy Bit Rates*, 1996.

Kartalopoulos, S., *Next Generation Sonet/SDH*, IEEE Press, Piscataway, NJ, 2004.

Mukherjee, B., *Optical Communication Networks*, McGraw-Hill, New York, 1997.

Sexton, M., and A. Reid, *Broadband Networking: ATM, SDH, and SONET*, Artech House, Norwood, MA, 1997.

T1.105-2001, *Synchronous Optical Network (SONET) Basic Description Including Multiplex Structure, Rates, and Formats*, 2001.

5

TCP/IP PROTOCOL SUITE

Network Infrastructure and Architecture: Designing High-Availability Networks,
By Krzysztof Iniewski, Carl McCrosky, and Daniel Minoli
Copyright © 2008 John Wiley & Sons, Inc.

5.1 INTRODUCTION

This book is based on the premise that the transmission control protocol/Internet protocol (TCP/IP) suite will be the dominant set of telecommunications protocols used by a wide range of endpoint devices in future evolution of the Internet. Given this premise, it is necessary to present the reader with a background in the TCP/IP protocol suite. Our primary interest in the TCP/IP suite is in how it affects the underlying transport network. With this interest in mind, we concentrate on the systemic impact of the TCP/IP suite but do not attempt to present all the details. A fuller presentation of the TCP/IP suite can be found in the excellent book *TCP/IP Illustrated* (Stevens, 1994, 1995). We begin the chapter with an overview of the entire protocol suite.

5.2 STRUCTURE OF THE PROTOCOL SUITE

The TCP/IP protocol suite uses a simplified layered structure of four layers instead of the seven-layer OSI stack. This reduction in layer count is accomplished mostly by paying less attention to the various higher layers of the OSI stack and lumping them into one layer called the *application layer*. The TCP/IP protocol stack is organized into layers as shown in Figure 5.1. From the bottom up, these layers and their primary roles or responsibilities are:

1. *Link layer.* The principal responsibility of the link layer is to move packets from point to point over direct links. The link layer consists of the physical means of signaling, the electrical or optical devices that drive the physical link, the definition of the data formats used on the link, the definition of the protocol's signaling and state transitions which control the use of the link, the hardware that controls the link [usually called a network interface

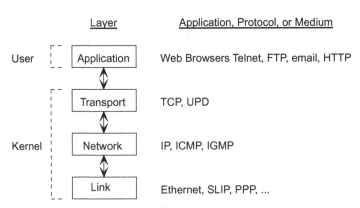

FIGURE 5.1   Four-layer protocol stack of the TCP/IP suite.

controller (NIC) card], and the operating system components and device drivers that interface to the NIC card. Fast Ethernet is an example of a link layer protocol.

2. *Network layer.* The principal responsibility of the network layer is to move packets across the network from source to destination. The network layer consists of protocol elements such as header formats, signaling, and endpoint state transitions which serve the purpose of moving packets through the network, from endpoint through multiple internal nodes to endpoint. The TCP/IP network layer provides a simple unreliable datagram service; that is, it neither recognizes endpoint-to-endpoint flows nor explicitly attempts to maintain a quality of service. Instead, it makes a best effort to deliver correctly each datagram that it is given by the transport layer. The Internet protocol (IP) is the dominant network layer of the modern Internet and the basis for much of the material in this book.

3. *Transport layer.* The principal responsibility of the transport layer is to establish and manage end-to-end communications flows and to provide a reliable service. The transport layer controls the movement of data between two endpoints. This is achieved by a distinct layer of protocol, which consists of header format definitions, signaling between endpoints, and endpoint state transitions. The transport layer views the channel provided by the network layer as an abstract pipe through a network "cloud." The layer has two primary responsibilities: to provide reliable service over the inherently unreliable network layer, and to regulate the flow of information between endpoints over that pipe such as to maximize throughput and to minimize congestion and packet loss in the pipe. TCP is the dominant transport layer of the modern Internet.

4. *Application layer.* The application layer consists of all and any software that makes use of the transport, network, and link layers to achieve endpoint-to-endpoint and process-to-process communications in support of some user purpose. The best known application layer entities are FTP (file transfer protocol) for moving files from endpoint to endpoint, Telnet for remote log-in services, and SMTP (simple mail transfer protocol) for e-mail services. Web browsers (e.g., Safari, Mozilla, Firefox, Explorer, and Netscape) are all applications that support a range of Web services that depend on a range of application layer protocols, such as the familiar "http" service.

It is important to emphasize that the meaning of the word *transport* differs in the terms *transport layer* and *transport networks.* The transport layer of the TCP/IP suite is as defined above, whereas transport networks are systems that act as a worldwide carrier for the TCP/IP protocol suite and appear to operate at the link layer of the TCP/IP protocol stack.

While the TCP/IP stack is the key service provided by the protocol suite, the remaining elements of the protocol suite provide other services and support the key TCP service as well as an alternative transport layer service. Figure

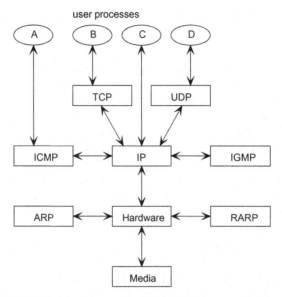

FIGURE 5.2   Protocols of the TCP/IP protocol suite.

5.2 illustrates the full TCP/IP protocol suite and positions it with respect to the physical media and endpoint hardware. The elements of the TCP/IP suite are as follows:

- *Media.* This layer consists of the physical medium (e.g., copper, fiber) that carries the signal.
- *Hardware.* This layer consists of the link terminating hardware, normally including the media interface and the digital hardware that implements the link layer interface.
- *IP.* The Internet protocol is the TCP/IP protocol suite's definition at the network layer. IP makes its best effort to deliver packetized datagrams from endpoint to endpoint, where endpoints are defined by the IP addressing scheme.
- *TCP.* The transmission control protocol is one of the TCP/IP protocol suite's definitions at the transport layer. TCP provides a reliable flow-controlled, end-to-end connection.
- *UDP.* Like TCP, the user datagram protocol runs on top of IP and provides a data service to user processes. However, UPD is a simplified layer that provides only an unreliable datagram service.
- *ICMP.* The Internet control message protocol supports IP by communicating error and control status among IP nodes in a network.
- *IGMP.* The Internet group management protocol supports the multicasting of UDP packets.

- *ARP.* The address resolution protocol supports a distributed mapping from IP addresses to physical link layer addresses.
- *RARP.* The reverse address resolution protocol supports a distributed mapping from physical link layer addresses to IP addresses.

The four user processes in Figure 5.2 represent the possible ways in which application processes can interface with the TCP/IP protocol stack:

- *Process B* uses the stack consisting of TCP/IP/hardware/media as a reliable but relatively expensive means of transferring data.
- *Process D* uses the stack consisting of UPD/IP/hardware/media as an unreliable but lighter-weight and lower-latency means of transferring data.
- *Process C* illustrates the rarely used mode of sending unreliable data directly over IP/hardware/media.
- *Process A* illustrates the fact that user processes can communicate directly with the ICMP protocol to assist in managing the overall communications system.

As user data pass from the application layer down the stack to the link layer, each protocol layer adds information in the form of new headers prepended to the user's data or in the form of new trailers postpended to the user's data. This incremental accretion of control information is illustrated in Figure 5.3. At the top of Figure 5.3, we suppose that some application needs to transfer 32 bytes of information via a TCP/IP protocol stack over an Ethernet link layer. The application immediately appends its own header, in this hypothetical case an 8-byte header which carries an indication of the data type and length and some form of control information indicating what is to be done with the application data.

The application hands the 40 bytes consisting of the application header and application data to the TCP layer. TCP typically adds a 20-byte header, the various roles of which we examine as we proceed through the chapter. The 60 bytes consisting of the TCP header and the application information are referred to as a *TCP segment*. TCP hands the TCP segment to the IP layer, as TCP depends on IP to provide end-to-end transport. The IP layer adds another header of 20 bytes. The aggregate 80 bytes is referred to as an IP segment. In this example, IP is dependent on Ethernet to provide the link layer services. Consequently, IP hands the IP segment to Ethernet software, which adds an Ethernet header and an Ethernet trailer before sending the complete Ethernet segment over the physical link.

All four of these layers are typically implemented in software, as their behaviors are too complex to be implemented directly in hardware. Software running on appropriate processors is capable of completing the various functions in a timely manner. The application layer is typically a user process; the

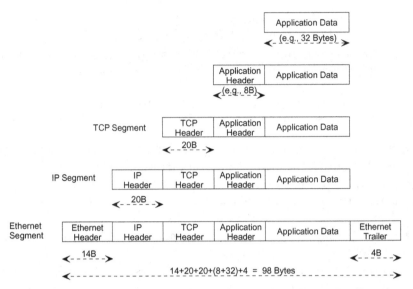

FIGURE 5.3   Progressive addition of protocol headers and trailers.

remaining layers (TCP, IP, and Ethernet) are typically system processes buried in an operating system or its device drivers.

Several key points should be noted about this layered aggregation of headers and trailers.

1. The segment grows substantially from its application layer needs to the form that is finally transferred on Ethernet. In this case the segment has grown from 8 + 32 bytes to 98 bytes, for an overall expansion ratio of 2.45. If the application layers needs were to transfer a 512-byte disk block, we would have seen 512 bytes grow by the much more modest factor of 1.11, to 570 bytes. In either case, though, the demands on the telecommunications systems are increased noticeably by the requirements of the telecommunications system itself.

2. The processing required for each layer is summarized in later sections of this chapter, but we can safely assert that it is not a trivial amount of work. Thus, the telecommunications system imposes processing loads on the TCP endpoints and on every network element that has reason to examine or process the segment as it makes its way to its destination endpoint.

3. The natural way to implement the various layers of the protocol stack is to have separate software modules for each layer. It would be easiest for each of these layers to accept a pointer to the incoming segment and to build a new, expanded version of the segment with added headers and/or trailers in a new memory space, copying the incoming segment in the process. However, this is

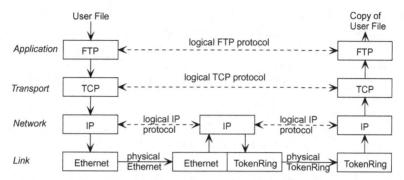

FIGURE 5.4   File transfer application using the TCP/IP protocol suite.

a moderately expensive operation, as modern processors are much slower at moving large blocks of memory than they are at arithmetic and logical calculations. Implementations are more efficient when the incoming segment does not have to be moved in memory, but new headers and/or trailers can be added in place. This is a worthwhile goal but is also a challenging problem for software design. Some of the better implementations achieve some of the possible memory-to-memory copy reduction, but in general it remains a significant cost. From the software point of view, this is another cost (or difficulty) imposed by the needs of the telecommunications system to layer its definition. However, modern software engineering tends toward comparably layered designs, so perhaps this cost should be attributed to the overall costs of defining complex systems.

Figure 5.4 shows another view of the layering of TCP/IP. On the left of the figure we see our four-layer protocol stack consuming some user file passed to it by the FTP application, which passes the file (in segments) to TCP, then to IP, then to Ethernet. At the right of the figure we see the stream of information rising from the link layer, up to IP, to TCP, to FTP, and finally, back as a copy of the user's file. Thus, we see our layered protocol implemented as stacks of (software) processing blocks carrying the file as a stream of segments or packets.

Notice the lateral correspondences between the left and right layers. Each pair of protocol layer implementations has a defined relationship with its peer at the same level at the other end of the connection. Thus, the two FTP entities appear to each other to share a logical FTP protocol with which they transfer the file from left to right. However, although these two FTP entities appear to be using the FTP protocol, they in fact send each other messages by appealing to the services provided by TCP. Thus, the two FTP entities have a logical relationship and a services relationship, the first with their application layer peer, the latter with their transport layer service module.

A similar relationship exists between the two transport layer entities. In this case they also receive requests from their application layers, but they share a logical TCP protocol with their same-layer peer and a service relationship with the IP layer that lies below. Of course, IP depends on the link layer in the same manner as it accomplishes the requests of the IP layer entities. The dataflow arrows in Figure 5.4 flow from the user file on the left to the copy on the right. This indicates the direction in which the user data move but does not reflect the full situation. In fact, most protocol layers require a duplex communication pattern in order to move the user's file successfully and reliably. We examine these duplex protocols as the chapter proceeds.

If we look more closely at Figure 5.4, however, we see more interesting features. The left protocol stack terminates with Ethernet at the link layer. The right protocol stack terminates with TokenRing at the link layer. There is a two-layer entity connected between them which accepts Ethernet on the left and TokenRing on the right. This intermediate entity takes segments from the left, moves them back up to the IP layer, then moves them down to the Token-Ring link layer. In moving up, the Ethernnet header and trailer are removed. In the IP layer, IP routing (to be discussed) is performed. In moving down to the tokenring link layer, TokenRing control information is added to the IP segment.

The entity in the middle of this network and link layer path is an IP router with both Ethernet and TokenRing ports. The router operates fundamentally at the IP layer but has protocol entities (and physical ports) to converse with both Ethernet and TokenRing. This simple example illustrates the source of a substantial part of the advantages of layered telecommunications architectures. Because the left and right entities (e.g., PCs) share FTP at the application layer, TCP at the transport layer, and IP at the network layer, they can communicate even though they have made different choices at the link layer. You might think of the left entity being your personal PC in your Ethernet-connected home, while the right entity is a file server buried in some service provider's infrastructure which is connected with TokenRing. Happily, you don't know or care that the service provider chose to use TokenRing. Because you've used shared standards (FTP, TCP, IP, and Ethernet), and because you are connected to an IP router with Ethernet ports, you need not be concerned that that router also has a TokenRing port connected to your TokenRing-using service provider.

The idea of link layer independence illustrated in Figure 5.4 is much more profound than that simple example shows. In fact, the entire concept of this book—transport networks that carry IP segments from user to user—is a conceptually simple extension of Figure 5.4.

Figure 5.5 shows the router at the link and network layers of Figure 5.4 replaced by a single entity: a transport network. Just as the Ethernet/router/TokenRing combination in Figure 5.4 carries IP services from place to place, the transport networks we examine in this book serve the same purpose. Real transport networks are vastly more complex, varied, and interesting than the

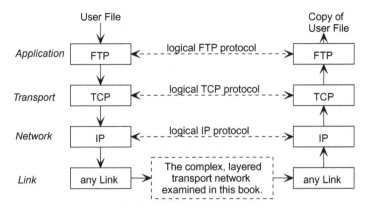

FIGURE 5.5 TCP/IP protocol suite over a transport network link layer cloud.

simple box shown in Figure 5.5. They consist of many complex interacting WANs which cover most of the world. We can, in fact, think of the chapters of this book that follow the current chapter as an attempt to describe the complexity that goes into building the link layer entity in Figure 5.5.

5.3   INTERNET PROTOCOL

The Internet protocol (IP) is the network layer solution for the TCP/IP protocol suite. All of the network traffic we are concerned with in this book is carried by IP. Ideally, this universality of IP might seem to make it the one and only protocol that we need to consider in studying the underlying transport network. However, we shall see that IP does not provide certain critical services on its own but that users must rely on the higher-layer protocols, especially TCP, to provide a full range of services. Thus, while IP is "universal," it is also incomplete in important ways. As we have seen, the network layer in general and IP in particular are responsible for moving varying-length datagrams (packets or segments) from one IP endpoint to another. The service offered by IP is unreliable in the sense that IP can drop a packet due to detected bit errors, queue overflow, or complete failure of network components. The IP layer makes no attempt to correct any such loss. It may or may not inform other layers or other parts of the system of such losses.

5.3.1   IP Addresses

One of the most important components of IP is the definition of endpoint addresses. These addresses uniquely define the endpoints between which IP segments are routed. IP (version 4) addresses consist of 32 bits. There are five formats of IP address, classes A through E (Figure 5.6). Classes A through C

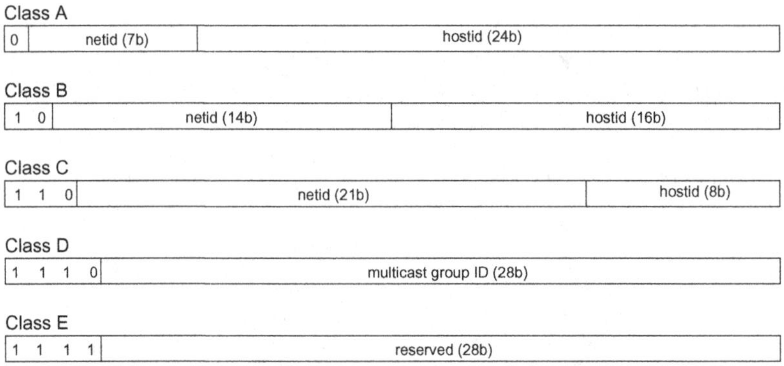

Class A

0	netid (7b)	hostid (24b)

Class B

1	0	netid (14b)	hostid (16b)

Class C

1	1	0	netid (21b)	hostid (8b)

Class D

1	1	1	0	multicast group ID (28b)

Class E

1	1	1	1	reserved (28b)

FIGURE 5.6   IP address formats.

are unicast addresses that specify one endpoint. The differences in the three classes have to do with the number of address bits used to describe the identity of the network on which the node resides (netid) and the number of address bits used to describe the identity of the host within its network (hostid). The three classes were provided to permit the description of a few networks with very large numbers of hosts each (class A), and many networks with few hosts each (class C). Class D is reserved for multicast and is discussed later in the book. Class E is reserved. IP addresses are written as four-decimal separated integers in the range 0 to 255, with each integer value specifying the contents of 8 bits of the address (e.g., 197.210.7.9).

5.3.2   IP Header Format and Function

The header appended to the user segment by the IP protocol is illustrated in Figure 5.7. The mandatory portions of the IP header consume 20 bytes, or five 4-byte words. The various fields are discussed in turn below.

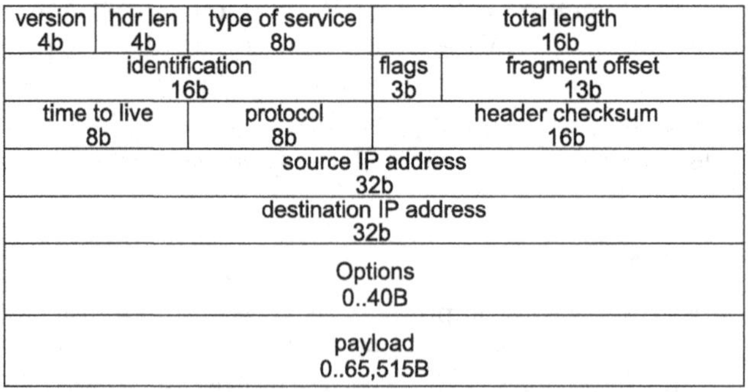

version 4b	hdr len 4b	type of service 8b		total length 16b	
identification 16b			flags 3b	fragment offset 13b	
time to live 8b		protocol 8b		header checksum 16b	
source IP address 32b					
destination IP address 32b					
Options 0..40B					
payload 0..65,515B					

FIGURE 5.7   IP header format.

- *Version* (word 0, bits 0 to 3). This field distinguishes IP version 4 from IP version 6. As version 4 remains the dominant standard, most of this chapter covers that version. Version 6 is described in Section 6.10.

- *Header length* (word 0, bits 4 to 7). This field encodes the number of 4-byte words in an IP header. For a minimum length header of 20 bytes, this field is the number five. As options (see below) are added to the minimal header in units of 4 bytes, this field is incremented toward its maximum value of 15, which indicates five words of nonoptional header plus 10 words (or 40 bytes) of options. Most IP segments do not carry any option information.

- *Type of service* (ToS; word 0, bits 8 to 15). This field carries a 3-bit priority field and a 4-bit packet treatment field (minimize delay, maximize throughput, maximize reliability, minimize cost). These fields were not used in most IP implementations for many years, so there are many IP routers and endpoints in the field that do not set or respect these bits. As we shall see, were these bits respected from end to end throughout the Internet, some serious problems of network quality of service could have been solved or at least mitigated. However, the lack of end-to-end support for the seven defined ToS bits (one bit is undefined) means that attempts to add quality of service based on IP ToS are doomed to failure as soon as the traffic passes outside a ToS-respecting domain.

- *Total length* (word 0, bits 16 to 31). This is the total length of the IP segment in bytes, including the header. The minimum value is 20, which occurs when no options and no payload are being sent. The maximum value is $2^{16} - 1 = 65,535$. Assuming only the minimum IP header of 20 bytes, the payload can range from 0 to 65,535 bytes. However, IP nodes are not required to accept IP segments greater than 576 bytes. Such an IP segment, with a minimal header, carries 556 bytes of payload.

- *Identification* (word 1, bits 0 to 15). This field is used by the IP protocol to identify each IP segment uniquely. Generally, an endpoint that generates IP segments has a 16-bit counter which is copied into each segment sent and incremented before the next segment is processed. Although the counter wraps around every 2^{16} segments, it is still possible for IP endpoints to assume that all packets within a suitably large window are labeled uniquely.

- *Flags* (word 1, bits 16 to 18). Two bits are used to support IP's segment fragmentation mechanisms (to meet a segment size constraint on some IP link). Fragmentation is described later in this chapter. One bit is turned on to indicate that the current segment is not the last fragment of a larger segment; the other bit is turned on to prohibit fragmentation of the segment.

- *Fragment offset* (word 1, bits 19 to 31). When an IP segment has been fragmented, this field gives the offset in bytes, from the beginning of the original segment, of the contents of the current segment.

- *Time to live* (TTL; word 2, bits 0 to 7). Initially this field, contains the maximal number of IP routers through which the segment should be propagated before being discarded. Each router visited is expected to decrement this count by 1, until the segment is discarded when the count goes to zero. The intent is to avoid infinitely looping packets congesting the network when there are errors in routing tables.
- *Protocol* (word 2, bits 8 to 15). This field contains an integer code for the transport layer protocol being carried by the IP segment (e.g., TCP or UDP).
- *Header checksum* (word 2, bits 16 to 31). This field is a checksum computed over the entire IP header used to detect transmission errors. Note that the checksum must be recomputed at each router as the TTL field changes.
- *Source IP address* (word 3, bits 0 to 31). This is the IP address of the node that generated the segment.
- *Destination IP address* (word 4, bits 0 to 31). This is the IP address of the node to which the segment is destined, or a multicast address (for class D).
- *Options* (words 5 to 14). The contents of the options field is a variable-length description possibly including security and handling conditions, a trace of the route taken by the segment to date, a sequence of timestamps from visited routers, a loose source routing (giving a list of IP address which may be transited by the segment), and strict source routing (giving a list of IP address which must be transited in sequence by the segment, with no other nodes being visited). Trace options are rarely used in modern networks, because the maximal length of 40 bytes is insufficient for use with most of the purposes in complex modern networks with many routers.

IP datagrams can be split into multiple smaller IP datagrams when in transit. This resegmentation occurs when an oversized IP datagram needs to transit a link that permits only small IP segments. The fragment offset and flags fields of the IP header support this process. Once an IP segment has been split, it will not be reassembled until it reaches its destination.

5.4   USER DATAGRAM PROTOCOL

The user datagram protocol (UDP) is a transport layer protocol carried by IP. UDP adds little to the basic unreliable IP datagram (see the UDP header format in Figure 5.8). The only new functions it adds are source and destination port numbers and a checksum over the payload. The source and destination port numbers, of 16 bits each, allow multiple processes or servers on

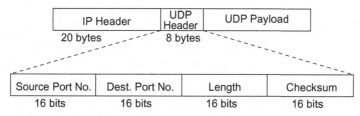

FIGURE 5.8 UDP header format.

endpoints to define independent UDP channels (e.g., server X might use UDP port P, while server Y on the same endpoint might use port Q).

With almost 2^{32} IP addresses for sources and sinks, and with 2^{16} UDP ports for sources and sinks, the Internet supports $2^{32} \times 2^{16} \rightarrow 2^{32} \times 2^{16} = 2^{96} \cong 8 \times 10^{28}$ possible UDP connection paths. Recall that the IP header checksum protects only the IP header and leaves the payload unprotected. The checksum in the UDP header covers the header and the UDP payload. Without this UDP checksum, the payload would not be protected.

The length field contained in the UDP header is redundant with respect to the IP length field. UDP's demands on the transport network are relatively light, as it requires only a simple best-effort attempt to get payloads to their destination, with no retransmission, strong quality-of-service conditions, or error notification. When UDP is used, the endpoint applications assume responsibility for dealing with a possibly poorly performing network. Thus, the challenges imposed by UDP on modern transport networks are modest. Instead, we focus on TCP, which provides no end of interesting challenges in its attempts to provide a higher level of services.

5.5 TRANSMISSION CONTROL PROTOCOL

The transmission control protocol exists at the transport layer of the TCP/IP stack. TCP adds substantial new functionality to the underlying IP protocol. The most important new feature is that TCP is intended to be a reliable service, whereas IP is not required to be reliable. This reliable service feature, more definite ideas regarding quality of service, and other features make TCP a very interesting challenge, both for the design of the protocol itself and for the transport networks that must carry TCP/IP segments while avoiding causing service failures in which TCP fails to achieve its goals of high-quality services to endpoint processes.

Each established TCP connection is bidirectional. Each direction of these duplex connections is modeled as a simple stream of bytes that are to be communicated in order. When a new connection is established, an initial sequence number (ISN) is exchanged between the two endpoints for each direction of

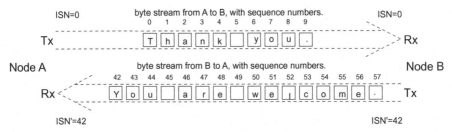

FIGURE 5.9   TCP byte stream concept.

flow. The bytes in each direction of flow are numbered from this ISN up to the last byte in the sequence. Sequence numbers occupy 32 bits, which allows them to range from over 0 to $2^{32} - 1$. Since sequence numbers have a finite range and can wrap around to zero, it is necessary that TCP accommodate these wraps without getting confused, by recognizing that $2^{32} - 1$ precedes zero. A random ISN is used as protection against the establishment of random TCP connections based on errors in the establishment protocol.

An example of two numbered byte streams which comprise a complete duplex TCP connection is illustrated in Figure 5.9. The direction from node A to node B has (by chance) started with an ISN of zero. The byte stream to be communicated is "Thank you.", the bytes of which are numbered 0 to 9. The direction from node B to node A has an ISN of 42. The byte stream communicated is "You are welcome.", the bytes of which are numbered 42 to 57. The bytes of the two byte streams will be packed in TCP/IP segments and transmitted to their destination endpoint. The logic of TCP and IP, coupled with the timing of arrival of the bytes from the application layer, will determine how many bytes are packed into each TCP/IP segment. Obviously, there is an advantage to packing the bytes into as few segments as possible, to save on header overheads. However, packing the byte streams into segments is dependent on the maximum segment size being used and the timing of the arrival of the bytes from the application layer in the source.

The example in Figure 5.9 seems to imply an interactive conversation between the two endpoint nodes. If this is the case, we expect that the A-to-B message ("Thank you.") appears all at once from the application layer in A, that the entire byte stream will fit in one TCP/IP segment, that the node B application layer will reply "You are welcome." all at once, and that that reply will be returned in one segment to A. However, there could be arbitrary temporal delays inserted by the endpoint application layers at any point in a more complex conversation.

Or, instead of an apparent conversation, the A-to-B byte stream could be the entire contents of a file (plus some control overheads), and the B-to-A byte stream could be nothing but some control information. This scenario brings to mind the FTP file transfer application. In this case, the only temporal delays in the byte stream will be caused by the process of reading the file and preparing it for transmission over the TCP/IP link. In this scenario, IP segments will tend to be as large as permitted.

The point of these two examples is that the TCP byte stream model allows for a huge range of byte stream sizes and temporal behaviors by the endpoints, some of which we would call *interactive*, some of which we would call *batch*. But the entire range of behaviors must be managed by a common TCP/IP mechanism. We shall see that TCP is defined as a way to make progress through the byte stream offered by the application layer, regardless of when it appears at the TCP interface. Regardless of these variations in the load model, the goal of TCP is to reliably reconstruct the byte stream at the receiver.

5.5.1 TCP Header Format and Function

We begin our study of TCP by examining its header, which nestles between the TCP payload and the IP header, as shown in Figure 5.10. Like the IP header, the basic TCP header consists of five 4-byte sections, for a total of 20 bytes. Also like IP, TCP has provision for optional extensions to the header. The fields of the TCP header are:

- *Source port number* (word 0, bits 0 to 15). This field contains an integer in the range 0 to $2^{16} - 1$, which identifies the logical port, and hence the software service, which originated the segment on the source endpoint.
- *Destination port number* (word 0, bits 16 to 31). This field contains an integer in the range 0 to $2^{16} - 1$, which identifies the logical port, and hence the software service, which will consume the segment on the destination endpoint.

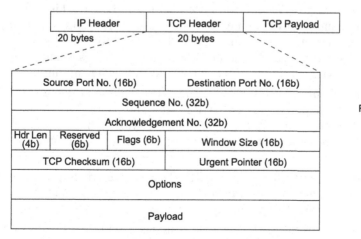

FIGURE 5.10 TCP header format.

- *Sequence number* (word 1, bits 0 to 31). This field identifies the logical position of the first byte of payload carried by this segment in the overall byte stream that constitutes the TCP connection. The range of this field is 0 to $2^{32} - 1$, which allows unique numbering for sequences of up to about 4 billion bytes. TCP manages the wraparound of sequence numbers.

- *Acknowledgment number* (word 2, bits 0 to 31). This field identifies the logical position of the first byte in the overall byte stream that constitutes the TCP connection which has not yet been received successfully by the endpoint that originated the TCP segment. This concept is described more fully in sections that follow.

- *Header length* (word 3, bits 0 to 3). This field encodes the number of 4-byte words in the header. For the minimum-length header of 20 bytes, this field is the number 5. As options (see below) are added to the minimal header in units of 4 bytes, this field is incremented toward its maximum value of 15, which would indicate five words of nonoptional header plus 10 words (or 40 bytes) of options. Most TCP segments do not carry any option information.

- *Reserved* (word 3, bits 4 to 9).

- *Flags* (word 3, bits 10 to 15). This field contains six single-bit Boolean flags: URG indicates that the urgent pointer is valid (unused in practice); ACK, indicates that the acknowledgment number is valid; PSH, indicates that the receiver endpoint should pass the payload to the user process as soon as possible (unused in practice); RST, indicates a request to reset the TCP connection; SYN, declares an attempt to synchronize TCP sequence numbers between the two endpoints; and FIN, states an intention to terminate a TCP connection (described more fully later).

- *Window size* (word 3, bits 16 to 31). This field plays a role in the sliding window transmission control system built into TCP. This scheme is described more fully later, but for now it is appropriate to state that this field specifies the number of bytes in the connection's byte stream that the receiver is prepared to accept. This number of bytes is understood to be those bytes starting with the current acknowledgment number.

- *TCP checksum* (word 4, bits 0 to 15). This field is an error detection checksum computed over the TCP header and payload. As IP does not protect its payload (the TCP header and payload), this TCP checksum provides an important protection service that would otherwise be missing.

- *Urgent pointer* (word 4, bits 16 to 31). This field describes a subset of the connection's byte stream which should be treated as urgent (when the URG flag is set). This feature is rarely used.

- *Options* (words 5 to 14). This field contains from 0 to 40 bytes of TCP option information. Unlike the IP options field, this field has at least one significant role: carrying the maximum segment size from receivers to senders, to avoid buffer overrun in TCP receivers. Like UDP, with almost 2^{32} IP addresses for sources and sinks, and with 2^{16} TCP ports for sources and sinks, the Internet supports $2^{32} \times 2^{16} \rightarrow 2^{32} \times 2^{16} = 2^{96} \cong 8 \times 10^{28}$ possible TCP connections.

5.5.2 Connection-Oriented Service

TCP is a connection-oriented service. Thus, connections must be established when needed and torn down when their purpose is completed. The details of establishment are not of great importance to the purposes of this book, so will not be dealt with in detail. A quick overview will suffice: One end proposes to the other that they establish a connection, the recipient of the proposal agrees with an acknowledgment message, and the original proposer acknowledges the acknowledgment. In the process, the two endpoints exchange the port numbers they intend to use and the maximum segment size that each will tolerate, and the starting sequence numbers that will be used to refer to progress in transmitting the two unidirectional byte streams that constitute the duplex interchange. A variant of this scheme occurs when both endpoints actively request a connection.

5.5.3 Receiver Window

At any time, some number of bytes in each of the two byte streams that make up a duplex TCP conversation have been transmitted from the source all the way to the destination. This sequence of bytes starts with the mutually agreed upon ISN and proceeds to the most recent acknowledgment number received by the transmitter. These bytes have been transmitted successfully and never need to be retransmitted by this particular connection. The acknowledgment number gives a definite indication of where this completion of transfer ends and where more work must be done to complete the transmission of the byte stream. From the acknowledgment number and onward, we may find that some bytes have already been transmitted but are in progress from the source to the sink, which has not yet seen TCP segments with those bytes. We may also find that some bytes have reached the destination and a suitable acknowledgment has been formulated and sent but has not yet reached the transmitter, which consequently cannot know that those bytes have been transmitted successfully.

In addition, we may find that some segments containing bytes from this byte stream have been lost while being transmitted from the source to the destination. Lost? Can networks lose segments? Yes, they can. When a segment with a bad checksum is received, it is thrown away (with possible notification back to the source). When a router has a full buffer and no place to store an incoming frame, it (or some lower-priority frame) is thrown away. So yes, segments

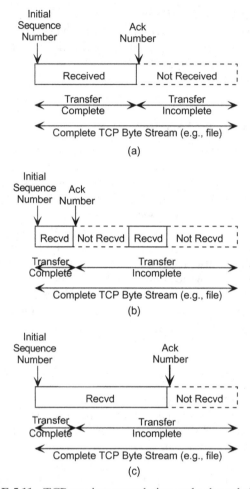

FIGURE 5.11   TCP receiver completion and acknowledgments.

regularly get "lost." The design of TCP is interesting in large part because it finds effective means to get around the problem of providing a reliable service in the presence of errors and overflows.

In Figure 5.11(a) we see a simple situation in which an initial segment of bytes from the unidirectional byte stream from source to destination has been transmitted and received successfully. The acknowledgment number is pointing to the next byte in sequence. We do not know from this diagram whether this next byte has been transmitted (in a TCP segment, presumably with other succeeding bytes), or whether a frame is presently returning from the far end with the Ack flag on and an acknowledgment number indicating that this next byte has been received by the far end. Or, of course, either the data-carrying segment or the Ack-carrying segment could have been lost.

Figure 5.11(b) shows a more complex situation. The receiver has received two segments of the byte stream, but they are separated by a segment of bytes that have not been received. The leftmost received segment of bytes starts from the beginning of the overall byte stream and is indicted as being completely received up to the byte before the acknowledgment number. Then the unreceived gap begins, followed by the second received segment. This situation could have arisen for a variety of reasons. For the moment, let's keep things simple and assume that the acknowledgment for the "unreceived" gap has merely been delayed in transit. A short time later, that delayed acknowledgment arrives at the sender. With that arrival, the sender understands that all the bytes from the beginning of the byte sequence through the end of the (formerly separated) second segment received have in fact been received. The TCP protocol implementation in the sender then advances its value of the acknowledgment number to just past the second of the earlier segments, as shown in Figure 5.11(c).

5.6   TCP FLOW CONTROL

At the network layer, IP offers no flow control mechanisms. IP sources transmit whenever they receive a segment from the transport layer. Networks and sinks can be overwhelmed by overly aggressive traffic loads, resulting in high loss rates and poor overall service. Some layer of the protocol stack must assume responsibility for flow control in order that networks can deliver reasonably efficient and reliable services. It is clear that networks cannot rely on the application layer for flow control discipline. Although we do not want to encourage the notion that the customer (i.e., the application layer) is the enemy of networks, we must accept the fact that customers will try to obtain the best/most services that they can, and their self-interest seems to be best served by submitting more traffic. If all customers pursue this narrow self-interest, networks will tend to congest, and all customers will suffer. What is needed is some combination of enlightened self-interest responding to some sense of communal best interest. Discipline to prevent network chaos must come from somewhere. We have only one remaining layer as a source of discipline: the transport layer. Happily, the TCP protocol at the transport layer does take up this challenge and provides several interacting mechanisms to enforce a reasonable level of flow control in the face of network limitations.

TCP's flow control mechanisms can be divided into two types: receiver-based flow control and transmitter-based flow control. *Receiver-based flow control*'s primary role is to prevent receiver buffer overflow. *Transmitter-based flow control*'s primary role is to further limit the flow in response to network conditions.

5.6.1   Receiver-Based Flow Control

We begin by examining the receiver-based mechanism. Each receiver (or each receiver portion of every endpoint, as TCP communications are bidirectional) has a window consisting of buffer space for incoming payloads. During the opening of a TCP connection, each receiver informs its far-end transmitter of the size of its receiver window. Normally, receiver windows are $W = 2^{16} - 1 =$ 65,535 bytes. During operation of the connection, a transmitter is never allowed to send payload bytes if it does not know for certain that the payload bytes will fit into the receiver window. This ensures the property that a receiver will never receive payload bytes that it cannot save because it has no buffer space available.

A naive flow control scheme is the *stop-and-wait scheme*, where a transmitter sends only one payload segment, then waits for an acknowledgment of that segment before sending the next payload segment. This scheme is simple to understand and requires only one segment buffer in the receiver. But it produces poor performance, and performance degrades with increasing time of flight between source and destination. Figure 5.12 illustrates this scheme and its relatively wasteful use of network facilities.

TCP avoids the inefficiencies of stop-and-wait flow control by defining a windowed flow control scheme. The basic idea of a windowed scheme is that the transmitter is free to send enough packets to fill the window in the receiver, but then must stop. The receiver must have at least one window's worth of segment buffers. As the receiver frees up used segment buffers (by moving their contents up to the application layer), it returns credits for further use of the receiver buffers to the transmitter. The transmitter than takes advantage of the increased credit by transmitting more segments. There are a number of (nearly) equivalent ways to manage a window-based flow control scheme. TCP

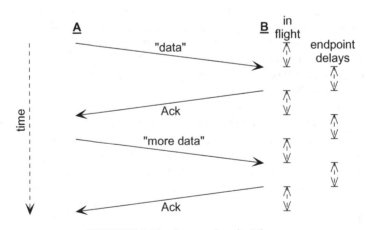

FIGURE 5.12   Stop-and-wait delays.

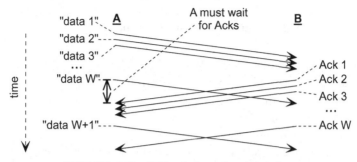

FIGURE 5.13   TCP windowed transmission.

uses a method know as a *sliding window without selective acknowledgments.*
We examine this scheme in some detail in the following paragraphs.

Figure 5.13 illustrates a windowed flow control scheme. The transmitter
begins with some allocated budget, W, of bytes that it can transmit before any
response from the receiver. When the transmitter has exhausted this window,
it must stop and wait for acknowledgments from the receiver. Each acknowledgment received informs the transmitter that one of its units of data got
through successfully and that the receiver is prepared to receive another such
unit. So the transmitter takes each acknowledgment as an authorization to
transmit another unit of data. If the window size is large enough with respect
to the round-trip delay from A to B to A, transmission can be a nearly continuous process. However, if the window is too small, more and more delays are
forced on the pair. When W becomes one segment, the window has degraded
to a stop-and-wait system.

Notice that this example has allowed the receiver, node B, to send acknowledgments that allow reuse of the receiver window immediately upon receiving
each packet. This differs from the details of TCP, which are explained below.
In anticipation of a fuller explanation, we can say that TCP acknowledgments
acknowledge proper reception of data but do not allow the transmitter to
reuse receiver buffer space. TCP has a related receiver window size, which is
also returned to the transmitter, which informs the transmitter when the messages have actually been removed from the receiver window by the receiver
application layer. Only at this time is the window space reusable.

It is clear that a new connection allows the transmitter to send W bytes in
a carefree manner, as the receiver has informed the transmitter that W bytes
are available in its receiver window. But how does the connection manage to
make progress past this point? The solution is to allow the receiver window
to "slide" forward in the connection's byte stream as progress is made. Recall
that each direction of communications in a TCP connection consists of a byte
stream of arbitrary length (let's assume that it is longer then W). The transmitter segments this byte stream into payload segments, adds TCP headers, and
hands them off to IP. Each such transmission subtracts a number of bytes

available in the known receiver window equal to the byte length of the payload segment. Normally, the receiver performs the following functions:

1. It receives these segments of payload in order.
2. It checks to ensure that no errors occurred.
3. It acknowledges their correct reception by an acknowledgment message back to the transmitter.
4. It moves the payload segments received out of the receiver window and up to the application layer where other buffers will hold the contents, as desired or required by the application.

The sequence of events in this receiver action list is critical. If on receiving the acknowledgment, the transmitter concluded immediately that another segment could be sent to fill the space of the segment acknowledged, that next segment could arrive in the receiver before the previous contents of the buffer were actually moved to the application layer. If this were to happen, the TCP receiver would be left with no option but to drop the new segment or overwrite the old segment. TCP deals with this potential problem by advertising a current receiver window size with each segment. The current window size is the number of bytes that the transmitter can fill, either with segments that have already been sent or with segments to be sent. The transmitter is not permitted to go beyond the end of the last receiver window size that it received from the far-end receiver. The management of this window is moderately complex. The window "closes" (becomes smaller) when a segment of data is acknowledged. The window "opens" (becomes larger) when the application layer entity in the receiver removes data from the receiver window's physical buffer. There is a third buffer operation, called "shrinking," which we do not discuss, as this option is discouraged by the TCP standards.

The operation of the TCP sliding window is described in reference to the example illustrated in Figure 5.14. In this example, the byte stream to be sent from the transmitter to the receiver is the sequence of bytes 0 to 5. For simplicity, these bytes will be sent with just 1 byte per TCP segment. Or, the reader can think of each of these integers as representing a larger amount of data: for example, one disk block each. The situation opens in Figure 5.14(a) just after a connection has been established and no data segments have been sent or acknowledged. The receiver window has been established by the opening of the connection with a length of 3. The example shows the bytes of the connection numbered from zero, but this is for convenience only. The opening of the connection will have established an ISN in the range 0 to $2^{32} - 1$.

Just before the case shown in Figure 5.14(b), the transmitter has used its open window of length 3 to send the first data byte, 0. As this byte has not yet been received and acknowledged, the window does not move. The transmitter

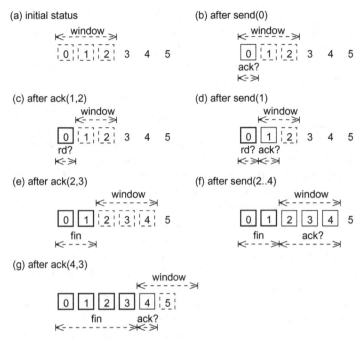

FIGURE 5.14 TCP receiver window progress.

keeps track of the fact that it has used this part of its window, but does not alter the window. Byte 0 is labeled with "ack?" to indicate that the transmitter is waiting for these data to be acknowledged.

The situation shown in Figure 5.14(c) occurs just after an acknowledgment has been received by the transmitter of the byte stream. This acknowledgment carries two pieces of information relevant to the sliding window: It carries the number of the first unacknowledged byte in sequence (its first parameter), and it carries the size of the currently advertised window (its second parameter). In Figure 5.14(c) the acknowledgment number is 1, indicating that the byte sequence has been received correctly up to byte 0. The advertised window size is 2, which is a reduction from the original window size of 3. This represents a "close" of the window. The reason for this reduction in window size is that the TCP receiver still has that data (the single byte, 0) in the receiver window and must inform the transmitter that it cannot assume use of that space at this time. Only when the receiver application layer has read and assumed responsibility for that part of the window can the occupied space be readvertised to the far-end transmitter. Byte 0 is labeled with "rd?" to indicate that TCP is waiting for these data to be read by the application layer.

Figure 5.14(d) is a snapshot immediately after data byte 1 has been sent in a TCP segment. The window size remains unchanged, but the transmitter has consumed part of its window by filling it with these data.

Figure 5.14(e) depicts the situation after an Ack(2,3) has been received, indicating acceptance through byte 1 and a current window size of 3. The left-hand edge of the window closes by 1, due to acknowledgment through byte 1. At the same time, the window size has increased back to its original value of 3, due to a counteracting opening of the window by 2. The reason the receiver indicated an opening of the window by 2 is because the receiver application layer has by this time read the first 2 bytes out of the receiver window, making this space again available for the transmitter. Thus, the receiver has decremented the window size once for the close due to the acknowledgment and has incremented the window size twice for the two spaces made available by the application layer reads of bytes 0 and 1 $[(W = 2) - 1 + 2 \rightarrow (W' = 3)]$.

The transmitter gets down to serious work before the stage shown in Figure 5.14(f), and sends bytes 2 through 4. The window remains unchanged (as there have been no new acknowledgments and no new reads by the receiver's application layer). However, the transmitter is aware that it has filled its current window with transmissions and is not free to transmit any more data. If it did, only lucky timing would prevent loss of data. The transmitter must wait for new window information from the receiver.

Good news arrives from the receiver just before the situation shown in Figure 5.14(g). An Ack(4,3) is received, indicating that the receiver has all the bytes through byte 3 and that the window size is back to 3. As the last acknowledgment was through byte 1 (the acknowledgment number being 2), we now know that two more bytes have been received (bytes 2 and 3). This closes the window by 2. At the same time, the receiver has advertised a current window size of 3. This implies that the receiver application layer consumed the next two bytes, bytes 2 and 3. Thus, this single acknowledgment indicates that bytes 2 and 3 have been acknowledged and consumed. The transmitter consequently shifts both the left and the right edges of the window to the right by 2. The transmitter knows that byte 4 is still within that window, awaiting an acknowledgment.

This discussion of Figure 5.14 has considered ideal progress of a TCP connection and its sliding receiver window. There are several possible complications:

- Segments carrying payloads are lost in transmission (dropped due to bit errors, dropped due to router buffer overflow, or dropped due to TTL expiry), never arrive at receivers, and thus never generate an acknowledgment.
- Acknowledgments are lost in transmission, as the packets that carry them are lost.
- Payload segments arrive out of order.
- An acknowledgment arrives out of order.

We consider the first two problem cases in Figure 5.15 and the following discussion. We begin by assuming that the segment carrying the payload from

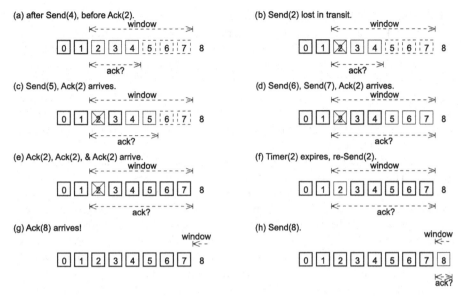

FIGURE 5.15 TCP window progress in response to lost segment.

the transmitter to the receiver has been lost in transmission. Obviously, the receiver will not generate a corresponding acknowledgment. For simplicity, in Figure 5.15 we assume that the receiver's application layer reads the received byte stream out of the receiver's TCP window as soon as the data arrive, so acknowledgments effectively reopen the receiver window immediately. In Figure 5.15(a), we see our first view of the status of the connection. The transmitter has sent 0 to 4 and is awaiting acknowledgments for 2 to 4. In part (b) we state that payload 2 has been lost. This fact is unknown to the transmitter or receiver, but obviously the receiver will never receive this byte of payload.

In Figure 5.15(c), progress around our lost segment has continued. Segment 5 has been transmitted. An acknowledgment arrives with a pointer of value 2. This acknowledgment was presumably generated by the correct arrival of segment 3, but since the receiver can acknowledge only just past the longest sequence of completely received information, it must send 2 as its acknowledgment. In part (d), transmitting progress continues, yet another acknowledgment for 2 arrives, as the fact that segment 2 is missing means that the receiver can only acknowledge up to that point, regardless of what higher number of bytes arrive. In part (e) more acknowledgments limited to 2 arrive. Notice that at time (e), no further transmissions can be made, as the receiver window is full.

Finally, by the time depicted in Figure 5.15(f), the transmit timer for segment 2 has expired. The transmitter realizes that it has failed to receive an acknowledgment for this segment. It acts by resending segment 2.

By the time depicted in Figure 5.15(g), the retransmission of segment 2 has arrived at the receiver, which has emitted an acknowledgment, which has arrived back at the transmitter. But this acknowledgment is not (explicitly) for segment 2. Instead, it is an acknowledgment carrying the number 8, as all of the segments from 0 to 7 have now been received correctly now that segment 2 has met with success. The transmitter now understands that the receiver's window has jumped to 8 and extends its full initial width (we assume that the receiver has advertised its full window to the transmitter). In Figure 5.15(h) we see that the transmitter is beginning to fill the newly positioned receiver window with segment 8 and that normal progress can be expected (unless or until some future segment is lost).

To this point we have discussed TCP's receiver window–based flow control scheme. The primary goal and accomplishment of this scheme is to prevent the transmitter from overrunning the buffers in the receiver. This is a critical component of the overall TCP flow control scheme because segment losses due to overrun in the receiver would have to be made up for by repeated transmission of these lost segments by the transmitter. The probability of loss would increase with overall traffic load, so the repeated retransmissions would make the situation worse by increasing the overall traffic.

5.6.2   Transmitter-Based Flow Control

We are far from done with flow control issues. If the network consisted only of a straight wire (fiber) from the transmitter to the receiver, and all segment loss occurred in receiver buffer overflow, the windowing scheme defined so far would solve the problem nicely. The transmitter would burst away as fast as the fiber and receiver window control permitted.

In fact, networks are much more complex than this single point-to-point fiber. Paths between communicating TCP endpoints may traverse several tens of routers and an equivalent number of interrouter links. Each router can be forced to drop incoming segments if it or its buffers become overwhelmed. Similarly, some routers can be forced to drop outgoing segments if their egress links and buffers are overwhelmed. Thus, each router is a potential point of loss at which TCP/IP frames can be dropped. The consequence is that a particular TCP connection may not be able to communicate at its full point-to-point, receiver window–limited rate. It may be necessary to communicate at a lower rate due to congestion in intervening routers caused by limited capabilities or by the crossing traffic experienced by a connection. If TCP transmitters did not respect these sorts of limitations and forced too many segments into the networks, excessive segment loss would occur within the network.

Thus, it is necessary for TCP to limit its transmission rate to the minimum of what the receiver buffers can accept and what the intervening network can effectively carry. As we have seen, it is relatively easy to maintain a good idea of what the receiver buffers can accept. TCP has an effective scheme for this form of flow control (Figure 5.16). But how can one estimate the limitations imposed by the intervening network? This is a difficult problem. Straightfor-

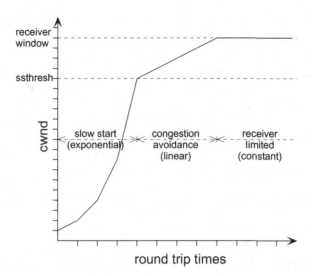

FIGURE 5.16 TCP transmitter flow control modes.

ward solutions are available when new connections are required to establish complete paths from source to sink for defined traffic flow. Call setup (or acceptance) can establish the path for a new connection: reserve buffers and/or time in the various links and switching and routing nodes. When this is done, the new connection knows how much network bandwidth it can use and knows that its use of this bandwidth is assured. However, TCP/IP has no such reservation mechanism. Instead, IP is designed for a fire-and-forget network. TCP does establish a connection between endpoints, but this establishment makes no attempt to reserve bandwidth or buffers on the path to be taken by the connection. Instead, TCP is left to do the best it can to provide a reliable service and to avoid destructive flooding of the network, with traffic doomed to be dropped. In the next few paragraphs, we examine the ideas that TCP uses to further limit the rate of transmission when the network cannot support the full receiver window–based rate.

TCP is given no explicit information to lead it to a sensible transmitter-controlled rate. There is no database that it can examine to determine how well a connection from A to Z will do at any particular time. Instead, TCP can experiment only by sending segments and seeing how well they do in the current network situation. In fact, TCP uses packet loss rates from its recent transmissions to guide it in selecting its rate of future transmissions. The full story of TCP transmitter flow control is not trivial. To simplify the presentation, we look at it one behavior at a time. A new TCP connection begins by sending one segment. Should that segment be received and an acknowledgment be returned, the TCP transmitter takes that returned acknowledgment as a sign of good network conditions and increments the number of segments it is willing to have outstanding at any time from one to two. It then fires off two more segments (receiver window permitting). One round-trip time later, two more acknowledgments are returned if the network remains uncongested. These are taken as reason to increase the number of outstanding segments from two to four (each acknowledgment received allows the transmitter to send another pair of new segments). Each round-trip time without packet loss results in a doubling of the number of segments sent in the next round-trip period. This process increases the transmission rate exponentially. Curiously, it is called *slow start* because it starts with only one segment being sent. But with an exponential growth in transmission rate, it does not remain slow for long.

Exponential growth stops when it gets to a threshold known as *ssthresh* (for "slow-start threshold") or when the first segment loss is encountered. We consider the *ssthresh* limit first. At that point the transmitter increases its transmission rate by only one segment per round trip time instead of doubling every round-trip time. Increasing by one segment per round trip time is a linear growth rate. This phase of increasing flow is called the *congestion avoidance phase*. The idea is that we've gotten the rate up pretty quickly, we've not hit a segment loss yet, but we've passed the *ssthresh*—it's time to become conservative. Yet there could be more network capacity above *ssthresh*, and we'd hate to waste it by refusing to transmit faster. So we compromise by continuing our rate increase but at a much slower rate. Eventually, the congestion avoidance phase may run into the receiver window control mechanism. At this point, the transmitter stops increasing its transmission rate, but obeys the receiver controls.

The slow-start and congestion avoidance behaviors described above have not considered what happens to the transmitter controls when a segment is known to have been lost somewhere in the network (presumably, due to congestion at some node). When a packet loss is discovered by the expiry of the timer that was set when the packet was first sent by the source, the transmitter responds by reducing its transmission rate. In particular, *ssthresh* is reset to a level one-half the current transmit level, and the transmitter restarts in slow start with one acknowledgment. Behaviors of this sort are illustrated in Figure 5.17. The changing value of *ssthresh* is shown as a dotted line.

This behavior allows the transmitter to respond to network congestion, which is the goal of TCP transmitter-based flow control. It is also quite robust. But the effects are not ideal. Transmission has more or less stopped while the packet acknowledgment timer expires (the events that occur as the timer expires can be quite complex, depending on the state of the sliding window). This delay is unacceptable for high-quality-of-service applications, and undesirable for all applications. Also, since packet losses tend to occur in batches as a queue overflows somewhere in the networks, many connections may enter this wasteful timer expiry period simultaneously, and the previously congested network paths will tend to drop to a state of little or no traffic. Such oscillations in traffic load are not good. Fortunately, later work in defining the TCP/IP

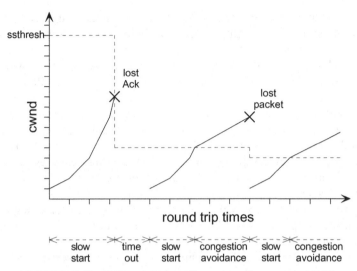

FIGURE 5.17 TCP transmitter flow control recovery modes.

protocol suite added improved recovery mechanisms, which are discussed below.

5.6.3 Fast Retransmit and Fast Recovery

When a network is forced to drop a packet, the receiver will never generate an acknowledgment for those data. The transmitter will eventually time out and realize that either the packet or its acknowledgment was lost due to queue overflow, or was corrupted and dropped. At that point, the transmitter will retransmit the packet in question. Although this action causes the connection to recover, we have had to suffer a timer expiry, during which no (or little) progress was made on the connection, the user has had to wait, and we have not been able to use the traffic source to keep network resources busy.

The *fast retransmit algorithm* is a mechanism to avoid waiting for the timer expiry. Fast retransmit is based on the fact that when a TCP receiver notices a missing packet (by observing sequence numbers), it must retransmit the acknowledgment sequence number just past its highest received byte in the connection's byte stream. Suppose that a transmitter sends a series of packets of length L, where the packets begin with sequence numbers $S_0, S_1, S_2, S_3, \ldots$. We would normally expect a sequence of acknowledgments with acknowledgment sequence numbers $S_0 + L + 1, S_1 + L + 1, S_2 + L + 1, S_3 + L + 1, \ldots$. But suppose that the second packet was lost in transit. Then the receiver would generate the sequence of acknowledgment numbers $S_0 + L + 1, S_0 + L + 1, S_0 + L + 1, \ldots$, as the arrival of the first packet was acknowledged but the arrivals of the third and fourth packets after the missing second packet then forced the receiver to continue to acknowledge sequence number $S_0 + L + 1$. Thus, the transmitter sees a duplicated acknowledgment number in response to the loss of a single packet.

It is possible for the IP network to mis-order two segments, in which case the transmitter would see one duplicated acknowledgment followed by an acknowledgment for two segments' worth of data. But when the same acknowledgment sequence number is repeated, it is highly likely that one packet is lost and that each succeeding successful segment is causing a repetition of the acknowledgment sequence number. In this situation (reception of three duplicated acknowledgment sequence numbers), the transmitter goes into fast retransmit mode. Its immediate response is to retransmit the segment just past the repeated acknowledgment. The transmitter does not wait for the acknowledgment time-out for the missing packet, but continues to emit segments at its currently appropriate rate. When only one segment has been lost, this is the fastest possible recovery.

After fast retransmission, the transmitter goes into its congestion avoidance mode (increasing the number of outstanding acknowledgments by one per round-trip time) instead of beginning again with slow start. This avoids dropping the transmission rate unnecessarily. This mechanism is called *fast recovery*. When the conditions caused by lost segments or acknowledgments are more complex than can be handled by fast retransmission/recovery, the normal time-out and slow-start mechanisms take over. Thus, the two fast mechanisms give the connection a chance to recover without serious loss of progress when the disruption caused by a lost segment somewhere in a network queue is minor.

5.6.4 Delayed Acknowledgment

Overanxious sending of acknowledgments by receivers can result in unnecessary work for the network. Consider what could happen when data are sent to a server and a reply is returned after a brief computation on the server. A naive solution would be to have the TCP receiver entity on the server send an acknowledgment as soon as the original data are received. Then, after the server has computed its response, the server will send another packet with the response (which will be acknowledged separately later). This situation corresponds to Figure 5.18(a), in which three TCP segments are sent (with lengths of 40, 44, and 45 bytes, for a total of 129 bytes).

A better solution is to delay the acknowledgment to the original data [Figure 5.18(b)]. In this example, the receiver may wait for 200 ms before it returns the acknowledgment. But before its 200-ms timer expires, the server has computed its response and has requested that a TCP segment be sent to the client. Since the acknowledgment has not yet been sent, it can be combined (piggybacked) with the server's reply. This means that only two TCP segments will be sent (with lengths of 44 and 45 bytes, for a total length of 89 bytes).

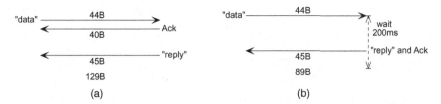

FIGURE 5.18 Delayed acknowledgment in TCP.

Thus, a delayed acknowledgment can significantly reduce network traffic. TCP couples this delayed acknowledgment solution with a preference for sending only one acknowledgment for every two payload segments received. This mechanism also reduces the number of small segments that must be carried by the networks. As many network elements are limited by their number of packets per second, this is a useful step.

5.6.5 Naigle's Algorithm

With some applications, the byte stream to be transmitted is generated in very small chunks separated by many breaks in time. A naive implementation of TCP would generate a new packet for each release of a few bytes from the source application layer. This would generate many inefficiently packed TCP/IP segments. To avoid this, *Naigle's algorithm* refuses to generate a new, small TCP/IP segment until the last segment sent has been acknowledged. This reduces the frequency of small packets in such applications to one per round-trip time, which is more acceptable than one packet per appearance of a byte at the source application interface.

5.7 IP ROUTING MECHANISMS

Once TCP or UDP segments have been injected into a pure IP network by their sources, switching decisions are made at the IP level by routers and host computers that enable routing functions. As we shall see as we progress through the book, there are other ways in which IP segments are carried by underlying mechanisms such as SONET, MPLS, and ATM (asynchronous transfer mode), but for the moment we assume a pure IP network without concern for link layer issues, as this view is useful for describing important aspects of the IP world.

Each router contains a routing table that contains information for the output links which should be followed by incoming IP segments. These routing tables, and a fairly simple lookup algorithm, are the source of all IP routing decisions. The discovery of the current topology of the Internet and construc-

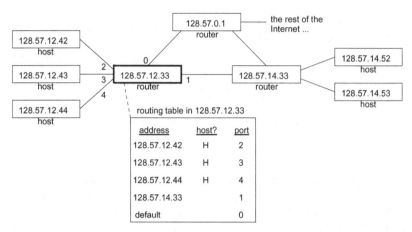

FIGURE 5.19   Routing table in a class B subnet.

tion of these routing tables is discussed in the next section. In this section we present a simplified view of how these tables are used to route IP datagrams. Figure 5.19 shows a simple network with five hosts (128.57.12.42, 128.57.12.43, 128.57.12.44, 128.57.14.52, and 128.57.14.53) and three routers (128.57.0.1, 128.57.12.33, and 128.57.14.33). In this example we are concerned by what router 128.57.12.33 should do with IP datagrams that it receives (or generates). Router 128.57.12.33 contains a routing table, as shown. With the destination IP address from each datagram to be routed, the router examines its routing table in three ways, in sequence, until a suitable route is found. The first pass on the table looks for a "host" entry (indicated by an "H" in the second column) and an exact match on the destination address (the first column). When such a match is found, the packet is forwarded out the "port" indicated (in the third column). For example, if the router in question received an IP datagram from host 128.57.12.42 destined for host 128.57.12.44, it would find an exact host match in the third entry in the table, so would forward that datagram out its egress port 4, which is connects to the desired destination host, 128.57.12.44.

   If no exact host match is found, another pass is made over the table to find an appropriate router to which the datagram should be forwarded. For example, if the router in question had a datagram intended for 128.57.14.52, it would not find an exact host match. But it would find that the fourth entry for router (nonhost) 128.57.14.33 matched the class B and network ID portions of the address. This partial match of a nonhost address indicates that the datagram should be forwarded to that router, so the datagram would be sent out port 1 to router 128.57.14.33.

   If neither of the search strategies described above is successful, the router in question examines the table for a default route. A default route is to be taken when no entries in the local table give specific instructions for forwarding of the packet. For example, if the router in question held an IP datagram with destination address 192.0.54.3, it would fail to find a local host match, it would fail to find a responsible local router, and it would select the "default" entry. In this case the default entry follows egress port 0 to router 128.57.0.1, which is connected to "the rest of the Internet." Router 128.57.0.1 may or may not have better routing information for 192.0.54.3. If it does, the datagram will follow that path. If it does not, the datagram may follow another default link. But eventually, the datagram will be directed to a router with knowledge of either the host 192.0.54.3, or at least of the network on which it resides.

The use of nonhost (router) routes and default routes is of enormous importance to practical IP datagram routing. If neither mechanism were available, each IP host or router would have to maintain a routing table for up to all of the 2^{32} (4 billion) IP addresses. The memory requirements of such a solution would be absurdly expensive. Probably more important, the costs of computing and distributing complete host tables would be overwhelming. The nonhost (router) and default mechanisms allow routing tables to be compressed. Entire local networks can be addressed via one responsible router, thereby reducing the tables in any connected host or router. More important, the use of "default" routes allows the routing tables in entire local networks to contain no knowledge of the rest of the network. As a consequence, these many routing tables can be much smaller and need not participate in detail in network route computations. Without these two address compression mechanisms, IP route calculation and route storage would not be feasible.

5.8 IP ROUTE CALCULATIONS

Given the IP routing strategy described in Section 5.7, a question arises: How are the contents of these routing table discovered? The enormous size of the Internet is an obvious challenge for route calculation. Less obvious, but more important, is the dynamic nature of the Internet. Hosts, routers, and links are constantly winking in and out of existence as they are installed, powered up, powered down for maintenance, fail, or are replaced. From the perspective of IP route calculation, the Internet is a huge sea of constant change, yet at all IP packets are expected to be routed properly. The issues involved in responding to this challenge are complex and constitute specialist knowledge well beyond the scope of this book. Consequently, in this section we present only an overview of the solutions in use.

There are two fundamental approaches to the route calculation problem which make the problem tractable:

1. The overall Internet is divided in to a collection of *autonomous systems* (ASs), each of which is a defined subset of the overall Internet, consisting of a set of hosts and routers and links, with connection to other ASs only via a

subset of the routers. This division into a set of nonoverlapping ASs allows the use of a divide-and-conquer approach: First, routing solutions are computed within each AS (in parallel); then the top-level routers of all the ASs interact to compute routes between the multiple ASs. Of course, traffic from one AS to another must pass through the routers that connect the two ASs, using default routes to get to their border router and hence into the other domain, where nondefault routes will eventually be found.

2. As it is not practical to recompute routes immediately in the face of network topology changes, nor to hold off topology changes to the process in a batch from time to time, it has been found necessary to continually recompute the routing tables. Each such recomputation in each AS notices or catches any topology changes that have occurred since the last route calculation. This uncoordinated batching of route calculations is more manageable, but does leave errors in routing tables at times (e.g., when a router or link goes down, or when some time passes before affected routes are recalculated).

In the remainder of this section we outline the most commonly used route computation algorithm for autonomous systems, the open shortest path first (OSPF) algorithm. A network of IP routers in an autonomous system can be thought of as an arbitrary mesh of routers with links between them. A useful routing algorithm must decide which outgoing link any router should use to forward an IP segment to some other router or attached host. For example, Figure 5.20 represents a small mesh of five routers and with six edges connecting them. From which edge should router A emit a segment bound for router B?

OSPF's goal is to find the lowest-cost path between each pair of routers. Costs are given by link distances. In Figure 5.20, the integers associated with each link represent the distance of that link. If we consider the pair A and B, it is clear that there are three possible paths that connect them: A–F–B, A–B, and A–C–D–B. If we add up the total lengths of these paths, we find that they are 6, 5, and 7, respectively. The lowest-cost path is A–B, at a total

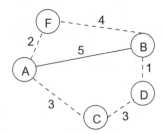

FIGURE 5.20 IP shortest-path routing.

cost of 5, so this path should be used for traffic from A to B (and the reverse path for traffic from B to A). In general, the costs of the various paths can be computed in a variety of ways, including the number of routers found on the path, the bandwidth of the path, and the dollar cost of the path, among others. But the key idea remains that one numerical cost is found for each path, and an IP routing algorithm choses the lowest-cost path for each routed connection.

OSPF is a combination of two algorithms: Dijkstra's shortest-path algorithm and the reliable flooding algorithm. Given full knowledge of the topology of the network (which routers are connected to which other routers, and the length of each link), *Dijkstra's algorithm* computes shortest paths. Dijkstra's algorithm runs in (an attached processor on) each router to compute the routing table to be used by that router. The *reliable flooding algorithm* also runs in each router: Each router emits a packet that describes its connections to other routers; then all routers cooperate to broadcast all such information packets to all other routers in a "reliable flood." Where there are N routers, each emits a link information packet that must be sent to all of the other $N - 1$ other routers, so $N(N - 1)$ packet deliveries must be (reliably) completed.

The information packets, called *link state packets* (LSPs), contain the identity of the source, a list of connected neighbors and their distances, a sequence number, and a time to live. The flooding algorithm can be described in terms of a series of questions that arise as a new LSP arrives:

- Do I have LSP information for the source of this packet? If not, save the information from the LSP and emit a copy of the LSP on all other links (flood to the rest of the network).
- Do I have an older copy of LSP information for the source of this packet? By *older*, we mean a smaller LSP packet sequence number. If so, replace the older information with the current LSP and emit the LPS on all other links.
- Do I have a newer or equal-age (sequence number) record of LSP information from this source? If so, drop the LSP without recording or forwarding (ending the flood).

Three related rules are required by the algorithm:

1. As a source of LSP packets, increment the LSP sequence number with each LSP emitted.
2. Whenever local or computed route changes result in new routing information, forward new LSP packets to all neighbors.
3. On startup, set the LSP sequence number to zero and forward an initial LSP to all neighbours.

This algorithm is not strictly reliable, in that sequences of links coming up and going down can prevent complete flooding. However, the failings are small and unlikely in practical networks. Also, the flooding process is repeated sufficiently often to repair any errors quickly.

Notice that the description of the reliable flooding algorithm seems to assume that sequence numbers increase forever. Of course, this is not the case, as sequence numbers wrap around from the maximum positive number to zero. It is vital that this not confuse the algorithm into thinking a new LSP (numbered zero) is older than its predecessor (with the maximum number). A standard solution to this problem recognizes that the sequence numbers are actually in a loop, and for each number in the loop recognizes a subset of numbers preceding it in the ring as being "younger" and another subset of numbers in the ring as being "older." If older numbers in the ring die away (by completing their flooding) in time, they are never confused with younger numbers.

So reliable flooding does a pretty good job of getting the critical information on connectivity to each router. From this information, each router can form its own view of the network, similar to the view given in Figure 5.20. Dijkstra's shortest-path algorithm makes use of this information to find a shortest path to each other node. The algorithm builds lists of triples of information as it computes its routes. The triples consist of <Node Address, Cost of Best Known Route, Egress Port to Begin Route>. Two such lists are maintained: a confirmed list and a tentative list. The *confirmed list* contains triples of node addresses, route costs, and starting egress ports for all known shortest paths. The goal of the algorithm is to find a confirmed entry for every other node in the system. The other list of triples is the *tentative list*, which describes best current estimates of shortest paths, but paths that have not yet been confirmed as being absolutely the shortest. Dijkstra's algorithm follows:

(1) initialize Confirmed ={<Myself, 0, ~>}; Tentative ={}
(2) Next = last node added to Confirmed
(3) **for each** Neighbor directly connected to Next
    Cost = dist(Myself, Next) + dist(Next, Neighbor)
    if (Neighbor not in Confirmed) and (Neighbor not in Tentative) **then**
        add <Neighbor, Cost, next-hop from Myself to Neighbor> to Tentative
    if (Neighbor in Tentative) and (Cost < cost of Neighbor in Tentative) **then**
        replace Neighbor in Tentative
            with <Neighbor, Cost, next-hop to Neighbor>
(4) **if** Tentative is empty **then** stop
    else move lowest-cost entry on Tentative to Confirmed and **go to** (2).

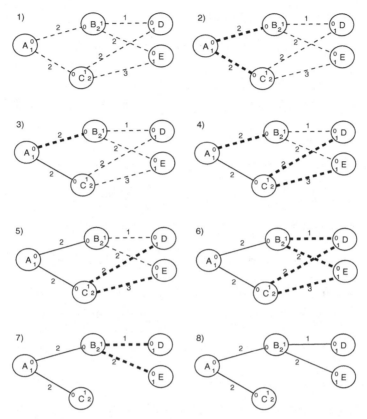

FIGURE 5.21 Example of Dijkstra's shortest-path algorithm.

Figure 5.21 illustrates a network of five nodes and the progress of Dijkstra's algorithm running in node A with current and correct information about the state of the network (from a recent completion of the reliable flooding algorithm). In the figure, the light dashed lines [such as all the lines in part (1)] represent known links with annotated distances, but make no assumptions about their use in any shortest paths. The heavy dashed lines represent tentative information about shortest paths, but information that must be confirmed before they are known to be shortest paths [e.g., in step (2) we have tentative routes from A to B and from A to C, but have not yet confirmed this information]. The solid lines represent confirmed shortest-path information [e.g., by step (3) we know that the shortest path from A to C takes the solid line which connects them directly].

The progress of Dijkstra's algorithm in the example of Figure 5.21 follows:

‖ Graph 1

(1) Confirmed = {<**A,0,**~>};   Tentative = {};

‖ Graph 2

(2) Next = **A**;

(3) Cost$_B$ = 0 + 2 = **2**;   Tentative = {<**B,2,0**>};

    Cost$_C$ = 0 + 2 = **2**;   Tentative = {<B,2,0>,<**C,2,1**>};

‖ Graph 3

(4) Confirmed = {<A,0,~>,<**C,2,1**>};   Tentative = {<B,2,0>};

‖ Graph 4

(2) Next = **C**;

(3) Cost$_A$ = 2 + 2 = **4**;   A already confirmed

    Cost$_D$ = 2 + 2 = **4**;   Tentative = {<B,2,0>,<**D,4,1**>};

    Cost$_E$ = 2 + 3 = **5**;   Tentative = {<B,2,0>,<D,4,1>, <**E,5,1**>};

‖ Graph 5

(4) Confirmed = {<A,0,~>,<C,2,1>, <**B,2,0**>};   Tentative = {<D,4,1>,<E,5,1>};

‖ Graph 6

(2) Next = **B**;

(3) Cost$_A$ = 2 + 2 = **4**;   A already confirmed

    Cost$_D$ = 2 + 1 = **3**;   Tentative = {<**D,3,0**>,<E,5,1>};

    Cost$_E$ = 2 + 2 = **4**;   Tentative = {<D,3,0>,<**E,4,0**>};

    . . .

‖ Graphs 7 and 8

(4) Confirmed = {<A,0,~>,<C,2,1>,<B,2,0>,<D,3,0>,<**E,4,0**>}

5.9   DIFFICULTIES WITH TCP AND IP

Although the TCP/IP protocol suite has allowed the emergence of the Internet, with all of its astonishing features and applications, nevertheless, these protocols leave transport network designers with serious problems. In this section we gather together the various serious problems that arise due to the nature and established use of the TCP/IP protocol suite.

5.9.1   One Shortest Route, Regardless of Load Conditions

IP's routing algorithm results in one path for all traffic between any two endpoints. Although the use of the shortest path makes sense in the case of a single connection, it can be a very poor choice in a congested network. The problem is that nearly-as-short paths are ignored and their carrying capacity is lost. Thus, too much traffic is forced onto the best link, resulting in congestion, while nearly-as-short links may be starved for traffic. In addition, any attempt to preserve quality of service (e.g., for VoIP) may be compromised by the congestion, while ideal channels exist for such traffic on the nearly-as-short links.

5.9.2 Deliberate Congestion and Backoff

Congestion is bad for networks, as it leads to packet loss and compromises quality of service for sensitive services. We would like to avoid congestion entirely, yet TCP deliberately seeks congestion. It ramps up its transmission rate (within the constraints of the receiver flow control system) until it discovers congestion by seeing packet loss. TCP uses this mechanisms to seek ever-higher transmission rates within available capacity, which is a good goal. But it has no way to avoid increased transmission rates before congestion happens. So TCP breeds congestion. Further, after congestion (unless fast retransmission applies), the transmission rate of the source drops dramatically. This tends to leave the pipe unfilled after congestion events that have affected numerous TCP flows. During such periods, the formerly congested pipe will be underloaded. This mechanism would probably be acceptable if it were not for the increasing demands to support higher quality of service for such applications as video links and voice over IP (VoIP). Such services cannot be useful if their TCP/IP connections are constantly subject to higher queuing delays, congestion, and packet loss.

5.9.3 Lack of Quality-of-Service Support

While IP was designed to support a modest system of quality-of-service (QoS) differentiation (via the type of service bits in the IP header), implementers ignored these features for many years. As a result, attempts to add QoS services at the IP level today are compromised: QoS requires end-to-end support, and the occasional new router cannot force all the other routers in the Internet to begin respecting the TOS bits. Consequently, it is essentially impossible to make use of IP's TOS facilities at this late point in time for Internet-wide applications. In addition, modern QoS standards require more support than IP's TOS facilities can provide.

5.9.4 Receiver Windows and Round-Trip Times

The size of the receiver window of a TCP connection limits the bandwidth possible on a connection: The maximal bandwidth of a connection decreases with increasing round-trip time of the connection. The reasons for this unfortunate result are explained below.

Each unidirectional TCP connection has a TCP receiver window which the TCP transmitter continually attempts to fill with new segments, and which the receiver application continually tries to empty. As we have seen, the TCP transmitter is not permitted to transmit a segment unless it knows that the receiver window has space for its segment. Once the receiver has transmitted a segment toward a receiver buffer, it cannot transmit another segment into that buffer space until two things have happened: The receiver has acknowledged the receipt of the segment, and the receiver application has moved the segment out of the buffer, thereby causing the receiver TCP module to include that buffer space in the window that it advertises to the transmitter. To simplify our discussion below, we assume that the receiver application always removes a newly received segment from the TCP receiver window as soon as it arrives. Thus, when the TCP receiver sends an acknowledgment for the packet, the acknowledgment already includes an advertisement for the window space corresponding to the segment acknowledged. If the receiver application layer were slower in moving the segment out of the TCP receiver window buffer, it would delay the time at which the TCP transmitter could send another segment to the buffer in question, so we are assuming optimal performance.

A receiver window's (normal) size is $W = 65,535$ bytes, or $W = 524,288$ bits. This space can be divided into separate segment buffers as desired without changing the nature of the negative performance result we are developing in this section. For the sake of simplicity, consider the entire receiver window to be one large segment window. With this arrangement, we can easily see that once a segment is transmitted, we must wait one full round-trip time of the link before we can reuse that single buffer segment: One end-to-end transmission is required to carry the segment from the transmitter to the receiver. A second end-to-end transmission is then required to carry the acknowledgment (with its advertised receiver window of W) back to the transmitter to allow it to then send the next segment. This implies that we can transmit only W bytes per round-trip time. If we divided the receiver window into W spaces for single-byte segments, the same property would apply: Each byte of the receiver window can support only 1 byte of transmission per round-trip time on the link, so we would still have an aggregate of W bytes of communications per round-trip time. Any intermediate segment size leads to the same result. Of course, the larger the segment size, the less we pay in header overheads for the connection.

Thus, we have the unfortunate property that progress in a connection depends on the length of the link (or actually, the time in flight of bits on the link, which is based on a link propagation speed of approximately 200,000 km/s for both copper and fiber). The bad news is that TCP connections, with constant receiver window sizes, get progressively less effective bandwidth out of a connection as it gets longer. In equational form, a link of length D with a medium that propagates at 200,000 km/s, carrying a TCP connection with a receiver window size of W bits, can carry at most an average bit rate of $W/[2(D/200,000\text{ km/s})]$. For a link from Toronto to Kingston, Ontario ($D = 250$ km), we get 209.7 Mb/s; for a link from Vancouver to Halifax ($D = 6000$ km), we get only 8.7 Mb/s. This is a serious performance issue with TCP (and other

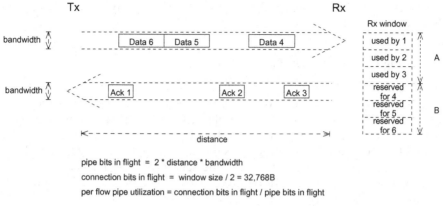

pipe bits in flight = 2 * distance * bandwidth

connection bits in flight = window size / 2 = 32,768B

per flow pipe utilization = connection bits in flight / pipe bits in flight

FIGURE 5.22 TCP connection round-trip effects.

similar protocols). Notice that the argument for decreasing bandwidth with increasing round-trip delay was made independent of the bit rate of the link. Obviously, if the link is slower than the receiver window maximal rate, the effective rate will be limited by the link's bit rate.

Also, it should be obvious that the delays are increased by the time it takes to transmit/receive a packet. It takes some finite time to launch a complete segment at the transmitter and to receive the complete packet at the receiver. No use will be made of the segment until it has been received and checked completely. So we must include in the round-trip delay the time required to launch (or receive) the segment (Figure 5.22). This factor increases the effective round-trip delay of the connection and thereby decreases the effective throughput.

5.9.5 Long, Fat TCP Pipes

Now consider the effects of the receiver window as transmission "pipes" get both longer and carry more bits per second (get "fatter"). By *fat pipes*, we mean high-bandwidth links. For example, a SONET OC-192 link, which carries approximately $B = 10 \text{Gb/s}$, is a fat pipe. An example of a *long pipe* is our Vancouver to Halifax ($D = 6000 \text{km}$) link. We have seen that a single unidirectional TCP data flow can utilize only 8.7 Mb/s of this pipe. This means that it takes 10 G/s/8.7 Mb/s = 1149 unidirectional TCP data flows to fill this 10-Gb/s pipe. If there are fewer TCP connections on the link at any time, it is not possible to utilize the entire link, even if the existing connections have a demand (at the application layer) for more than 10 Gb/s.

Both of these factors—limited bandwidth per TCP connection and inability to fully utilize large pipes—make it difficult to build satisfactory transport networks for TCP. Although TCP undoubtedly brings many advantages, it also brings serious liabilities to the Internet.

5.9.6 Big Packets on Thin Pipes

This issue is not related directly to transport networks, but is included for completeness. It is normally to have *thin* (low-bandwidth) *pipes* at the very edges of the Internet. The best example of this is an DSL connection to a subscriber's home. The problem is that big IP packets can take a considerable amount of time to transmit on these slow pipes. If real-time services such as VoIP are being supported at the same time, sending one big packet on the thin pipe may block the small VoIP packets, which must arrive in time to prevent failures of voice quality.

Avoiding this problem (without dedicating another thin pipe for real-time applications) requires that all IP datagrams be segmented to a small enough size that they can be transmitted safely without destroying real-time application needs. This could be done by segmenting IP datagrams to a small enough size, but as IP has no mechanism for reassembling larger datagrams from fragments while they are in flight, the entire network is then burdened with many inefficiently small IP datagrams. A better solution has been to use ATM cells to carry voice over IP and fragments of larger IP datagrams over DSL links. The ATM connection is terminated at each end of the thin link, so the many small fragments are easily reassembled into more efficient IP datagrams for the Internet.

5.10 IPV6: THE FUTURE?

Various limitations of Internet protocol version 4 (IPv4) have been recognized for many years. Much work has gone into the definition of a replacement network layer protocol known as IPv6 (Internet protocol version 6) or IP-NG (Internet protocol–next generation). IPv6 is a well-defined protocol with several significant advantages over IPv4, which we discuss briefly in this section. However, a protocol as firmly established as IPv4 is very difficult to replace. Today's Internet is utterly dependent on IPv4. Customers are unwilling to shut the Internet down for an hour (or a day?, a month?, a year?) to replace one nearly invisible layer with another nearly invisible layer. Simply stated, the Internet has become far too important to the world to ever be shut down intentionally. Consequently, given that IPv4 is the Internet's heartbeat and blood flow, it will be exceedingly difficult to replace. However, before we conclude that all change is impossible, let us examine some of the advantages of IPv6.

1. *Expanded address space.* IPv4 permits only 2^{32} (roughly 4 billion) endpoint addresses. Surprisingly, this has become far too few addresses. Although

the number of connected devices has not yet risen to this number, it is not far off. One IP-addressable computer, PDA, cell phone, refrigerator, or vehicle per person on this planet, and we have overspent our IP endpoint budget. In addition, it is impossible to manage IP address allocations such that they are all used efficiently. There are several factors that lead to a lumpy, inefficient allocation of these addresses. IPv6 offers 2^{128} endpoint addresses. This is a preposterously large number: 340 billion billion billion billon, or 3.4×10^{38}. Clearly, this well of IP addresses is unlikely to run dry.

2. *Simplified management.* Considerable complexity (outside the scope of this book) is required to manage IPv4. It is difficult to allocate pools of IPv4 addresses, to allocate individual addresses within these pools, and most difficult of all, to reallocate pools of addresses. IPv6 uses its enormous address space to simplify these difficulties. By freely utilizing enormous portions of this 2^{128} space, IPv6 comes up with much simpler management solutions.

3. *Simpler header organization.* The header of IPv6 consumes 40 bytes, due to the use of such large source and destination addresses. However, it is a much simpler structure. Several issues that are dealt with directly in the IPv4 header are pushed out of the IPv6 header. For example, all variable-length options in IPv6 are pushed out into a separate header when required. The effect of this and other simplifications is that the IPv6 header is a fixed-length, fixed-format structure that is simple and inexpensive to process in hardware. Hardware header processing permits the higher segment-processing rates required in the core of the network.

4. *Integration of security.* For a wide variety of reasons, security of communications has become a critical issue in the Internet. Security issues are not dealt with in IPv4. This was, for a time, an advantage in that it permitted a range of semiexperimental solutions to be adopted. But the Internet has matured to the point at which a solid standard solution of the various security problems has become vital. IPv6 addresses this need. It integrates strong security mechanisms in an additional level of header that is invoked when required (somewhat like the options header).

5. *QoS support.* IPv4 included a type of service (TOS) field, which could have been used to provide at least rudimentary QoS support. However, this field was ignored (not implemented) for years. As a result, an IP segment crossing today's Internet will find very few nodes that respect any coding placed in its TOS field. As QoS must be respected from end to end to be of any value, it is impossible to obtain QoS without replacing or updating all the routers on any path that QoS sensitive segments may take. This is completely impractical for large or public networks. Consequently, it may be necessary to wait for a next generation to overcome or bypass the mistakes of the past. IPv6 has reasonable QoS support. It is hoped that IPv6 will progressively carry its QoS capabilities into the Internet, gradually replacing IPv4's failure to support QoS.

6. *Peer-to-peer and multicast support.* IPv6 includes stronger support for peer-to-peer applications and for multicast. Both classes of features have been poorly supported by IPv4.

So what are the prospects for serious adaptation of IPv6? There is no doubt but that the widespread use of IPv4, coupled with the telecommunication industry's unwillingness to reinvest to replace functioning and profitable systems and with the technical challenges of either quickly or slowly replacing IPv4 with IPv6, means that IPv4 will remain in service for quite some time.

Yet the advantages of IPv6 are compelling. For instance, to give every phone and IP address, those addresses must be IPv6, not IPv4. A partial solution is available. Certain applications (e.g., IP telephones) can use IPv6 within their own domains. Then when telephony segments are to be sent across the Internet, the IPv6 carrying the conversations is "tunneled" through IPv4. Although tunneling is a complex topic in its own right, it is simply a form of carrying IPv6 segments as payloads to IPv4 segments—thus, IPv4 acts to carry IPv6 segments across the older Internet IPv4 technology. It is generally expected that this is how IPv6 will make its best progress against IPv4.

It is difficult to believe that eventually, IPv6 (or some successor) will not replace IPv4. But technology and business lessons of the last decade make it essentially impossible to predict such events.

5.11 CONCLUSIONS

TCP/IP has been one of the most important foundations of the Internet and the continuing growth of the Information Age. It has provided decent functionality and an enormous communality of technology which has enabled a huge range of new applications. To say the very least, TCP/IP has been a technological home run! These protocols have won a beachhead in our lives which it is essentially impossible to dislodge. There are far too many TCP/IP conversant bits of technology for us to throw away willingly, because we want to replace TCP/IP with something better. Any evolution to solutions better than TCP/IP can only take place in an incremental manner which remains compatible in important functionalities with the existing TCP/IP solutions. Yet we know that TCP has serious flaws—flaws that have already prevented interesting and potentially useful solutions with today's networks.

KEY POINTS

- The TCP/IP protocol suite is the standard interface by which users of the Internet make use of the facilities of transport networks.

- The TCP/IP protocol suite establishes a four-layer protocol model. The layers are: application, transport, network, and link. TCP is the primary transport layer solution. IP is the primary network layer solution.
- IP establishes a set of 32 bit endpoint addresses, divided primarily into classes A, B, and C for large, medium, and small networks, respectively.
- IP defines a 20 byte header for each IP frame. TCP defines a 20 byte header for each TCP frame. TCP/IP traffic has a copy of each header type. Both headers can be expanded by options.
- IP makes a best-effort attempt to deliver IP frames to their intended endpoints, usually in order.
- TCP assumes responsibility for end-to-end flow. This includes the provision of reliable transport, and two flow control mechanisms: receiver-based flow control and transmitter-based flow control.
- The primary role of receiver-based flow control is to prevent buffer overflow in the receiving device.
- The primary role of transmitter-based flow control is to competitively seek higher transmission rates for the user, without destructively congesting the overall network.
- Network-wide algorithms compute IP routing tables, which are stored in each IP router and used to determine how each IP frame should be forwarded.
- TCP/IP presents a set of systematic challenges to the underlying transport network. Much of the rest of the book will deal with the resolution or partial resolution of these problems.

REFERENCES

Internet Protocol Darpa Internet Program Protocol Specification, Request for Comments 791, September 1981.

Internet Protocol, Version 6 (IPv6) Specification, Network Working Group Request for Comments 2460, December 1998.

Stevens, W. R., *TCP/IP Illustrated*, Vol. 1, The Protocols. Addison-Wesley, Reading, MA, 1994.

Stevens, W. R., *TCP/IP Illustrated*, Vol. 2, The Implementation. Addison-Wesley, Reading, MA, 1995.

6

PROTOCOL STACKS

6.1 INTRODUCTION

In this chapter we examine a number of protocol stack solutions for carrying the TCP/IP protocol suite over the Internet's wide area networks (WANs). We begin by reviewing the problems imposed by TCP/IP. Although it might seem attractive to remedy these problems by redefining the TCP/IP suite, we must accept that the commercial success of the suite has precluded any more

Network Infrastructure and Architecture: Designing High-Availability Networks,
By Krzysztof Iniewski, Carl McCrosky, and Daniel Minoli
Copyright © 2008 John Wiley & Sons, Inc.

than evolutionary change to the TCP/IP protocols. Even the change from IPV4 to IPV6, which does not remedy all of the problems, has been going on for years, has made only limited progress, and may never be complete. A solution to the difficulties imposed by TCP/IP must be sought somewhere in the protocol stack, which lies between IP and the underlying fiber. Somehow, we must utilize protocols in this transport network protocol gap to remedy the shortcomings of TCP/IP without disturbing the millions of TCP/IP endpoints.

In this chapter we introduce some key protocol layers which are used to fill the requirements between IP and fiber and which attempt to improve on the service possible with TCP/IP alone. The protocols, presented briefly, are as follows:

1. *Asynchronous transfer mode protocol* (ATM) was at one time widely expected to replace TCP/IP as the protocol of choice to deliver data, voice, and video to the desktop. This revolution did not occur, due to shortcomings in ATM and to the entrenched technology and market position of TCP/IP. Instead, ATM has become a carrier for TCP/IP in two situations where improved protection of quality of service (QoS) for TCP/IP streams is required. We shall examine ATM's role in QoS protection.

2. *Generic framing procedure* (GFP) is a simple protocol that is used to patch over small but important failures of adequate foresight (or generality) in underlying protocol layers.

3. *Multiprotocol label switching* (MPLS) is one of several initiatives to enable QoS delivery in IP-based layer 3 networks. By combining the attributes of layer 2 switching and layer 3 routing into a single entity, it provides several benefits to traffic engineering. MPLS can potentially eliminate an IP-over-ATM overlay model and its associated management overhead.

4. *Gigabit Ethernet* (GE) is the evolution of the familiar Ethernet standard to operate at 1 Gb/s. GE is used in the Internet is several ways, as shown in this chapter.

5. *Resilient packet ring* (RPR) is a standard for frame transport over fiber-optic rings. The basic concept uses Ethernet framing in SONET-style rings. The goal of RPR is to define a high-performance, high-availability optical transport suitable for carrier networks in metropolitan service areas. RPR networks are optimized to transport data traffic rather than circuit-based traffic. This frame transport, with bandwidth consumed only between source and destination nodes, is more efficient than a time-division multiplexing (TDM) transport such as SONET/SDH.

6. *Digiwrapper* is a layer 1 technology that can be considered as an extension and/or a replacement of SONET/SDH. It has the capability to carry any data traffic, be it SONET/SDH, ATM, IP, Ethernet, or fiber channel. It is standardized under the name G.709. It uses forward error correction (FEC) techniques to obtain higher link reliability from poorer-quality optical links. It is

intended for use in mostly optical and all-optical portions of the core network, as they emerge with advancing technology and bandwidth requirements.

It is economically unattractive to scrap the enormous SONET infrastructure in place. Consequently, it is necessary to find some solution that is capable of providing improved TCP/IP service over SONET. Any such solution must be provided by layer(s) of protocol between TCP/IP and SONET, which results in enhanced network performance. In time, SONET may disappear slowly, so any solution for IP over SONET should not preclude IP over the digital wrapper technology.

In this chapter we discuss several solutions for the transport of TCP/IP:

- Two legacy solutions, which have carried most of the TCP/IP traffic in recent years but which are limited in their future applicability for a variety of reasons
- A solution that appears to be the best present-day course and appears to be being followed by a significant portion of network providers
- Future optical network solutions

In addition, two other important protocol stacks are discussed:

- A solution that is appropriate for direct connection of private Ethernet domains across the WAN
- A solution that is appropriate for carrying storage area networking (SAN) traffic over the WAN

6.2 DIFFICULTIES WITH THE TCP/IP PROTOCOL SUITE

We begin this examination with a simplified vision of WAN implementations. Let us assume that WANs are built with no technology other than IP routers connected by the simplest link layer solution available. All packets are routed at the IP layer through this network of routers. Routing tables are maintained by the IP routing algorithms (OSPF and BGP). The simplest link layer is undoubtedly Gigabit Ethernet (or 10-Gb Ethernet). Although Ethernet has not been examined in this book, we accept it for present purposes as a straightforward way to move packets from point to point over fiber. For present purposes we ignore certain distance limitations of Ethernet. This "routed Ethernet" WAN is attractive in its simplicity. In the spirit of Occam's razor, we must ask why we should "unnecessarily complicate" this network by adding layers of protocols. If we cannot find strong reasons, perhaps the "routed Ethernet WAN" model should prevail in the world of real networks.

Another way to look at this chapter is to think of the gap between light over fiber and the TCP/IP protocol suite. How big is this gap? What functions need to be provided? What's the simplest of available satisfactory solutions? Is it perhaps some form of Ethernet, or is more complexity necessary to achieve the required network solution? The key to understanding what must be used to fill this gap is to understand the shortcomings of the TCP/IP protocol suite, which were identified in Chapter 5. Those shortcoming raise the essential issues with which this chapter must deal. Consequently, this chapter includes a review of the shortcomings of the TCP/IP suite. It quickly becomes apparent that more than the simplest possible link layer solution (Ethernet) must be employed to remedy the shortcomings of TCP/IP.

The difficulties identified with the TCP/IP protocol suite in Chapter 5 are:

1. *One shortest route, regardless of load conditions.* IP's routing algorithm results in one path for all traffic between any two endpoints. Although the use of the shortest path makes sense in the case of a single connection, it can be a poor choice in a congested network. The problem is that nearly-as-short paths are ignored, and their carrying capacity cannot be used to alleviate congested shortest paths.

2. *Deliberate congestion and backoff.* TCP deliberately seeks congestion. It ramps up its transmission rate (within the constraints of the receiver flow control system) until it discovers congestion by seeing packet loss. TCP uses this mechanisms to seek ever-higher transmission rates within available capacity, which is a good goal. But it has no way to avoid increased transmission rates before congestion happens. So TCP breeds congestion. This mechanism would probably be acceptable if it were not for the increasing demands to support higher QoS for applications such as video links and VoIP. Such services cannot be useful if their TCP/IP connections are subject to higher queuing delays, congestion, and packet loss.

3. *Lack of QoS support.* While IP was designed to support a modest system of QoS differentiation (via the type of service bits in the IP header), these features were ignored by implementers for many years. As a result, attempts to add QoS services at the IP level today are compromised: QoS requires end-to-end support, and the occasional new router cannot force all the other routers in the Internet to begin respecting the TOS bits. Consequently, it is essentially impossible to make use of IP's TOS facilities at this late point in time for Internet-wide applications. In addition, modern QoS standards require more support than IP's TOS facilities can provide.

4. *Limitations due to receiver windows and round-trip times.* The size of the receiver window of a TCP connection limits the bandwidth possible on a connection; the maximal bandwidth of a connection decreases with increasing round-trip time of the connection.

5. *Long, fat TCP pipes.* Given that it is difficult for any one TCP/IP connection to utilize a large amount of bandwidth on a long link, it necessarily takes many such connections to fill a long, high-bandwidth pipe. This difficulty makes it more difficult to engineer appropriate networks.

6. *Cost of routing.* Each time an IP packet arrives at a router, an IP address lookup must be performed to determine the next hop for the packet. This is a moderately expensive function given the worst-case peak arrival rate of IP minigrams.

These failings make it difficult or impossible for the TCP/IP suite alone to satisfy the needs of modern networks. Some other solutions must be applied to achieve the following goals in the face of the TCP/IP difficulties:

1. *Consistent provision of QoS* (at least supporting good-quality voice and video connections, and preferentially supporting differential levels of data services for higher- and lower-cost services). It is not possible to support differentiated QoS with TCP/IP adequately when all classes of traffic must be routed over the same path, and IP's TOS field cannot be used with any reliability over legacy networks.

2. *Efficient use of network resources.* It is not possible to make use of secondary (slightly longer) routes when IP routing algorithms send all traffic over the shortest route.

3. *Network survivability in the presence of hardware failures.* New IP routes cannot be determined quickly enough in the face of link or node failures, as the distributed IP routing algorithms are necessarily slow. Consequently, IP services must suffer severely in the presence of hardware failures unless other remedies are found and applied.

4. *Network manageability by its operators.* While it might be possible to build network OA&M (organization, administration, and maintenance) facilities in a "routed Ethernet WAN," such capabilities do not exist, so today's and probably tomorrow's real networks must have other solutions for OA&M requirements.

For the reasons cited above, a simple "routed TCP/IP over Ethernet" solution cannot meet the various requirements of modern networks. At the same time, it is practically impossible to replace TCP/IP at this time.

6.3 SUPPORTING PROTOCOLS

Curiously, the solution to the problems encountered by carrying the TCP/IP protocol suite of the Internet's WANs is found by adding more protocols to the mix! It is interesting that, at least to some extent, adding a new protocol between two other protocol layers in a protocol stack can remedy existing problems. In the present section we briefly introduce a number of protocols that have been proposed as partial solutions to TCP/IP's problems in the WAN.

6.3.1   ATM

ATM is a cell-based protocol. All data streams, whether fundamentally constant bit rate services such as telephony or bursty data services such as TCP/IP, are segmented into 48-byte chunks. Each such chunk is given a 5-byte ATM cell header to form a complete 53-byte ATM cell. ATM was designed by telephony interests to allow one universal network to carry traditional telephony calls in the same environment as a number of new services, including video, compressed telephony, and data. A fixed cell format was adopted to allow efficient, high-bandwidth switches to be feasible. The cell format was made quite small to minimize latency as new cells were filled and delivered cells were played back. (One-byte voice samples are taken every 125 µs. It takes about 6 ms to fill or play back a single cell. This consumed all of the latency budget that could be allocated to the purpose. Larger cells would have consumed so much latency that echo cancellation equipment would have been required. This would have been prohibitively expensive.)

ATM was designed to be a universal solution. This included both the ability to support advanced QoS standards and the flexibility to become the networking protocol interface of choice for desktop computers. As a result of these two goals, ATM became a sophisticated and complex protocol. However, two market forces crippled ATM: Ethernet and TCP/IP. The development of Ethernet focused on low-cost connectivity to the desktop. As the market for Ethernet grew, the cost of connectivity decreased rapidly. The fact that the original 10-Mb/s Ethernet used a daisy-chain configuration further reduced costs to each desktop client. By the time ATM was attempting to connect directly to each desktop, its more expensive star connectivity (with a ATM switch at the center of the star), and its small market size resulted in noncompetitive costs to desktop customers.

In addition to having higher hardware costs, any mass movement to ATM was crippled by the widespread use of TCP/IP. By this time, TCP/IP had successfully connected millions of desktop computers. Its QoS was poor, but it had enabled a number of highly popular applications such as FTP (file transfer protocol) and e-mail. Two options were available to ATM proponents: Replace the TCP/IP network interface in the important applications with an ATM interface, or support TCP (or TCP and IP) over ATM. The first approach was infeasible due to the large number of applications and their enormously widespread distribution. The second approach was not attractive to customers, as it added complexity and seemed to add very little of value (at least to the initial applications, which did not require strict QoS).

As a result of these issues and forces, ATM has not "won the desktop." However, ATM has found two important roles in today's networks, and it is these roles that we focus on in this section. Before describing these roles, let's take a quick tour of some of the technical features of ATM. Only the briefest summary is given below, for the dual reasons that ATM's success has been too limited to merit a large amount of space in this book, and because many of the better ideas in ATM have been adopted by other protocols and proposals (e.g., MPLS).

The most relevant features of ATM are:

1. There are two cell header formats: UNI (user–network interface), which is used between client end stations and network switches, and NNI (network–network interface), which is used between ATM switches. Since ATM has failed to dominate the desktop, the UNI interfaces is of reduced importance.

2. Cells are routed by virtual path indicators (VPIs) and virtual channel indicators (VCIs). These small tags are used to establish connections and switch cells. A new client connection is allocated a VCI. VCIs are packaged inside VPIs, which tend to be more permanent entities used to bundle many VCIs through the network core, where they are switched as a collection.

3. The ATM community put great effort in establishing a strong technical basis for differential QoS services. Support was provided for constant-bit-rate (CBR) services and several forms of variable-bit-rate (VBR) services. Individual traffic flows were policed to ensure that they remained within their agreed traffic parameters and to ensure that their agreed QoS parameters could be supported by the network.

4. New connections stated their desired QoS parameters to their network interface. The network then attempted to meet these requirements and to find an appropriate route through available links and network switches. If appropriate resources were available, the new call/connection was established. The establishment involved setting VPI/VCI routing and swapping tables in each ATM switch on the established path.

5. Once a new connection was established with its stated QoS, it was highly probable that the network would successfully maintain the agreed-upon QoS parameters.

6. ATM provided several ATM adaptation layers (AALs), which supported the carrying of a variety of traffic classes within the ATM cells. These AALs managed the required segmentation and some forms of error recovery.

7. The small, fixed ATM cell size allowed efficient, high-aggregate bandwidth switches to be designed and built. On the other hand, ATM suffered a "cell tax" of 5 (or more, due to the AAL layer) bytes of overhead for each 53 bytes transmitted. This reduced overall network efficiency.

In our judgment, ATM did offer the possibility of one integrated network with strong QoS and efficient multiplexing of different traffic types. Had the world invested sufficiently in ATM, today's telecommunications systems would have been stronger and better integrated. The cell tax issue was of minor consequence, as bandwidth is relatively cheap, while the appropriate management of bandwidth remains illusive. However, ATM failed due to the advantages of being first to market (Ethernet and TCP/IP) and of low-cost entry (ATM would certainly have been more expensive to develop than the lower-cost solutions we have today; it is far from clear that would have been cheaper in the long term, but the initial costs were an obstacle to ATM.) Commercial forces and inertia, coupled with the explosive growth of data services over TCP/IP, have dictated the direction that networks have taken. Network engineers of all sorts continue to struggle with the difficulties of providing acceptable services (QoS) over hybrid TCP/IP networks. One could say that this entire book is directed toward this goal.

As mentioned above, ATM has found two successes in today's networks. The first success of ATM is slightly outside the scope of this book, but we shall mention it briefly in the interest of breadth. Many homes are served by *DSL links. These links provide "wideband" communications rates on the order of a few million bits per second. If a large TCP/IP frame (say, 1 k bytes) were sent down a 1-Mb/s link, the link would be committed to the transfer of that frame for 8 ms. This is too long a time if any real-time service such as telephony or video is sharing the link. Starving a voice connection for this time could lead to starvation at the playback device and an audible click in the sound received. To avoid this, ATM is used on such links. The large TCP/IP frame is segmented into ATM cells. Those cells then compete with higher-priority voice cells on a cell-by-cell basis. With the higher priority assigned to the voice traffic, the longest delay of a voice cell would be only 424 μs. The data traffic occupies all ATM cell times when the voice connection has no data to transfer. The costs of this technique are increased overheads (the ATM cell tax plus various segmentation costs); the benefit is that multiple services such as voice, video, and data can coexist on these common *DSL links.

The second success of ATM is more relevant to the focus of this book. If networks segment TCP/IP traffic into ATM cells, the sophisticated bandwidth allocation and flow control techniques of ATM can be employed to ensure fairness, maximal QoS, and enforcement of SLAs (service level agreements) on shared broadband Internet links. Further, ATM allows all major consumers of telecom services (i.e., phone, data, video, storage area networks, etc.) to share major network trunks efficiently. ATM is employed in this manner in a large portion of today's Internet core.

6.3.2 Generic Framing Procedure

The generic framing procedure (GFP) provides several features that bridge between the functions of SONET and the functions of the carried frames. The most important services provided by GFP are:

- Identification of the starts of frames within the carrying byte stream (e.g., SONET)
- The ability to carry more than one frame type within the byte stream, and to distinguish among the multiple frame types, should they be mixed in any one logical channel
- The ability to carry management frames within the GFP channel

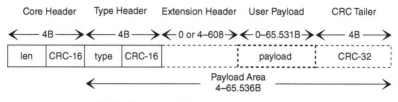

FIGURE 6.1   GFP headers and trailers.

- Strong CRC-based protection for all GFP header information and payloads

Figure 6.1 illustrates the three headers, the user payload, and one trailer of the GFP protocol.

1. *Core header.* The core header consists of the first 4 bytes of each GFP frame. It contains only the length of the GFP payload area and a CRC-16 protecting that length information. The core header provides the fundamental function of identifying frame boundaries in the carrying byte stream (e.g., SONET). GFP frame alignment is found initially by a search for a sequence of length fields, followed by correct CRC-16s computed from that length, with the length field pointing to the next such length field. Searching for this core header alignment is carried out before any frames are transferred.

2. *Type header.* Once the core header has identified the frame boundaries, it is necessary to describe the type of frame being carried. The focus of this book is on the use of GFP to carry MPLS frames, but one of the strong advantages of GFP is that it can be used to carry a wide variety of frame types. The type field of the type header describes all of the following information:

- Whether the frame is a data or a management frame
- The payload type, which can be Ethernet, MPLS, RPR, some SAN frame, or other options
- Whether the GFP frame carries any extension header information
- Whether a CRC-32 trailer is appended to the GFP frame

3. *Extension header.* This optional, variable-length field is provided for future expansions of functionality. At this time, it is unused.

4. *User payload.* The complete user payload frame (e.g., an MPLS frame) is carried in this variable-length field.

5. *CRC-32 trailer.* This optional 4-byte CRC-32 is computed over the GFP frame.

For a user payload type that carries its own protection, the normal GFP overhead consists of the 8 bytes of the core and type headers. Where the user payload requires CRC protection, the normal GFP overhead also includes the

CRC-32 trailer, which brings the total GFP field length up to 12 bytes. GFP thus provides several critical features that were not included in SONET/SDH, that was originally intended primarily for telephony services.

6.3.3 Multiprotocol Label Switching

The multiprotocol label switching (MPLS) protocol was introduced to remedy a number of the shortcomings of the TCP/IP protocol suite. MPLS is a bridge between the data link layer (layer 2) and the network layer (layer 3, i.e., IP). MPLS is a versatile solution used to address many problems faced by present-day networks: speed, scalability, QoS management, class of service (CoS), and traffic engineering. MPLS has emerged as an elegant solution to meet the bandwidth-management and service requirements for next-generation Internet protocol (IP) backbone networks. MPLS addresses issues related to scalability and routing based on QoS and service quality metrics.

MPLS is an Internet Engineering Task Force (IETF) standard that provides for the efficient designation, routing, forwarding, and switching of traffic flows through the network. MPLS performs the following functions:

- Specifies mechanisms to manage traffic flows
- Remains independent of the layer 2 and layer 3 protocols
- Provides means to map IP addresses to simple fixed-length labels
- Interfaces to existing routing protocols
- Supports IP, ATM, and other layer 2 protocols
- Reduces the costs of IP routing

Networks of MPLS routers form MPLS domains. Traffic can enter or depart these MPLS domains in special routers on the edge of the MPLS domains that are called *label edge routers* (LERs). LERs classify incoming traffic into one of many forwarding equivalence classes (FECs). This classification process examines the arriving frame's (or cell's) destination, priority, and possibly details of the encapsulated protocol or application. The goal of classification is to separate traffic into FECs that require different levels of service as they are forwarded by the interior of the MPLS network to their eventual exit LER. MPLS adds a small header to each frame forwarded through its interior network. This header contains little more than a label tag that is used to determine its treatment by the next internal router at which it arrives. Internal routers called *label swap routers* (LSRs) examine the MPLS header of arriving frames by performing a simple table lookup using the arriving header as an index. The results of the table lookup determine to which of an LSR's egress ports the frame should be sent and what replacement label value should be used in the MPLS header when the frame is emitted. The action of the LSR is thus very simple. Although it is referred to as a router, it is in fact only a switch, with a simple translation from incoming label to outgoing link and

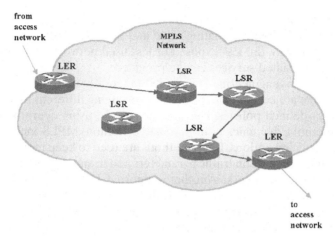

FIGURE 6.2 MPLS network.

replacement label. The cost of LSRs can be less than the cost of IP routers, due to this simpler process. (However, each LSR usually implements a complete IP router as a separate control path for a variety of reasons, including the carrying of the messages that establish the MPLS functionality.)

Eventually, the MPLS frame arrives at its egress LER. At this point the label is used to determine only to which (non-MPLS) egress port the frame should be directed. After this determination, the label is stripped off the frame, and the frame transported is emitted in its native protocol (e.g., IP). It is necessary to understand how an FEC of frames is routed and switched to the proper exit LER. A sequence of LSRs to be visited is determined for each FEC by automated management software. Figure 6.2 represents one such path, for one FEC, from an entry LER on the left to an exit LER on the right. This intended path is encoded into the next-port and next-label tables in each router. The tag of an arriving frame (and all frames in its FEC, with the same label on the same link) is used to determine which MPLS router should be visited next. Thus, any particular path through the MPLS network can be programmed by entering values into these lookup tables. These encoded paths, called *label-switched paths* (LSPs), are established for each FEC. These FEC/LSP pairs can be set up statically (preprovisioned) or dynamically. There can be as many separate LSPs from one LER to another as desired. Normally, there is at least one per class of traffic, but there may be more if the MPLS bandwidth management control finds it necessary to divide one class of traffic into more than one FEC/LSP in order to make use of multiple paths between the two LERs.

Traffic engineering is a process that enhances overall network utilization by attempting to create a uniform or differentiated distribution of traffic throughout the network. An important result of this process is the avoidance of congestion on any one path. It is important to note that traffic engineering does not necessarily select the shortest path between two devices. It is possible that for two frame data flows, the frames may traverse completely different paths even though their originating node and the final destination node are the same. In this way, the less exposed or less used network segments can be used and differentiated services can be provided.

In MPLS, traffic engineering is provided inherently using explicitly routed paths. The LSPs are created independently, specifying different paths that are based on user-defined policies. They can be created using operator intervention or preferably, dynamically in automated fashion. MPLS includes traffic shaping and policing options. These methods are used to keep individual traffic sources within agreed-upon traffic parameters and to smooth the flow of traffic through the MPLS network. Another supported feature of MPLS is *label merging*, a mechanism that allows an LSR to merge multiple LSPs into one new, larger egress LSP. This mechanism makes aggregation of flows straightforward. However, this mechanism does not support the demultiplexing of the multiplexed flows. A more useful mechanism that supports both multiplexing and demultiplexing is described next.

MPLS supports the tunneling of one MPLS network's flows through another MPLS network. This feature is useful for "carrier of carriers" networks. When an MPLS frame enters a carrier-of-carriers network (from the carrier network), the original MPLS label is retained and another MPLS label is pushed on in front of the frame. This is referred to as *label stacking*. A carrier-of-carriers network uses the second, outer label for routing until the intended LER is reached. At that point the carrier-of-carriers MPLS label is stripped off and the resulting frame (carrier's MPLS label and payload) is delivered to the carrier's MPLS network. Labels can be stacked in this manner to any depth, thereby facilitating any depth of tunneling of networks.

MPLS labels consist of 4 bytes, containing the following fields:

- *Label* (20 bits): the label that will be used when the frame arrives at its next LSR or LER to look up a new exit port and a new label value.
- *Class of service* (CoS, 3 bits): the MPLS class to which the frame belongs. This field will be used to determine queuing, policing, and dropping policies in each MPLS router.
- Bottom of stack indictor (S, 1 bit): when set to 1, the bottom label in an MPLS label stack, or the first label created; when set to zero, this label is stacked on top of the first label.
- *Time to live* (TTL, 8 bits): a count of the number of MPLS hops that remain before the frame will be assumed to be miss-routed and thus dropped.

It would have been possible to retain one label value on all hops in each of the LSPs, avoiding the need to replace labels. But this would have left only 2^{20} labels for all LSPs in an MPLS network. The fact that labels are swapped for new labels at each LSR means that there are 2^{20} labels for each port out of each LSR. Thus, doing the minor bit of work required to swap labels makes labels on all links independent and makes the label allocation problem easy even with huge numbers of LSPs.

MPLS addresses many network backbone requirements by providing a standards-based solution that has many advantages over a traditional IP-over-ATM overlay approach. First, it improves frame-forwarding performance in the network compared to routing IP frames. Second, it enhances and simplifies frame forwarding through routers using layer 2 switching. Third, it supports QoS and CoS for service differentiation. Finally, MPLS helps build scalable virtual private networks (VPNs) with traffic-engineering capability. In that light, MPLS can be viewed as a ATM replacement technology. But it might also be considered as a bridge technology between access IP and core ATM, effectively joining two disparate networks. To what extent MPLS will "replace" ATM and to what extent it will "collaborate" with ATM remain to be seen.

6.3.4 Ethernet over the Internet

Ethernet is a family of closely related packet-oriented protocols that operate at 10 Mb/s, 100 Mb/s, 1 Gb/s, and 10 Gb/s. The lower two rates are the dominant technology for the implementation of local area networks (LANs). The 1-Gb/s rate is used in more advanced LANs and to connect LANs. The 10-Gb/s rate is emerging to offer greater bandwidth in connecting LANs. The LAN use of Ethernet is not of direct interest to this book on transport networks. Although Ethernet-based LANs are a source of much of the traffic carried by transport networks, in general the Ethernet protocol has been terminated and the transport network is used to carry the contained IP frames. However, the widespread popularity of Ethernet in the LAN has led to demands for Ethernet support in the WAN or the Internet. There are two reasons: Ethernet has a reputation for low cost which customers would like to extend into the WAN, and customers with multiple Ethernet LAN sites would like to be able to connect these LANs without getting involved in more complex protocol stacks—hence they want Ethernet-over-the-WAN service. It is in this role that we are primarily interested in Ethernet.

There are three approaches to the support of Ethernet over the WAN. The first is called *transparent support* of Ethernet. In this solution, complete Ethernet frames are transported in point-to-point channels that carry only Ethernet frames. This can be accomplished by carrying Ethernet frames directly in optical fiber (or in optical colors in DWDM systems). With this approach, there is very little but Ethernet present; all the overheads and OA&M services normally associated with WAN links are absent. These solutions can often be implemented at lower costs than other solutions, but the lack of effective OA&M services limits the application of this option. An alternative is to embed Ethernet frames in SONET/SDH SPEs of varying sizes. For example, a SONET STS-24c of about 1.268 Gb/s is appropriate to carry 1-Gb/s Ethernet. With this option, bandwidth is lost to overheads, but standard OA&M services are available at the SONET/SDH level. Of course, Ethernet itself is essentially unmanaged.

The second basic approach for carrying Ethernet over the WAN is the use of switched Ethernet. With this option the WAN is provisioned with Ethernet switches, which can direct Ethernet frames based on their Ethernet endpoint addresses. This option supports more complex networks than the transparent mechanisms, which support only point-to-point services. A switched Ethernet WAN can be as large and as complex as desired as long as some mechanism is provided to manage the hardcoded Ethernet endpoint addresses. The Ethernet pipes between the Ethernet switches can be based on any of the transparent Ethernet solutions (e.g., embedded in SONET/SDH SPEs).

The final class of solution for extending Ethernet over the WAN is the use of routed Ethernet networks. In this class of solution, the routing is at the IP layer. Packets enter or depart routers on Ethernet links but are routed in the normal manner for IP frames. Links between routers may or may not be Ethernet. In other words, this solution is indistinguishable in essence from normal IP routed layers. The use of Ethernet for access and perhaps for interrouter links may reduce costs. This solution is the only one of the three that offers universal IP access to the data frames, as only in this solution is the traffic not isolated to an Ethernet island of isolated traffic.

6.3.5 Resilient Packet Rings

Resilient packet rings (RPRs) are intended for use in metropolitan area networks (MANs). The dominant MAN protocol has been SONET/SDH operating in fault-tolerant rings. To understand RPR, it is necessary to understand the features and shortcomings of SONET/SDH in this application.

- Protection of traffic is certainly not a shortcoming of SONET/SDH, which has led the industry in providing reliable and fast replacement of failed links and/or components of network elements. It is a goal of RPR to provide as good a level of connectivity protection as SONET/SDH has provided.
- SONET/SDH was designed originally primarily for telephony services. As packetized data services take over more and more applications and resulting network traffic, SONET/SDH becomes less appropriate in that its telephony features are unused (as telephony moves to VoIP) and it lacks some important data features (e.g., the features added by GFP). Thus perhaps it makes sense to replace SONET/SDH in MAN rings with a more packet-friendly service.
- SONET/SDH is not good at carrying packet multicast services.
- The standard SONET/SDH ring protection mechanisms are wasteful of ring bandwidth or at least make it more difficult to make use of otherwise idle protection capacity.

RPR uses ring architectures much like SONET/SDH. But instead of imposing a TDM division of bandwidth over the available physical channel, RPR makes use of a single simple channel. This single channel (per fiber direction in the ring) is used to transport RPR frames. Each node on the ring is no longer (just) a TDM switch but is a ring-oriented frame switch. The RPR logical ring layer can run on an Ethernet-like physical layer or within a standard SONET channel (e.g., OC-48C). Frame traffic is added to the ring though a frame multiplexer that mixes new traffic in with traffic received on the ring. Frame traffic is dropped from the ring when it has reached its (last) destination. Thus, traffic travels exactly as far as necessary around the ring, but no further. Communications capability (bandwidth on fibers) is consumed only where necessary, leaving the remaining capacity for other frames on other parts of the ring.

Traffic is prioritized, so QoS can be maintained for the more critical services. When a failure occurs, the RPR nodes on either side of the failure (which may be a link or an intervening RPR node) double the ring back in the other direction, much as SONET/SDH achieves ring repair. When a fault is isolated by this doubling-back process, less aggregate capacity remains on the ring for the transport of frames. If necessary, the lower-priority frames are the ones to be delayed or lost.

RPR Operation   RPR technology uses a dual counterrotating fiber ring topology. Both rings, inner and outer, are used to transport working traffic between nodes (Figure 6.3). In the example shown the inner ring carries the data flow in clockwise direction, while the outer ring carries the data in a counterclockwise direction. By utilizing both fibers, instead of keeping a spare fiber for protection as SONET/SDH does, RPR utilizes the total available ring

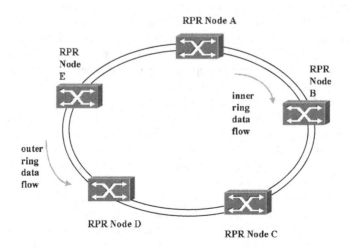

FIGURE 6.3   Resilient packet ring.

bandwidth. The fibers are also used to carry control signals for purposes such as topology updates or bandwidth control messages. Control messages flow in the opposite direction of the traffic that they represent. For the example used in Figure 6.3, the inner would carry control signals in a counterclockwise direction.

By using bandwidth control messages, a RPR node can dynamically negotiate for bandwidth with the other nodes on the ring. RPR has the ability to differentiate between low- and high-priority frames. Just like other QoS-aware systems, nodes have the ability to transmit high-priority frames before those of low priority. In addition, RPR nodes also have a transit path, through which flow frames destined to downstream nodes on the ring. With a transiting buffer capable of holding multiple frames, RPR nodes have the ability to transmit higher-priority frames while temporarily holding other lower-priority frames in the transit buffer. Nodes with smaller transit buffers can use bandwidth-control messages to ensure that bandwidth reserved for high-priority services stays available.

Media Access Controller for RPR   RPR is layer 2 protocol. RPR frames can be transported over both SONET/SDH and Ethernet physical layers. Thus, RPR is largely independent of whatever layer 1 technology is used. As a layer 2 network protocol, the media access controller (MAC) layer contains much of the functionality for the RPR network. The RPR MAC is responsible for providing access to the fiber media. The RPR MAC can receive, transit, and transmit frames.

Let us discuss a receiving path first. Every station has a 48-bit MAC address, similar to Ethernet. The MAC will receive any frames with a matching destination address. The MAC can receive both unicast and multicast frames. Multicast frames are copied to the host and allowed to continue through the transit path. Matching unicast frames are stripped from the ring and do not consume bandwidth on downstream spans. Now consider a transmit path. Nodes with a nonmatching destination address are allowed to continue circulating frames around the ring. Unlike point-to-point protocols such as Ethernet, RPR frames undergo minimal processing per hop on a ring. RPR frames are inspected only for a matching address and header errors.

The RPR MAC can transmit both high- and low-priority frames. The bandwidth algorithm controls whether a node is within its negotiated bandwidth allotment for low-priority frames. High-priority frames are not subject to the bandwidth-control algorithm. Thanks to this mechanism, RPR can provide QoS attributes in the RPR network. RPR uses a bandwidth-control algorithm that dynamically negotiates available bandwidth between the nodes on the ring. This applies only to the low-priority service. It ensures that nodes will not be disadvantaged because of ring location or changing traffic patterns. The algorithm manages congestion, enabling nodes to maximize the use of any spare capacity. Nodes can be inserted or removed from the ring without any bandwidth provisioning by the host.

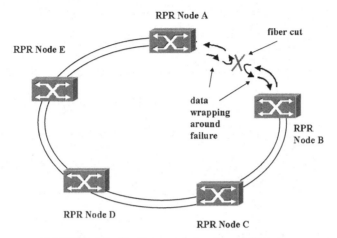

FIGURE 6.4 Traffic restoration in an RPR ring.

RPR Topology Discovery and Protection Mechanisms RPR has a topology discovery mechanism that allows nodes on the ring to be inserted or removed without manual management intervention. After a node joins a ring, it will circulate a topology discovery message to learn the MAC addresses of the other stations. Each node that receives a topology message appends its MAC address and passes it to its neighbor. Eventually, the frame returns to its source with a topology map of the ring. RPR has the ability to protect the network from single span failures. When a failure occurs, protection messages are quickly dispatched and traffic is wrapped around the failure point as illustrated in Figure 6.4. Within 50 ms the ring will be protected and the data traffic can resume safely. RPR features as fast a restoration of service as that of SONET/ SDH. RPR restoration is clearly superior to the slow restoration of Ethernet-based networks.

RPR Benefits RPR technology has several benefits over existing solutions. First, it provides an efficient multicast function. One RPR multicast frame can be transmitted around the ring and can be received by multiple nodes. Second, RPR is more efficient than SONET/SDH, as it provides spatial reuse. RPR unicast frames are stripped at their destination. Unlike SONET/SDH networks, where circuits consume bandwidth around the entire ring, RPR allows bandwidth to be used on multiple idle spans. Spatial reuse is illustrated in Figure 6.5. Third, RPR supports resiliency and topology discovery. RPR ring nodes are added and removed from the topology map automatically. Fourth, RPR provides some quality of service. High-priority frames are delivered with minimal jitter and latency. Finally, RPR offers a lossless transport, as nodes on the ring will not drop frames. Compare this to Ethernet, which can lose frames in congested conditions.

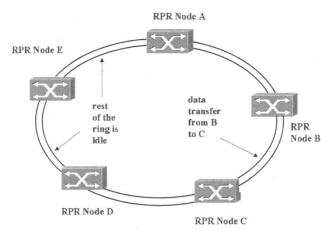

FIGURE 6.5 Spatial reuse in an RPR ring.

6.3.6 G.709: Digital Wrapper Technology

Creators of digital wrapper technology recognized some of weaknesses of
SONET/SDH and acknowledged the diversity of protocols used in real networks. Digital wrapper technology was formalized in 2001 as the ITU G.709
recommendation and is frequently referred to as G.709 technology. Its purpose
was to build a new standard based on experience in SONET/SDH for next-
generation networks that can include other forms of gigabit network protocols
such as Ethernet or Fiber Channel.

Like SONET/SDH, G.709 has a layered structure that includes protection,
performance monitoring, and other management services. G.709 provides
interfaces for an optical transport network (OTN). The OTN structure shown
in Figure 6.6 is quite similar to that of a SONET transport network. It consists
of an optical transmission section (OTS), optical multiplexing section (OMS),
and optical channel section (OCh). It is clear that OTS resembles the section
segment in SONET, OMS the path section, and OCh the line section. OTS is
a transmission span between two optical amplifiers. OTS contains data payload
and OTS overhead. OMS is responsible for multiplexing in WDM terminals.
OMS adds its own overhead. Finally, OCh is responsible for end-to-end channel
connection and adds OCh overhead. The relationship between OTS, OMS, and
OCh closely mimics that of SONET.

Forward Error Correction   One interesting addition in G.709 is forward error
correction (FEC) for improvements in transmitted bit error rates (BERs)
(Figure 6.7). Whereas FEC can always be added in SONET or any other protocols, it is mandatory and standardized in G.709, which ensures that some
degree of interoperability between hardware vendors. Due to its extra man-

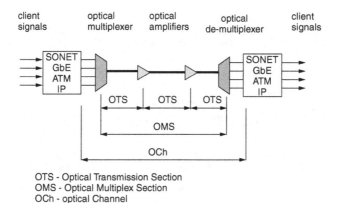

FIGURE 6.6 Optical transport network.

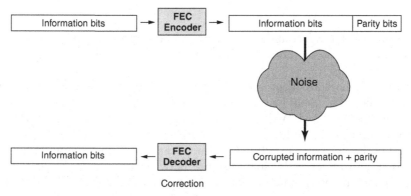

FIGURE 6.7 Forward error correction concept.

agement functionality and FEC addition, G.709 has extra overhead compared to SONET/SDH. But what exactly is *forward error correction*? It is a technique that enables a high degree of link reliability, even with the presence of noise. FEC increases the signal-to-noise ratio (SNR) for the optical link. For an input BER of 10^{-4} (a seriously degraded link), the output BER might become 10^{-12} (a good link) when FEC is used. For SONET/SDH, an acceptable BER is 10^{-12}.

How is higher FEC performance accomplished? The basic concept is illustrated in Figure 6.8. A number of parity bits are added to the information bits being sent through a communication channel. Even in the presence of noise, the receiver is capable of correcting a certain number and type of errors, due to higher redundancy in the data structure. The degree of FEC gain depends

FEC code parameters:
- k = number of information symbols
- n = number of data symbols after FEC code is added
- latency

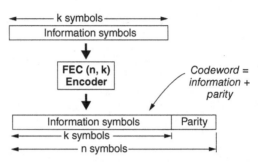

FIGURE 6.8   Basic FEC definitions.

on the particular algorithm used and the amount of extra parity bits added. More bits increase the error correction power, but also increase latency and data overhead.

Although there are many ways to increase performance, and the SNR ratio in particular, FEC might be a cheaper alternative than other means of accomplishing the same goal. With FEC, for the same BER, you can relax SNR at the receiver. In the optical domain, that translates to relaxing the power budget at the optical transmitter. Or, equivalently, you can keep optical power the same and increase the span distance between terminating/regenerating/amplification points. Improved noise tolerance means that in DWDM systems, the wavelength spacing can be tighter. Clearly, using FEC allows for many interesting optical network optimization trade-offs.

The FEC implementation in G.709 uses a well-known Reed–Solomon code called RS (255, 239). This notation means that for every 239 bytes of data, 16 bytes are added for error correction (239 + 16 = 255). RS (255, 239) code can correct up to eight bit errors in a code word when used in error correction mode. If used in error detection mode, up 16 errors can be detected. Digital wrapper hardware can typically be configured either to detect only or to correct as well.

Digital Wrapper Frame   The digital wrapper frame consist of data payload, overhead, and FEC. Data payload could be SONET, Ethernet, fiber channel, or any other frame. A digital wrapper does not care what sits inside its payload, so it can carry any protocol. The digital wrapper frame has four rows of 4080 bytes, for a total of 16,320 bytes (Figure 6.9). The interesting question is how FEC data are incorporated into the frame. As mentioned previously, a digital wrapper requires RS(255, 239) FEC implementation. This FEC code requires

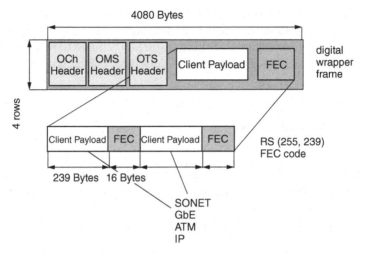

FIGURE 6.9   Data structure in OTN.

that after every 239 bytes of client payload, 16 extra FEC bytes are added. As a result, data payload is "chopped" to create this 239 + 16 scheme as shown at the bottom of Figure 6.9.

Although FEC addition improves bit error rates, there is a price that has to be paid for this feature. The addition of 16 extra FEC bytes on top of the existing 239 payload bytes implies additional overhead of 6.7% (= 16/239). As a result, a hardware processing digital wrapper protocol has to operate at rates that are 6.7% higher than nominal. For example, to achieve effective 10-Gb/s throughput through the G.709 network, hardware has to run at 10.67 Gb/s. Fortunately, in practice, 6.7% is small enough that this extra requirement is not a large obstacle.

Digital Wrapper Benefits   Digital wrapper technology has many benefits. First, it can carry any type of data, whether it is SONET/SDH, Ethernet, ATM, IP, or fiber channel. As the structure of OTN suggests, the digital wrapper was built with WDM technology in mind. The question might therefore arise whether digital wrapper can be used in a non-WDM environment. The answer is yes, it can, in which case OTS and OMS layers can be bypassed completely.

Second, a digital wrapper can be used at gigabit rates. The OTN hierarchy was built specifically for gigabit networks, so 2.5-, 10-, and 40-Gb/s rates are supported. The standard does not support lower-bandwidth SONET rates at 155 and 622 Mb/s, although it would have been straightforward to do so if there had been a need for this addition.

Third, G.709 offers prompt provisioning of optical channels, which is a large improvement over slow service provisioning in SONET. With prompt provisioning, carriers can offer service-level agreements on any protocol or service. With G.709 they can provide end-to-end optical channel monitoring and management. Despite its interesting features, it remains to be seen whether digital wrapper technology be widely deployed commercially in optical networks.

6.4 LEGACY SOLUTIONS

We now turn our attention from the problems of TCP/IP and the nature of our protocol tools to the issue of how these protocol tools are used to at least partially provide solutions to the limitations of TCP/IP in a world that requires QoS, high throughputs, and automated OA&M. To begin with, two important legacy solutions must be considered.

6.4.1 IP over SONET

The earliest solution to carrying IP frames over SONET was the simplest. IP frames were embedded in SONET byte streams with either packet over SONET (POS) or point-to-point protocol (PPP) serving to embed the frames in the SONET byte stream. Both POS and PPP are based on the HDLC protocol, which suffers one very serious weakness: The means used to embed control information in the byte stream meant that packets carrying a particular byte code would require twice as much space in POS (or PPP) as they did at their native IP layer. This pernicious expansion of bandwidth made it impossible to make strong QoS assurances without dedicating approximately twice the bandwidth that would otherwise be required—the network always had to be prepared for a burst of the bandwidth-doubling data code. Although severe performance losses due to random occurrences of this code are extremely unlikely, they provide an opportunity for network opponents to damage overall network behavior, and the practice remains subject to accidental damage by applications that unintentionally transfer many bytes of the offending control code.

A second weakness of IP over SONET was that it provided no ability to distinguish differential QoS for different TCP/IP traffic classes. A third weakness of these protocols was that it was not possible to interrupt a lower-priority frame with a higher-priority QoS-critical frame such as voice over IP. This issue became important on slow channels, as the critical information could be delayed too long by large frames on these slow pipes.

6.4.2 IP over ATM over SONET

This solution carved the IP frames into segments that could be carried by ATM cells, with 48 bytes of payload. An ATM adaptation layer (AAL) carried the segments of the IP frames within the byte stream provided by a sequence of

ATM cells. Although ATM and the AAL layer had heavy overheads, three advantages resulted: (1) there was no possibility of doubling the required bandwidth as in HDLC-derived protocols; (2) as each ATM cell was scheduled independently, it was possible for higher-priority traffic to interrupt lower-priority traffic; and (3) independent ATM cell flows, via the VPI/VCI tags, allowed differential QoS for different services and for different service-level agreements (SLAs).

ATM has been used extensively in some service providers' networks and in the ADSL links to customer homes. Were it not for the difficulties of managing an ATM WAN, and for the high overheads of ATM cells, ATM might have provided a network-wide solution with benefits comparable to the MPLS-based solution discussed in much of this book.

6.5 NEW PROTOCOL STACK SOLUTIONS

6.5.1 Using MPLS

Figure 6.10 illustrates the overall networks protocol solution that the authors feel is most appropriate for future general-purpose TCP/IP networks. Significant energy is being expended pursuing network solutions based on these ideas, or ideas very like them. However, the future of the Internet has proven to be difficult to predict, so we must confess that this solution may or may not—in the fullness of time—come to dominate the public Internet. That said, it is clear that the protocol stack solution discussed in this section has some compelling advantages.

Reading Figure 6.10 from the top, we see that TCP and UDP over IP are preserved at the network and transmission layers. Thus, all the existing TCP/IP protocol suite implementations and applications can continue to be compatible with this proposed network. At the bottom two layers of Figure 6.10, we

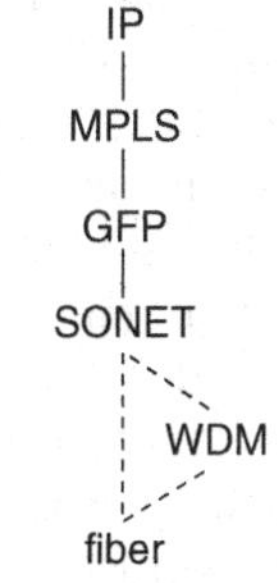

```
TCP    UDP
  \    /
   IP
   |
  MPLS
   |
  GFP
   |
 SONET
   :
   :  WDM
   :
  fiber
```

FIGURE 6.10 A protocol stack for the next-generation Internet.

see that traffic is still carried in SONET pipes over fiber, with or without the use of WDM technologies. In the case where SONET is carried over WDM, over fiber, SONET requires additional OA&M and protection solutions to deal with the issue of carrying independent SONET pipes in the same physical path.

It is the two interior layers of this protocol stack that provide the features required to bridge from the TCP/IP protocol suite to the SONET/fiber layers while remedying or overcoming the shortcomings of the TCP/IP layers. Most of the solution is provided by MPLS. When IP frames enter an MPLS label edge router (LER), the LER provides the image of an IP-routed network to the IP layer. But it immediately adds behaviors unbeknownst to the IP layer, which solve problems created by IP. The LER examines the IP packet at the IP layer and at contained layers. Any such information can be used to classify the packet for transmission in one of MPLS's forwarding equivalence classes (FECs). Thus, a LER may classify all VoIP traffic to a particular destination into one FEC, while classifying all data traffic to the same destination into one of two other FECs, depending on the priority of the data. Then the MPLS system of a series of Label switch routers (LSRs) and a destination LER treat all three FECs differently by sending them through independently allocated and controlled channels. The IP layer at the source and destination LERs is unaware of this differentiation of service provided by MPLS. Yet none of this added sophistication disrupts any assumptions made by the TCP/IP layers.

Let us now examine how well MPLS provides solutions for the standard TCP/IP problems:

1. *One shortest route, regardless of load conditions.* The MPLS layer is free to support multiple paths between two LERs. It can use independent FECs to separate traffic onto the multiple paths. In dividing TCP/IP traffic over multiple paths, it is desirable to keep each TCP/IP flow (defined by source and destination IP address and IP port number combinations) on one path to avoid out-of-order delivery within any one flow. Yet a large degree of freedom for the management of traffic over multiple paths is introduced by MPLS.

2. *Deliberate congestion and backoff.* Arbitrary TCP/IP and UDP/IP endpoints will continue to attempt to fill the channels they use until limited by receiver buffers, or congestion and packet drop occurs. Packaging this behavior within MPLS mechanisms will not change this basic IP protocol suite behavior. MPLS can, however, mitigate the problem by managing queues of IP packets in improved ways: for instance, by applying the random early discard (RED) packet discard mechanism. Recall that RED prevents synchronized TCP/IP backoff behaviors when many TCP/IP flows experience packet drop at the same time. Instead, RED deliberately inflicts random packet drops as the queue nears fullness, thereby desynchronizing the backoff behaviors of the various TCP/IP flows. Nevertheless, we must expect TCP/IP data flows such as ftp to continue to experience packet loss. However, more disciplined flows such as VoIP and videoconferencing can have much better outcomes when

used with MPLS. These applications require high QoS to be useful, but they also offer a great benefit to transport networks. Their communications requirements are bounded: A VoIP connection requires only an (approximately) fixed amount of bandwidth. Sending such a flow through a later pipe will not cause it to expand its use of the pipe until congestion occurs. Instead, the flow is inherently limited. Thus, such flows can be added to QoS-protected flows by a call acceptance control (CAC) procedure which refuses to add calls beyond what can be conservatively supported by the allocated TDM or physical channel. MPLS is a mechanism that can be used to gather such traffic into high-QoS flows and to protect these higher-priority real-time flows from opportunistic data flows.

3. *Lack of QoS support.* MPLS provides a powerful mechanism to classify IP traffic into multiple QoS classes, and to provision channels separately for each class across the MPLS network. As MPLS networks are "entirely new," end-to-end provision of QoS support is possible.

4. *Long fat, TCP pipes.* As MPLS takes over the packing of TCP flows into MPLS LSPs, there are several avenues toward solution of this problem. The most useful solution is that "fat" SONET pipes will probably be divided into multiple MPLS channels to provide differential QoS support and differential channels for different source–destination combinations. So there will be less tendency to have "one big pipe" to pack with TCP/IP frames.

The layer below MPLS in the protocol stack of Figure 6.10 is GFP. GFP is provided for a variety of reasons, which were introduced in Section 6.4:

- Identification of the starts of frames within the carrying byte stream (e.g., SONET)
- An ability to carry more than one frame type within the byte stream, and to distinguish among the multiple frame types, should they be mixed in any one logical channel
- An ability to carry management frames within the GFP channel
- Strong CRC-based protection for all GFP header information and payloads

Although the GFP layer seems "thin" in functionality it is technically necessary and provides important opportunities for offering non-TCP/IP services over the same core network (i.e., other services supported by GFP/SONET/ fiber).

6.5.2 Future All- or Mostly Optical Networks

Optical fiber technology has enabled the explosive growth in networks which has culminated in today's Internet. Without this technology, costs would be higher, available bandwidth would be lower, and error rates would be higher.

The race to transmit more and more bits per color and per fiber has produced an excess of technological capability. SONET/SDH channels have grown from OC-3 (155.52-Mb/s) and OC-12 (622.08-Mb/s) to OC-192 (9,953.28-Mb/s) and OC-768 (39,813.12-Mb/s) in a remarkably short period of time. OC-192 links are in common use, but very little of the network has needed to move to OC-768 links even years after this 40-Gb/s capability was first developed. So, although optical technology has enabled our present-day networks, the same technology has gone far beyond today's requirements.

Researchers continue to push the capabilities of various optical technologies: more colors per DWDM fiber, more bandwidth per color, more distance per color, lower-cost amplifiers, practical optical switches based on MEMs, and the beginnings of color-changing switches that permit data streams to be switched from an ingress color to a new egress color. As much research is still in progress, it is not yet possible to predict all of the capabilities that optical technologies will eventually provide.

Some researchers are working toward the goal of an all-optical network. The authors of this book are skeptical that such a network will ever make sense, or at least that significant unseen developments must occur before all-optical networks will make sense. Our conservative conclusion is based on several key ideas:

1. Light is exceptionally good for communications because it has so little tendency for loss and interference. Until a photon stream is destroyed in a photodetector, thereby creating energetic free electrons, it has very little tendency to interact with other EM fields. Electrical signals behave in the opposite manner: They have a strong tendency to interact. As a consequence, electrical signals are relatively very poor for long-distance communications. However, a telecommunications system requires more than just communications. Information (e.g., packets) must be used as data in various algorithms (e.g., routing decisions) and must be stored (e.g., in queues, awaiting egress links). Neither light nor electricity is good at all three of these requirements. Light is good at communicating; electricity is good for computing and storage. In the opinion of the authors, this simple logic will preclude all-optical networks just as the advantages of light in fiber has precluded all-electrical networks.

2. The sensible and practical bandwidths in optical channels are enormous compared to the needs of most applications. Only a very small number of network applications require as much as 10-Gb/s (OC-192) or even 622-Mb/s (OC-12). Given that a bit stream of 10-Mb/s can support about as good a multimedia experience as a human can consume, it seems likely that requirements for vastly more bandwidth per end station are unlikely (at least until some astonishing new way to consume bandwidth has been invented). There are many network elements at the core of the network that regularly use many 10-Gb/s channels. But these are relatively rare compared to the many end stations that generate and receive most traffic. More important, the large core network elements and the larger SONET channels that connect them almost

always are used for heavily multiplexed streams ultimately generated by the many end stations. A very high bandwidth (e.g., 10-Gb/s) flow from one end-point to another is possible, but highly atypical of Internet traffic. The obvious conclusion is that the amounts of bandwidth offered by light in fiber appear to be far too generous for most applications.

3. We expect that much of network traffic will continued to be packetized. Certainly, there has been a strong tendency to move from logical or physical circuits to packetized communications. We expect that this tendency will continue. Packetized traffic has two consequences: The physical media (e.g., light) must be switched as quickly as packet boundaries occur, and the bursty nature of network traffic implies that packets will always have to be queued, awaiting some output channel or processing service. Both of these requirements drive us back to electrical solutions from optical solutions. Switching on packet boundaries in the optical domain implies that packets can be sensed by the routing algorithm and that optical methods can compute the required path. We observed above that light is difficult to sense (due to its weak interactions); this makes the entire process of packet routing a very difficult challenge. We anticipate that the ease of computing with electrical methods will continue to prevail. The second requirement just discussed was the ability to queue packets. Light has no simple, economical storage technologies. Light doesn't stop, so the best practical solutions for light storage are looping light paths. These are inefficient and extremely difficult to control. In comparison with electricity, today's methods for light-based storage are both primitive and expensive.

For the reasons stated above, we authors seriously doubt that all optical networks will replace today's optical/electrical hybrid networking solutions. Communications will continue to be done in light; computation and storage will continue to be done in electricity. Today's "optical computers" are not nearly capable of taking on the challenges regularly solved by CMOS-based computation and memory. The discussion above argues against any all-optical network, but certainly does not preclude having light-based solutions take over more functionality as better light-based technological solutions are invented and developed. The physical circuit layer of the Internet is certainly open for more and better optical solutions. We expect developments to continue in this area, at least when overall bandwidth demands are much higher than those handled by today's core physical layer network.

Even this enhanced optical core physical layer network requires many new technological "silver bullets" before it can be used in practice. The key issues appear to be improved ways to manage optical power in the different colors gathered together in a single DWDM fiber and optical switching solutions that permit physical channels to be switched both from port to port and from color to color. Without this capability, the overall utilization of any core network must remain limited due to the need to have unique colors from end to end through the network.

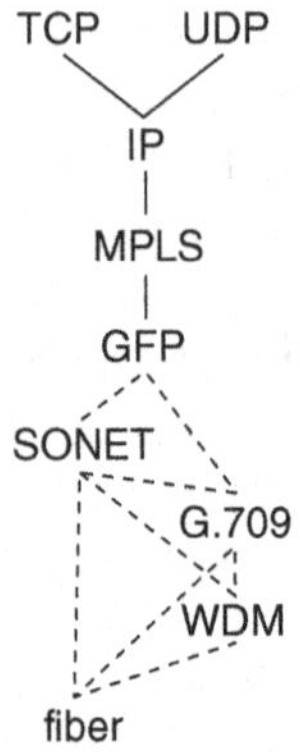

FIGURE 6.11   Alternatives with G.709.

If we are right in our prognostications above, a question must be asked: Should G.709 be used if we are not moving to an all-optical network any time soon? In our opinion, the advantages of G.709 are independent of the limits of optical networks, and it probably makes sense to at least use G.709 in any future extensions of core network capabilities. The reasons are: (1) G.709 allows the problem of protection of multiple colors in one fiber to be managed, and (2) G.709's FEC allows effectively better error rates on possibly lower-quality physical channels. These are compelling advantages that we believe will continue to drive G.709 into the core network.

We do not anticipate that G.709 will suddenly replace SONET/SDH. Rather, the more complex, staged development illustrated in Figure 6.11 seems more likely to occur. SONET will continue to run over fiber and WDM; SONET will begin to be carried over G.709; eventually, the upper layers of the protocol stack (e.g., GFP) will begin to be carried over G.709 without SONET. This evolution could continue for many years, leaving much legacy SONET-over-fiber in place until it requires replacement.

6.5.3   Gigabit Ethernet over the Internet

Although the focus of this book in on the transport of TCP/IP in MPLS frames, there are other important services provided by transport networks. The two most important such services are the transport of Ethernet frames and the transport of storage area network (SAN) frames such as fiber channel (FC). The transportation of Ethernet frames (over GFP, over SONET) allows distant Ethernet networks to be connected at the Ethernet level. This service is highly attractive to multisited corporations that want the ease and security of more-or-less direct site-to-site connection of Ethernet to support their data and communications operations. Many such Ethernet frames carry IP traffic, so this vision of TCP/IP/Ethernet/GFP/SONET is in direct competition with the more general Internet vision of TCP/IP/MPLS/GFP/SONET discussed in

Section 6.5.1. The first, Ethernet-based solution is simpler to manage and is certainly appropriate when the level of private traffic justifies the establishment of a permanent (or nearly permanent) pipe over SONET. But it does not provide strong capabilities for the mixing of traffic from multiple sources (i.e., outside the scope of a signal corporation); nor does it provide QoS support (without independent support provided in cooperation between the two ends of the Ethernet/GFP/SONET pipe). The TCP/IP/MPLS/GFP/SONET solution is intended as a much more general solution in which multiple traffic streams can be multiplexed and the various QoS and security requirements are supported in a general fashion.

6.5.4 Storage Area Network Protocols over the Internet

Although quite modest at the present time, use of the Internet to carry SAN frames such as FC is growing. SANs are used to connect storage devices (disks) to file servers and to corporate computing infrastructures. In general, SANs were developed to operate on one campus; they were not designed to operate well over extended distances. Yet multisited corporations have strong needs to integrate all of the data services into one all-encompassing network. This need is particularly strongly felt in corporation that use any of the forms of data mining to determine corporate strategies.

Fiber Channel was designed for short and medium reach, up to 10-km. Although this reach is fine for most applications, there has been increasing need to go further for data storage mirroring and disaster recovery applications. Imagine, for example, a large U.S. enterprise with offices on the west (San Francisco) and east (New York) coasts. The only practical way to connect two FC islands separated by tens or hundreds of kilometers is to send FC traffic over a wide area network (WAN). Sending FC data over WAN is easier said than done, as WAN or MAN networks do not recognize FC protocol. In fact, as discussed earlier in the book, WAN typically uses SONET as layer 1, ATM as layer 2, and IP as layer 3 protocol. As a result there are multiple ways to address FC transmission over WAN. In this section we briefly review some of these emerging technologies.

FC over SONET One idea is to map or encapsulate FC data into SONET. This seems to be the most straightforward approach, as SONET is so prevalent in WANs and MANs. It turns out that this approach is anything but simple. To accomplish mapping from FC to SONET, the general framing procedure (GFP) has be used. Using GFP protocol, FC frames can be packed into SONET frames as illustrated in Figure 6.12. Assume that data leaving a storage island in New York is in FC format. By *island* we mean any FC topology, such as arbitrated-loop or switch fabric. FC format is unpacked (unframed) and packed (framed) again into SONET format. This operation would typically be done at a next-generation SONET ADM called a multiservice provisioning platform (MSPP) at the edge of a WAN. As the data traverses the WAN or

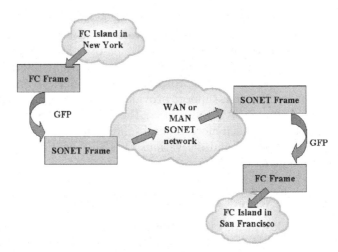

FIGURE 6.12   Sending FC over a SONET network using GFP mapping.

MAN, it is being recognized as a regular SONET data. Once it reaches its destination in San Francisco, the entire process is reversed.

One of the problems with this approach is that SONET is only a layer 1 technology, whereas FC reaches much higher in the protocol stack. As a result, many constructs present in FC (e.g., classes of service) are not supported in SONET. Second, the entire procedure involves frequent unframing and framing, a computationally intense process that wastes valuable hardware resources and increases latency in the data transfer.

The alternative approach to sending FC frames over SONET is to use digital wrapper technology. *Digital wrapping* involves packing FC frames into a superframe to avoid the unframing–framing process, as illustrated in Figure 6.13. Unfortunately, the digital wrapper approach involves larger frame overhead, decreasing the efficiency of data transport. In addition, there is very little hardware equipment at the moment that recognizes digital wrapper format. Nevertheless, this approach might prevail in the future, as the content of the FC frames is not affected.

FC over WDM   One way to avoid the unframing–framing process is to send FC data directly over the WDM infrastructure, as illustrated in Figure 6.14. As long as WDM equipment is available all the way from the storage island in New York to the storage island in San Francisco, this elegant solution avoids SONET framing complications. However, we have to keep in mind that signal amplification from New York to San Francisco can only be done in the optical domain by using EDFA or Raman amplifiers. If O-E-O regenerators were used in a WDM network, electronic hardware would have to be aware of FC formatting issues, and this approach would lose most of its advantages.

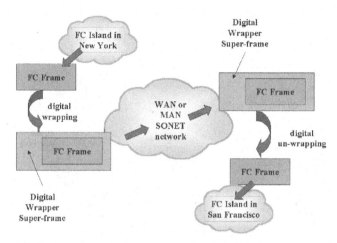

FIGURE 6.13 FC over a SONET network using digital wrapper technology.

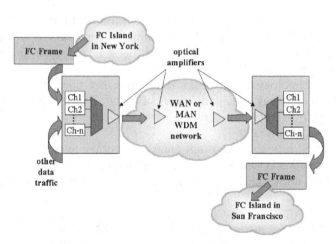

FIGURE 6.14 FC over a WAN network using WDM.

It is important to note that in WDM systems, FC traffic can be "mixed" with other types of traffic. For example, WDM terminals in Figure 6.15 can transmit FC, SONET, or Ethernet data as long as optical transparency is maintained on the entire link as each data stream is send on a separate wavelength.

FC over IP Another type of approach involves layer 3: Internet protocol (IP). Since the IP layer is really not discussed fully in this book, at this point we can only very briefly mention the names of emerging technologies. The first

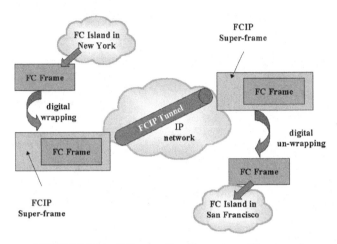

FIGURE 6.15   FC tunneling through an IP network.

in this class, FC over IP (FCIP), involves FC frames tunneling through IP network as illustrated in Figure 6.15. The tunneling process involves encapsulating an FC frame into an IP frame, so the process is somewhat similar to that of digital wrapper technology.

Another technology in this class is the Internet fiber channel protocol (iFCP). This technology essentially merges FC and IP protocol stacks. It relies on layer 4 TCP protocol for error recovery and offers a migration path to native IP storage networks.

Finally, iSCSI is a very important emerging storage technology that eliminates FC altogether. Using this protocol, all storage networking can be done using IP-based networks. Description of the iSCSI technology is beyond the scope of this book.

FC over MPLS   An attractive alternative for SAN applications is to make use of a general-purpose Internet at a lower layer. Probably the most attractive solution is to use GFP/MPLS to carry FC frames but to provide separate MPLS paths for the SAN traffic. This allows the QoS issues of TCP/IP to be avoided in today's systems while making maximal reuse of the QoS protection offered by MPLS. We expect that this mechanism will support rapid growth of SAN traffic in the coming years. Figure 6.16 illustrates the two ways in which SAN traffic can be carried on the common protocol stack envisioned in this book.

FC over WAN Summary   The extent to which enterprises will demand support for SAN traffic over the Internet or WANs remains unclear. The authors of this book anticipate large growth in this area, with the result that machines talking to machines (e.g., SAN traffic) will become progressively

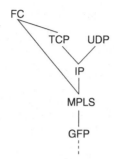

FIGURE 6.16 Carrying FC (SAN) traffic.

more important as information technologies mature. Existing SAN protocols (e.g., FC) were designed without consideration of the distances or the nature of WAN networks. As a result, SAN protocols require very low latencies and high qualities of service, both of which are difficult to provide in the WAN. Thus, there are significant challenges in providing satisfactory solutions for SAN over the WAN. In this chapter we examined a number of approaches to addressing this problem, but none of them are ideal. It is entirely possible that SAN protocols will have to change significantly before they can be used widely over the Internet.

KEY POINTS

Digital wrapper:

- A digital wrapper (G.709) has a layered structure that provides infra-structure for an optical transport network.
- The optical transport network consists of an optical transmission section, optical multiplexing section, and optical channel section.
- Forward error correction (FEC) is used in a digital wrapper to increase the reliability of the optical link. FEC improves the signal-to-noise ratio by adding extra parity bits.
- A digital wrapper frame consists of data payload, overhead, and FEC. Data payload could be SONET, Ethernet, fiber channel, or any other frame. A digital wrapper does not care what sits inside its payload, so it can carry any protocol.

MPLS:

- Multiprotocol label switching (MPLS) protocol layer represents a bridge between the data link layer 2 and network layer 3, with tight coupling to the network layer, represented by Internet protocol.

- MPLS is a networking standard that provides for the efficient designation, routing, forwarding, and switching of traffic flows through the network.
- For traffic engineering purposes MPLS is adding labels as the data enter the MPLS network. The label is removed when the data leave the network.
- An MPLS label identifies the path a frame should traverse across an MPLS network. The label is of fixed length (20 bits) and is encapsulated in a layer 2 header along with the frame.
- MPLS networking devices can be classified into label edge routers and label switching routers.

RPR:

- Resilient packet ring (RPR) technology uses a dual-counter rotating fiber ring topology.
- RPR is a layer 2 protocol. RPR frames can be transported over both SONET/SDH and Ethernet physical layers.
- An RPR media access controller (MAC) layer contains much of the functionality for the RPR network. MAC is responsible for providing access to the fiber media as it receives, transits, and transmits RPR frames.
- RPR uses bandwidth-control algorithm that dynamically negotiates available bandwidth between the nodes on the ring.
- RPR has a topology discovery mechanism that allows nodes on the ring to be inserted/removed without manual management intervention.
- RPR has the ability to protect the network from single-span (node or fiber) failures. RPR has two protection mechanisms: wrapping and steering. Regardless of the protection mechanism used, the ring will be protected within 50 ms.
- RPR benefits: very efficient multicast function, spatial reuse functionality, high resiliency, automatic topology discovery, and quality of service features.

FC extensions over WAN:

- The fiber channel (FC) over SONET approach involves frame translation using generic frame procedure or packing FC frames into digital wrapper frames. Frame translation requires intense hardware operations.
- The FC over WDM approach involves sending raw FC data using WDM technology and does not require any frame translation. This technology requires WDM equipment that exists all the way between both remote locations.

- The FC over IP approaches require utilization of the layer 3 IP protocol in different forms (FCIP, iFCP, iSCSI). These technologies have difficulty in maintaining quality of service characteristics of the FC protocol.

REFERENCES

Agilent Technologies, An Overview of ITU-T G.709, Application Note 1379, http://www.home.agilent.com/agilent/redirector.jspx?action=ref&cc=DE&lc=ger&ckey=67594&cname=AGILENT_EDITORIAL.

Davie, B. S., and Y. Rekhter, *MPLS: Technology and Applications* (Morgan Kaufmann Series in Networking), Academic Press, San Diego, CA, 2000.

Fiber Channel Industry Association, http://www.fiberchannel.org/.

IEEE, *Resilient Packet Ring,* IEEE 802.17 Working Group Documents, http://www.ieee802.org/17/documents.htm.

ITU (International Telecommunication Union), *Generic Framing Procedure,* Recommendation G-7041, http://www.itu.int/itudoc/itu-t/aap/sg15aap/history/g.7041/g7041.html.

Kembel, R. W., *Fiber Channel Switched Fabric*, Northwest Learning Associates, Tucson, AZ, 2001.

Kembel, R. W., *Fiber Channel: A Comprehensive Introduction*, Northwest Learning Associates, Tucson, AZ, 2002.

Kembel, R. W., and H. L. Truestedt, *Fiber Channel Arbitrated Loop*, Northwest Learning Associates, Tucson, AZ, 2000.

MPLS Resource Center, http://www.mplsrc.com/standards.shtml.

Perros., H. G., *An Introduction to ATM Networks*, Wiley, Hoboken, NJ, 2001.

7

VLSI INTEGRATED CIRCUITS

Network Infrastructure and Architecture: Designing High-Availability Networks,
By Krzysztof Iniewski, Carl McCrosky, and Daniel Minoli
Copyright © 2008 John Wiley & Sons, Inc.

7.1 INTRODUCTION

7.1.1 Integrated Circuits, VLSI, and CMOS

Integrated circuits (ICs) or VLSI chips, as they are frequently called, control almost everything in our external environment, from IP routers that process Internet traffic, PCs, cell phones, and car engines to household appliances. They are complex electronic systems embedded in a small volume of highly processed silicon. Although the cost of designing and developing them can be very high, when spread across millions of production units, the individual IC cost can be very low. ICs have migrated consistently to smaller feature sizes over the years, allowing more circuitry to be packed on each chip. This increased capacity per unit area can be used to decrease cost and/or increase functionality. It has been observed that the number of transistors in an IC doubles every two years or so. This famous observation is called *Moore's law*, after Gordon Moore, the first person to make that observation.[†]

Very-large-scale integration (VLSI) is the process of creating ICs by combining millions of transistors into a single chip. The VLSI era began in the 1970s when silicon chips achieved substantial complexity. Depending on the actual number of transistors, ICs can be of large-scale integration (LSI), very large-scale integration (VLSI), or ultralarge-scale integration (ULSI), but we will not bother with these distinctions; instead, we refer to all IC here as VLSI chips. The current complexity of VLSI is well represented by Intel's Montecito Itanium server chip, which contains over 1 billion transistors and is manufactured in the extremely advanced 65-nm manufacturing process, where 65 nm represents the smallest physical feature used on the chip.

VLSI chips typically use CMOS (complementary metal-oxide-semiconductor) processing technology to define basic transistors and to connect them using on-chip metal lines. CMOS represents a major portion of ICs sold worldwide (over 90%). The "MOS" in CMOS represents the basic sandwich structure of a MOS transistor, consisting of a metal gate formed on top of an oxide that in turn is grown on top of semiconductor (silicon) substrate. The word *complementary* in CMOS refers to the fact that both n-channel (carrying electrons) and p-channel (carrying holes) MOS transistors are available.

CMOS logic uses a combination of p- and n-type MOS transistors to implement logic gates and other digital circuits. CMOS digital gates feature the very interesting property that they do not dissipate any power in static conditions. As a result, CMOS circuits are power efficient, as power is dissipated only during transistor switching.[‡]

[†]Gordon Moore, Cramming more components onto integrated circuits, *Electronics Magazine*, April 19, 1965.
[‡]In very advanced CMOS circuits there is some finite amount of static power dissipation due to transistor leakage currents, so this particular advantage of CMOS technology is slowly diminishing as transistors continue to shrink.

7.1.2 Classification of Integrated Circuits

Integrated circuits can be classified into analog, digital, and mixed signal (where both analog and digital functions are performed by the same circuit). *Digital integrated circuits* can contain NAND/NOR logic gates, flip-flops, latches, multiplexers, counters, and higher-complexity blocks. They effectively process information in terms of 1 and 0 signals. *Analog integrated circuits*, such as operational amplifiers, comparators, references, or power management circuits, work by processing signals that have a continuous scale from 0 to 1. They perform functions such as amplification, filtering, modulation, or mixing. *Mixed-signal circuits* combine analog and digital functionality, and typical examples of this class include analog-to-digital (ADC) and digital-to-analog (DAC) converters.

Complex ICs, frequently referred to as a *system-on-chip* (SOC), can contain numerous classes of analog, digital, and mixed-signal circuits as shown schematically in Figure 7.1. To generate internal high-speed clocking, the analog clock generator is needed, typically designed using a phase-locked loop (PLL). To communicate with other ICs using traces on the same printed circuit board (PCB) or through copper/coaxial cables, high-speed I/O interface circuitry is required. Digital signal processing is performed using logic gates, microprocessor cores, on-chip field-programmable gate arrays, or dedicated digital signal processing (DSP) blocks. The data are stored internally in static random access memories (SRAMs) and/or dynamic random access memories (DRAMs). Various support circuits, such as a low drop-out (LDO) voltage regulator, might be required as well. Finally, circuits that support chip testing—built-in self-testing (BIST) and boundary scan (JTAG)—are also required.

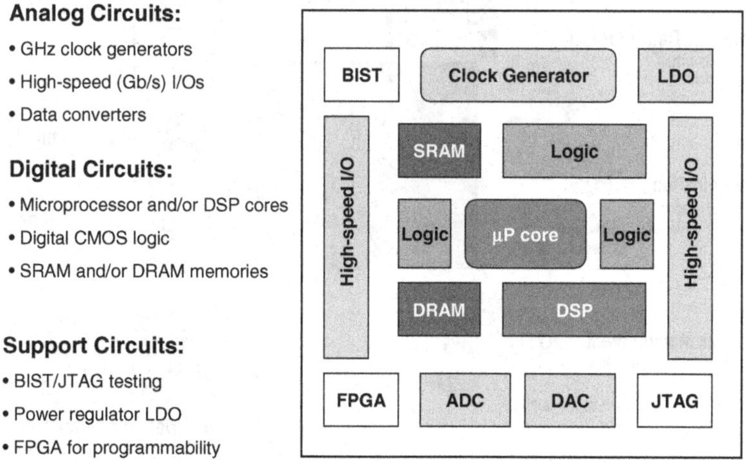

FIGURE 7.1 Schematic representation of a system-on-chip containing various analog, digital, and mixed-signal blocks.

ICs can be classified depending on the flexibility of their use. ICs designed for a specific application are called *application-specific integrated circuits* (ASICs). An example of the IC customized for a particular use could be a SONET framer that is being designed to process SONET frames in data networking equipment. We discuss the architecture of a SONET framer in more detail in Section 7.4.2.

More general ICs, which can be used in multiple applications, are called *standard products*. Examples of such ICs are ADC data converters or temperature sensor chips. Both ASICs and standard products have a fixed functionality that cannot be changed. We can say that in this case the functionality is permanently hardwired. One type of ASIC, known as a *gate array*, provides some flexibility of IC functionality at the manufacturing phase. A gate array circuit is a prefabricated silicon chip with no particular function in which transistors, standard NAND/NOR logic gates, flip-flops, and other circuits are placed at regular predefined positions and manufactured on a wafer, usually called the *master slice*. Creation of a circuit with a specified function is accomplished by adding a final surface layer metal that is interconnected to the chips on the master slice late in the manufacturing process, joining these elements to allow the function of the chip to be customized as desired.

Gate arrays therefore provide certain flexibility for the ASIC manufacturer, but still grant no flexibility to users. User programmability can be accomplished in hardware by using field-programmable gate arrays (FPGAs). FPGAs contain custom logic blocks and programmable interconnects that can be enabled (programmed) using built-in fuse technology, as shown in Figure 7.2. The

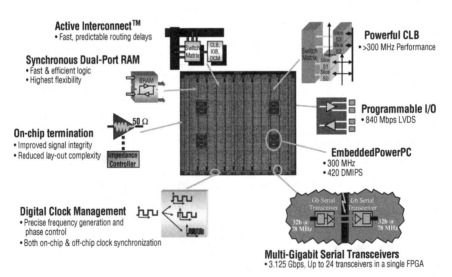

FIGURE 7.2 Characteristics of a field-programmable gate array chip. (Courtesy of Xilinx Inc.)

programmable logic components can be programmed to duplicate the functionality of basic logic gates or more complex combinational functions. Complex FPGAs contain various flip-flop configurations, memory blocks, processor cores, and programmable inputs/outputs (I/Os) as well. These characteristics allow the same FPGA chip to be used in many different applications.

FPGAs are generally slower than their ASIC counterparts, might not be able to handle complex designs, and will dissipate more power. However, they have several advantages, such as a shorter time to market, ability to program in the field to change functionality, and lower nonrecurring engineering costs. FPGAs could also be simpler, faster, and more effective to use than general-purpose microprocessors. For these reasons their use is becoming increasing popular, in particular when prototyping new systems.

7.1.3 Looking Ahead

This chapter is organized as follows: First, in Section 7.2 we discuss the application of IC in data networking. We divide ICs into layer 1, 2, and 3 devices and see how they correspond to the OSI open interconnect model and networking protocols discussed in Chapters 5 and 6. In Section 7.3 we cover chip I/O interfaces. We will show how networking ICs interface to each other, to memory chips, and to microprocessors. In Section 7.4 we discuss some architectural design examples, including time slicing, SONET framer, and network processor architectures.

The design of ICs is a complex process. Over the last 40 years it has changed so dramatically that today's techniques of modern IC design would probably seem strange even for the most visionary IC designers from the past. Because the subject matter is the focus of numerous books, and due to space limitation, we only touch on the most important topics in Section 7.5.

By the end of the chapter you should have some idea of what a VLSI chip is, how it is used in data networking, and how one goes about designing it. You will not have become a VLSI designer, though. If you would like to learn about silicon chips and methods to design them, we have provide a list of selected references at the end of the chapter.

7.2 INTEGRATED CIRCUITS FOR DATA NETWORKING

Data networking ICs are VLSI chips that enable processing and transmission of voice, video, and data, in either electrical or optical form, from one location to another across a broadband Internet network. They typically are responsible for the following functions:

- Modulation and amplification of electrical signals for transmission through a physical medium (an optical fiber, a twisted copper pair, or a coaxial cable). Similarly, in the receive direction: demodulation and equalization for reception of electrical signals. These devices are typically referred to as physical medium devices (PMDs) or physical layer devices (PHYs). Both PMD and PHY ICs are considered to be layer 1 devices, using OSI as a frame of reference.

- Formatting of data into frames or cells using predefined protocols (ATM, Ethernet, fiber channel, SONET/SDH). These devices, typically referred to as *framers* or *mappers*, are considered to be layer 2 devices.
- Processing of data packets. Processing functions include protocol conversion, packet forwarding, policing, lookup, classification, encryption, and traffic management. These devices are typically referred to as network processors, classification engines, or data link devices, and are considered to be layer 3 (occasionally, layer 4) devices.

Various classes of ICs for data networking are compared in Figure 7.3. Layer 1 devices are responsible for physical aspects of data transmission such as signal timing and frame formatting. Their functionality is defined rigorously by various standards established to create networking equipment interoperability. As a result, they typically have a fixed functionality and are implemented as ASICs. On the other hand, layer 3 devices require a good deal of flexibility, as they deal with tasks such as IP packet processing and forwarding that could be specific to a particular use of the networking equipment. As a result, they are implemented in ICs that provide means of programmability such as a general-purpose central processing unit or network

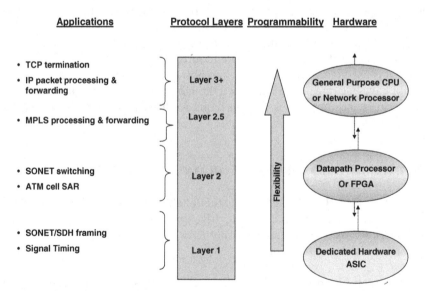

FIGURE 7.3 Comparison of data networking ICs, indicating their application space and position within the OSI model.

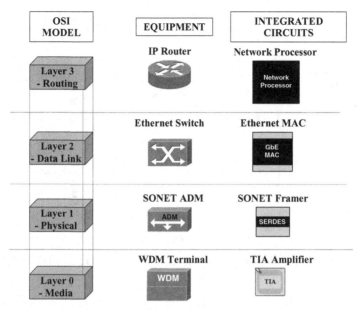

FIGURE 7.4   Conceptual relationship between ICs, data networking equipment, and the OSI reference model.

processor. Layer 2 devices fall somewhere in between and can be implemented as datapath processors (DSPs) or FPGAs. A conceptual relationship between data networking ICs, networking equipment, and the OSI reference model is shown in Figure 7.4.

7.2.1   PMD and PHY Devices (Layer 1)

Physical medium devices (PMDs) are used for I/O interfacing to physical media such as an optical fiber, a copper wire, or a coaxial cable. Typically, these ICs are data protocol independent, as they deal only with physical effects and electrical signals. PMDs include devices such as amplifiers for photodetectors or drivers for lasers. PMD devices require analog expertise to design and are discussed in more detail in Chapter 8.

Physical layer devices (PHYs) are responsible for defining how the traffic will be transported from one location to another. PHY devices include SONET serializers and deserializers (SERDES devices), Ethernet transceivers, cable modem chips, and xDSL transceiverPHY devices deal with issues such as clock generation and clock extraction. As a result, one of the important properties for these devices is signal *jitter*, which expresses uncertainty in the timing position of the signal. PHY devices have to comply with elaborate specifications for the amount of jitter being generated (intrinsic jitter), the amount of jitter the device can tolerate at its input (jitter tolerance), and the transfer

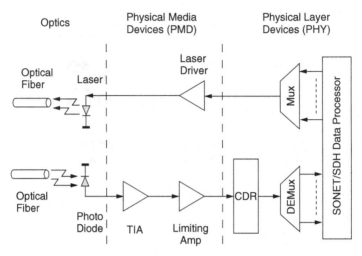

FIGURE 7.5 SONET link indicating the roles of PMD and PHY devices.

characteristics of the output jitter vs. applied input jitter (jitter transfer). Different jitter specifications exist for different transport technologies (SONET/SDH or Ethernet).

The interaction between PMD and PHY devices is illustrated using a SONET link as shown in Figure 7.5. We use that example to summarize basic functions performed by PMD devices.

Amplification   An optical signal is typically received by a photodiode and converted to an electrical signal. Since the photodiode produces current I whereas most electronic devices require voltage signal V, I is converted to V by a PMD device known as a *trans-impedance amplifier* (TIA). A TIA is a rather sensitive device, as it needs to clean and amplify small signals generated by the photodiode. The small voltage signal from a TIA frequently needs to be amplified further, with the limiting amplifier performing this function. The term *limited* comes from the fact that the limited amplifier output provides the same signal amplitude regardless of the strength of the optical signal being converted. A limiting amplifier is another example of a PMD device.

Clock and Data Recovery   The signal from the limiting amplifier is received by the clock and data recovery (CDR) circuit. The CDR recovers the clock from the NRZ data stream using analog phase-locked loop techniques. A detailed description of this process is given in Chapter 8. A CDR jitter tolerance parameter is typically the most challenging parameter to be met in design. After recovering the clock and "cleaning up," the signal received needs to be deserialized to a parallel stream of lower-data-rate signals. The chip that performs this function is called a *demultiplexer* or *deserializer*. Using the clock

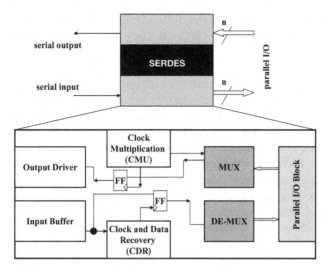

FIGURE 7.6 SERDES chip containing data multiplexing (mux and demux) and clock generation and recovery (CMU and CDR) functionality.

recovered and lower speed, further data processing can be done in the upstream framer device.

Clock Multiplexing Similar processing occurs in the reverse, egress direction. The data generated by the framer device is serialized using a multiplexer or a serializer device. The clock multiplication unit (CMU) is used to generate a high-purity reference clock for data transmission. The data are transmitted to a PMD device known as a *laser driver*, which in turn modulates the laser, effectively converting an electrical signal into an optical signal. Deserializers and serializers are frequently combined on the same chip, creating a *SERDES device*, shown in Figure 7.6. SERDES devices convert a high-speed serial data stream into a parallel combination of lower-rate signals. SERDES devices are used in numerous applications beyond data networking, as serial-to-parallel signal conversions are frequently desired in hardware design. More details of SERDES and CDR operation are given in Chapter 8. While Figure 7.5 shows all processing functions in separate IC blocks, many of the functions can be integrated on one chip. SERDES, CMU, and CRU are frequently integrated into devices called *transceivers*; the chip might sometimes also include a laser driver or limiting amplifies. At lower data transmission rates, integration of framer and transceiver functions is also possible.

7.2.2 Framers and Mappers (Layer 2)

Layer 2 devices perform various data coding, framing, and mapping functions. These tasks are layer 2 protocol specific, and as a result, the devices might have various names, such as DS3 mapper, STS-3 framer, or Ethernet MAC. Intrinsically, they all perform the same function: They process packets/frames/cells at the data link layer.

Data Coding Data coding translates user data into a format that is suitable for transmission. Various coding schemes are used in communication protocols. The most popular ones for broadband are scrambling, used in SONET, and 8b/10b, used in Ethernet. The 8b/10b scheme codes 8 bits of data into 10-bit code words. The expanded code space is used to maintain sufficient transitions for data recovery, to detect transmission errors, and to control emitted energy. In addition, special codes can be used to delineate frames. 8b/10b is a well-known coding scheme used in Ethernet protocol, although due to the high 25% overhead, similar, more efficient schemes such as 64b/65b or 64b/66b have been proposed for high-speed (10 Gb/s and above) transmissions.

The scrambling scheme transmits the exclusive-or of the data and the output of a pseudorandom source. The primary reason for scrambling is to introduce some transitions into long streams of zeros or ones. Scrambling is easy to implement and has no coding overhead. However, scrambling success is only probabilistic, and once in awhile very long sequence of zeros or ones will be produced, possibly causing bit error (although the probability of this happening is very low).

The most important feature of coding is transition density, as coding has to ensure that there are enough zero–one transitions to allow the receiver to recover the data sent by the transmitter. In that respect, 8b/10b has a higher transition density than PRBS23-type SONET traffic at the expense of additional overhead.

Framing Framing divides transmitted data into blocks, typically of fixed lengths. ATM cells, SONET frames, or Ethernet packets are results of this framing process. Framing allows the receiver to align with the transmitter and detect transmission errors. Framing devices are usually protocol specific, although multiprotocol devices have started appearing in the marketplace. An example of the multiframe device is an IP/ATM framer, shown in Figure 7.7, which can be configured to process IP packets over SONET or ATM cells.

Packet-over-SONET (POS) is an established technology for carrying IP and other data traffic over a SONET backbone. Variable-length data packets are mapped directly into the SONET synchronous payload envelope (SPE). POS provides reliable, high-capacity, point-to-point data links using SONET physical layer transmission standards. The POS/ATM framer from Figure 7.7 locates and locks onto the boundaries of each payload in the coming signal. As it checks for transmission errors it can detect the loss of a signal or frame. The framer might also perform some overhead processing functions. These functions include extracting key bits and initiating automatic protection switching (APS) when the fatal error has occurred. Another function of the framer device is section, line, and path termination. These extra bytes are added by

POS/ATM framer IC

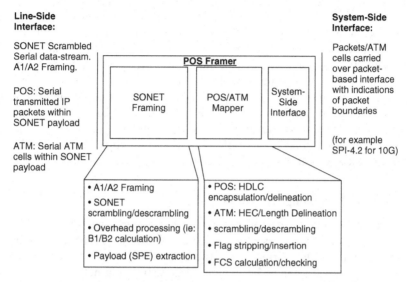

FIGURE 7.7 POS/ATM framer integrated circuit.

SONET protocol and need to be inserted or extracted as required by networking functionality.

Mapping Mapping processes involve squeezing different payloads into various frame types, such as a SONET envelope. The device performing mapping, the *mapper*, contains an elastic store to accommodate varying payload lengths. It can also insert or drop traffic, as needed. A related function is pointer processing, which keeps track of the location of each payload within a SONET envelope. The pointer processor also tries to align different tributary rates into a synchronous optical channel.

Forward Error Correction A critical factor for high-speed transmission is the bit error rate (BER). Due to the high throughput used, the consequences of sending "wrong" bits are serious, as many customers might be affected. To improve BER coding techniques one should use forward error correction (FEC). ICs that employ FEC algorithms greatly reduce the bit error rates in long-haul systems. The concept behind FEC is straightforward; extra bits are added to the data frame to introduce redundancy. These extra bits are used to check whether the transmission error has occurred and to correct it. FEC techniques are not unique to broadband communication; they are used in cell phones, compact disks, memory, and satellite systems.

For high-rate-data (10 Gb/s and above) transmission, system engineering of FEC is a must. The high-speed SONET link shown in Figure 7.5 might require one extra IC that performs the FEC function and would be positioned between the high-speed transceiver and the digital framer. There are two types of FEC techniques for broadband ICs: in-band and out-of-band. *In-band FEC* uses certain locations in the SONET frame for extra FEC bits; *out-of-band FEC* appends these extra bits after the frame. As expected, out-of-band FEC is more powerful than in-band FEC. One typical Reed–Solomon code implementation for OC-192 can improve BER from 1×10^{-4} to a value better than 5×10^{-15}. This is an improvement by over 10 orders of magnitude! The price for this BER improvement is the additional bit overhead required; in this example it is 7%, so the data have to be transferred at the 10.65-Gb/s rate. That requirement creates some extra demand on analog circuitry that has to perform clock and data recovery function at 10.65 GHz instead of 9.95 GHz.

7.2.3   Packet Processing Devices (Layer 3)

As expected, layer 3 ICs perform in hardware networking functions assigned in the OSI model. Examples of this include IP routing, ATM layer policing, and fiber channel interworking with the SCSI protocol in storage area networks. Layer 3 devices can be implemented as ASICs or as network processors. FPGA can also be used if the required gate count is low. Network processors (NPs) provide flexibility at the cost of complex firmware. As we will see later, a major decision in NP architecture design is to determine the amount of internal and external memory required (both SRAM and DRAM). A network processor consists of multiple CPUs, standardized high-speed interfaces, and data buffering capabilities. It is a software-based, highly flexible solution for networking processing.

Network Processor   The concept of the network processor is based on the microprocessors used in the computing industry. NPs are frequently advertised as universal devices for layer 3–7 packet processing. Although NP flexibility is a big bonus, it comes at the price of compromised performance. Complex firmware development is another disadvantage. Network processors are microprocessors designed specifically for packet processing, a core element of high-speed communication routers. Similar to the microprocessors used in a PC, a network processor is a programmable chip, but its instruction set has been optimized to support the operations that are needed in networking, especially for packet processing. The success of the microprocessor for the computing platform is based on the fact that its architecture is optimized according to the software characteristics of the application intended. For example, a desktop CPU is optimized for a broad range of window applications, while a graphic processor is optimized for two- and three-dimensional rendering, a DSP for signal processing, and a high-end RISC for scientific computations.

Packet Processing   Although commercial vendors are advertising NP as a "do it all" solution, it is quite possible that there will be no unique NP platform at all, but instead, that NPs will be application specific to some degree, as microprocessors for computing are. To define NP one needs to consider the commonality of various target applications. No matter what protocol is involved, packet processing requires the following tasks: (1) header processing, lookups, and classifications; (2) scheduling, policing, and queuing; and (3) encrypting and security. Due to these requirements, memory bandwidth becomes a common bottleneck for NPs in networking applications. In particular, lookups, packet classification, and queuing are memory-intensive tasks. As a result, NPs typically contain several external SRAM/DRAM I/O interfaces. The challenge for NP architects is to define memory architecture in such a way that the overall NP performance is close to the optimized hardware solution in which both external and internal memories are optimized for the given application.

Another key architecture design issue is the organization of parallel processing. There are multiple approaches: single-instruction multiple data; single-program multiple data; multiple-program multiple data; simultaneous multithreading, superscalar, as used in Intel's Pentium; super pipelining, as used by MIPS; or very long instruction word, as used in Intel's Itanium. It remains to be seen which parallel-processing scheme is the most efficient for packet processing. This area will probably be an interesting research subject for years to come.

In closing this network processor description, we would like to mention that there are examples of network processors that are designed for dedicated applications. They include classification processors, which can be used to lighten the processing load; security processors, which can be used to lighten the processing load by handling security algorithms; or traffic/policy managers, which can handle queuing, load balancing, and quality of service. However, the differences between these devices are beyond the scope of complexity that we can afford in this chapter.

7.3   CHIP I/O INTERFACES

In the world of communication the word *interface* may have several meanings, depending on the context within which it is used. For example, the open system interconnection (OSI) reference model outlines a seven-layer abstract description of the communication protocol that includes interfaces between different layers so that they can talk to the ones above or below. In this section, the meaning of *interface* is limited to the external interface, which allows communication of one IC device with another. As in any type of communication, this can be accomplished by means of certain predefined protocol. These protocols define the data and control signals as well as their timing parameters. Many of these interfaces are the subsets of certain standards that also define information about transmission mediums, type of connectors, and similar physical level information.

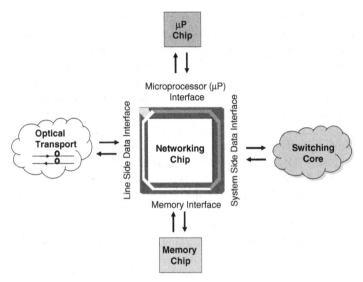

FIGURE 7.8 Networking IC with its external I/O interfaces.

Digital VLSI chips associated with data networking are responsible for processing data paths: extracting, processing, and inserting information from and to the data stream as it passes through the networking node. For that reason, networking chips typically have system- and line-side interfaces. The system side is described as the interface connecting the data path to the system side of a node within which the chip is placed. The line side is described as the side closest to the optical fiber transmit/receive facility.

A schematic representation of a networking IC with its data interfaces is shown in Figure 7.8. The chip contains two data interfaces, one on the line side and another on the system side. This is the path for the data flow. In addition, there is one interface dedicated to communication with a control microprocessor and another one to store data in the external memory. The microprocessor path is used for control purposes, and the memory path is used for storage. Needless to say, if demands for control or storage are small, they can be incorporated on-chip, resulting in no external microprocessor or memory interfaces being present.

In this section we first consider the physical nature of the I/O scheme: serial or parallel. Later we discuss some typical networking standards, followed by a description of data networking interfaces. Finally, we briefly examine both microprocessor and external memory interfaces.

7.3.1 Serial vs. Parallel I/O

Chip I/O interfaces can be divided into serial and parallel interfaces. In *serial interfaces*, each bit of data is allocated a time slot defined by a clock signal and

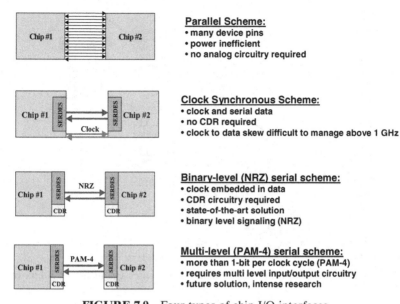

FIGURE 7.9 Four types of chip I/O interfaces.

transmitted one after another. Sometimes, the serial interface includes one or more control signals, the function of which is defined by a particular protocol.

Parallel interfaces require more device pins as the data are presented in bytes or words of a desired width. Parallel interfaces typically run at slower clock speeds, but the data throughput is multiplied by the number of bits transmitted or received in one time slot. In addition to data and clock signals, parallel interfaces frequently include control or indicator signals that help further processing of data.

A summary of various chip I/O interfaces is shown in Figure 7.9. The first, at the top, is a standard, parallel interface which is used frequently to connect low-speed ICs. It uses many device pins but its architecture is uncomplicated and requires no analog circuitry. Due to its large number of parallel signals, this scheme is power inefficient.

A clock synchronous interface uses the SERDES device described earlier to serialize parallel data for transmission, with clock information being sent separately. This scheme works well for medium speeds; however, above 1 Gb/s, significant issues arise due to a clock skew between the clock and the data lines that are not identical. For very high speeds (above 1 Gb/s), an NRZ (no return to zero) scheme with embedded clock is preferred. As the clock edge is embedded in the data stream, additional clock and data recovery (CDR) circuitry is required. As a result, the interface block has to contain both SERDES and CDR functionality, but this is the price that has to be paid for very fast data transmission.

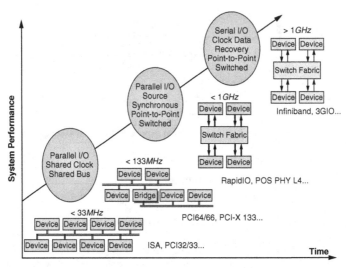

FIGURE 7.10   Time evolution of I/O signaling schemes. (Courtesy of Xilinx Inc.)

An NRZ scheme uses binary-level signaling (0 and 1 levels). More efficient schemes utilize multibit encoding to sent more than 1 bit of data in a given clock cycle. For example, PAM-4 uses four levels (1, 3/4, 1/2, 1/4, 0) to encode 2 bits of information. Although it would seem that PAM-4 would be twice as efficient as NRZ, significant difficulties exist in the implementation of high-speed analog circuitry that can resolve four voltage levels at gigahertz speed. As a result, the multilevel scheme is a subject of university research but is not used in commercial ICs.

As the required throughput of I/O communication lines increases, the serial I/O scheme is gaining more widespread adoption. Figure 7.10 illustrates this historical evolution. High-speed serial links are prevalent today in most point-to-point applications and frequently use differential signaling (see Figure 7.11). Although single-ended solution will save some device pins, typically it will not reduce dissipated power. Single-ended links are also problematic from a signal integrity and electromagnetic compatibility point of view.

From an electrical signal point of view there are numerous I/O electrical standards and proprietary schemes used for high-speed serial links. Low-voltage differential signaling, current mode logic, and pseudo-emitter-coupled logic are examples of electrical standards for I/O signals frequently used to connect networking ICs. We discuss some circuit aspects of high-speed I/Os in Chapter 8.

7.3.2   Networking I/O Standards

Data networking VLSI devices are integrated in the Internet system that spans the entire globe. As a result, they process data that in many instances

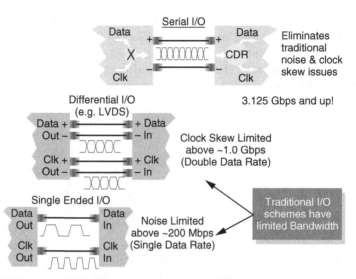

FIGURE 7.11 Serial I/O with clock recovery capability vs. source synchronous single-ended and differential I/O schemes. (Courtesy of Xilinx Inc.)

originated thousands of miles away. Therefore, it is easy to imagine that there is a need for standardization of data formats, data interfaces, and many other aspects of the communication system which have to be considered while designing an integrated circuit for communications. The networking standards make it easier for VLSI designers to derive specifications for the chip they design. This means that the specification of the chip begins from some common, internationally predetermined conditions within which the device has to operate. For example, transmitting data in certain formats dictates the frequency at which the data are transferred. The interfaces are agreed upon in different standard bodies that consist of many engineers from around the world, on many occasions from competing companies. Because of this standardization effort, various IC components can talk to each other regardless of who designed them. To illustrate the standardization concept, we discuss some of the networking I/O interfaces developed by the Optical Internetworking Forum (OIF). Although we use OIF standards here almost exclusively, keep in mind that many other standards and proprietary schemes do exist.

Optical Internetworking Forum Standards[†] The mission of the OIF is to foster the development and deployment of interoperable data switching and routing products and services that use optical networking technologies. To promote multivendor interoperability at the chip-to-chip and module-to-module level, the physical and link layer (PLL) working group within the OIF

[†]The text in this section is adapted directly from the OIF Forum Web site: http://www.oiforum. com/.

has defined electrical interfaces at two different points within a synchronous optical network/synchronous digital hierarchy (SONET/SDH) based communication system. These interfaces, the system packet interface (SPI) and the SERDES framer interface (SFI), are depicted in the reference model shown in Figure 7.12. SPI and SFI I/O physical hardware correspondence to the OSI model is also shown in Figure 7.13.

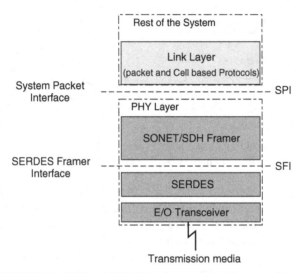

FIGURE 7.12   SPI and SFI I/O reference model adopted by OIF.

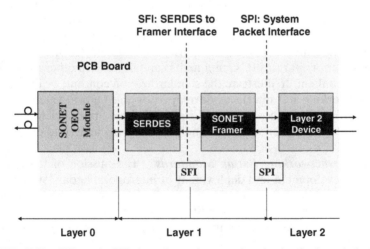

FIGURE 7.13   SPI and SFI interfaces in a printed circuit board hardware implementation.

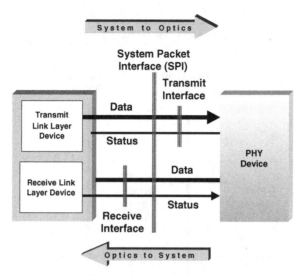

FIGURE 7.14 System packet interface between a framer and a link layer device adopted by OIF.

The SPI is found between the physical layer (PHY) device(s) and the rest of the SONET/SDH system (i.e. between the SONET/SDH framer and the link layer), as shown in Figure 7.14. This interface separates the synchronous PHY layer from the asynchronous packet-based processing performed by the higher layers. As such, the SPI supports transmitting and receiving data transfers at clock rates independent of the actual line bit rate. It is designed for efficient transfer of both variable-sized packets and fixed-sized cell data.

The SERDES framer interface (SFI) defines an electrical interface between a SONET/SDH framer and a high-speed parallel-to-serial/serial-to-parallel (SERDES) logic, as shown in Figure 7.15. This permits the SERDES and framer to be implemented in different speed technologies, allowing a cost-effective multiple-chip solution for the SONET/SDH PHY. The SFI interface can also be used as an interface for an FEC chip inserted between the SERDES and the framer as illustrated in Figure 7.16.

To keep up with evolving transmission speeds and technology enhancements, the OIF has defined several different versions of these electrical interfaces. Each version is tailored for a specific application environment and time frame. The OIF has adopted the following naming conventions to identify the various PLL interfaces and the application environments they are designed for:

- SxI-3 OC-48/STM-16 and below (2.488 Gb/s range)
- SxI-4 OC-192/STM-64 (10 Gb/s range)
- SxI-5 OC-768/STM-256 (40 Gb/s range)

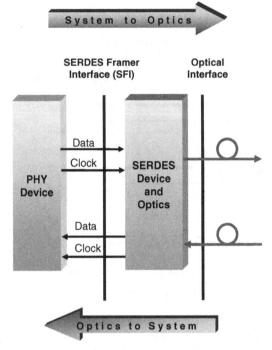

FIGURE 7.15   SERDES framer interface adopted by OIF.

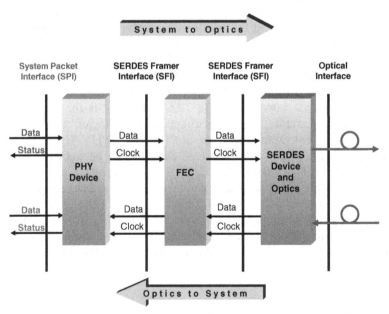

FIGURE 7.16   Forward error correction chip inserted between the SERDES and the framer.

where x can be either F (framer) or P (packet) in SxI. Both SONET and SDH terminology are adopted in the following manner: OC, optical carrier (SONET terminology); and STM, synchronous transport module (SDH terminology). To illustrate some attributes of SPI and SFI interfaces, we discuss briefly their 10-Gb/s implementations.

SPI-4 OC-192 Interface SPI-4 is an interface for packet and cell transfer between a PHY device and a link layer device that runs at a minimum of 10 Gb/s and supports the aggregate bandwidths required of ATM and packet over SONET/SDH (POS) applications. SPI-4 specifies independently its transmitting and receiving interfaces that allow more flexibility in the design of higher-layer devices. SPI-4 is well positioned as a versatile, general-purpose interface for exchanging packets anywhere within a communications system. The SPI-4 interface, illustrated in Figure 7.17, has the following attributes:

- Point-to-point connection (i.e., between single PHY and single link layer devices). Variable-length packets and fixed-sized cells. Transmit/receive data path that is 16 bits wide.
- Source-synchronous double-edge clocking with a 311 MHz minimum. 622 Mbps minimum data rate per line. Low-voltage differential signaling (LVDS) I/O (IEEE 1596.3–1996, ANSI/TIA/EIA-644–1995).
- Control word extension supported. In-band PHY port addressing. Support for 256 ports (suitable for STS-1 granularity in SONET/SDH applications (192 ports) and fast Ethernet granularity in Ethernet applications (100 ports)). Extended addressing supported for highly channelized applications. In-band start/end-of-packet indication, error-control code.

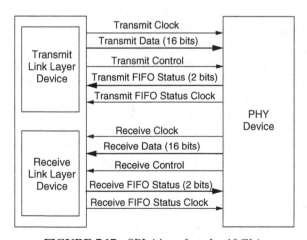

FIGURE 7.17 SPI-4 interface for 10 Gb/s.

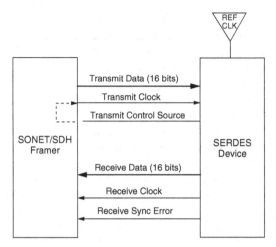

FIGURE 7.18   SFI-4 interface for 10 Gb/s.

- Transmitter/receives FIFO status interface; 2-bit parallel FIFO status bus; in-band start-of-FIFO-status signal; Source-synchronous clocking.

SFI-4 OC-192 Interface   The SFI-4 interface supports transmitting and receiving data transfers at clock rates locked to the actual line bit rate of OC-192. It is optimized for the pure transfer of data. There is no protocol or framing overhead. Information passed over the interface is serialized by the SERDES and transmitted on the external link. The SFI-4 interface, illustrated in Figure 7.18, has the following attributes:

- Point-to-point connection (i.e., between single framer and single SERDES devices). Transmitter/receives data path that is 16 bits wide. Source-synchronous clocking at 622.08 MHz.
- 622 Mbps minimum data rate per line. An aggregate of 9953.28 Mbps is transferred in each direction. Timing specifications allow operation up to 10.66-Gbps LVDS I/O (IEEE Std 1596.3–1996) in order to accommodate 7% FEC overhead.
- The 622.08-MHz framer transmit clock sourced from the SERDES. Uses a 622.08-MHz LV-PECL reference clock input. A low-speed receive loss of synchronization error is signaled when the receive clock and receive data are not derived from the optical signal received.

7.3.3   Design of Data Networking I/O Interfaces

As described above, many serial interface standards are used in networking devices. In this section we sketch how one might go about designing the simple

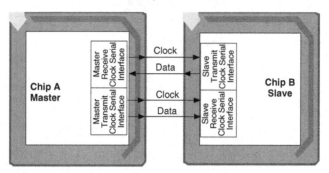

FIGURE 7.19   Master–slave source synchronous serial interface.

clocked serial interface shown in Figure 7.19. That serial interface will facilitate data communication between master and slave devices and requires a clock and data pair of signals in each direction (transmitting and receiving). As the name indicates, a master device controls the timing of the data transfer. Being the master in this case means initiating the clock used for data transfer.

Let us consider the task of designing a simple slave receive clock serial interface. Let us assume that we have already gone through the specification phase and know the block diagram surrounding the interface we are going to design. We also know the interfacing signal behavior as well as the timing of the interface signals. With these assumptions we can place the block we are designing, let us call it RxSerial, within the networking chip as illustrated in Figure 7.20. Note that all clock signals are distributed by a specialized module called a clock distribution block, which ensures the ease of the clock network design.

In the networking architecture, data are usually processed in bytes or words rather than serially. This is why the serial interface converts the serial bit stream of data into parallel sequences of bytes or words, which are presented to the rest of the device at much lower speed. Its function typically includes synchronization of the data to the internal system clock. This synchronization task is performed by means of a first-in first-out (FIFO) block, which is included in the interface circuitry.

In our example the system clock (SysClock) is decoupled from the timing of the data received; therefore, the only synchronization requirement that we have is related to data bandwidth [i.e., the frequency of the SysClock has to be higher than one-eighth that of the frequency of the RxClock]. This requirement is a result of the serial data being converted to octets of data passed to the logic interfacing RxSerial. Of course, due to other requirements of the chip, we may need to run SysClock at a higher frequency; therefore, together with SysData[7:0], we need a SysDataEn signal which indicates to the rest of the logic when the output data are valid. Figure 7.21 is an RxSerial implementation block diagram. The circuitry operates as follows. The serial data stream RxData

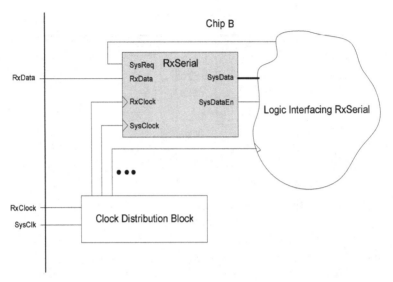

FIGURE 7.20   RxSerial design block.

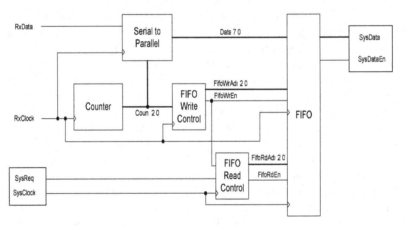

FIGURE 7.21   RxSerial implementation block diagram.

is clocked by the RxClock signal. A free-running counter counts each bit as it is collected in the serial-to-parallel converter. The most common implementation of this converter is a shift register. The counter value is used to control both the serial-to-parallel converter and FIFO write control.

When each time octet of Data[7:0] is ready, the Fifo Write control issues a FifoWrEn indicator and increments FifoRdAdr[2:0].[†] The write frequency to

†In our example the FIFO depth is 8. In general, FIFO depth depends on system requirements.

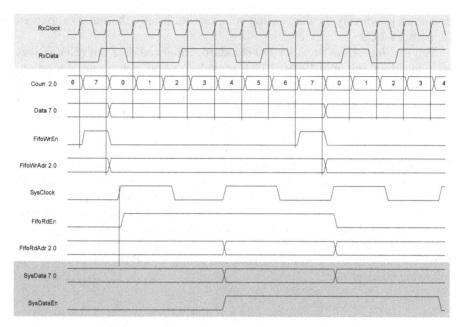

FIGURE 7.22 Timing diagram for the RxSerial operation.

the FIFO is predetermined by the RxData bit rate and has priority over FIFO read operations. FIFO is read each time there is a request from the logic interfacing RxSerial and there is something to read from it (i.e., FIFO is not empty). The FIFO not-empty condition is determined in the FIFO read control, which issues a FifoRdEn signal and generates FifoRdAdr[2:0]. The SysData read from the FIFO is passed to the rest of the core logic. The behavior of the RxSerial circuit, with individual signal behavior, is presented in the timing diagram of Figure 7.22. The shaded waveforms at the top and bottom belong to the external signals of the RxSerial block, while the remaining waveforms are internally generated signals.

7.3.4 Memory I/O Interfaces

To perform frame/packet processing the networking ICs need to store the processed data in memory elements. If these storage requirements are not too severe, the storage operation can be performed using internal on-chip memories. However, if the memory requirements are large, the external memory has to be used. As a result, many networking ICs need to interface to the external memory chips. External memories can be SRAMs or DRAMs. SRAM memories are faster but offer lower density and cost more than DRAMs. Memory interfaces are somewhat different than the data interfaces discussed previously. In this section we examine briefly the basic properties of memory I/Os.

The first important memory I/O standard, enhanced data out (EDO), was introduced in the mid-1990s. EDO provided an asynchronous operation, meaning that the output data were not aligned with a column access signal. The frequency of operation was less than 50 MHz. The I/O was specified to operate as a 5-V logic, had nonterminated signaling, and represented input capacitance of each memory pin to be CIN = 5 pF.

To improve I/O speed the next-generation synchronous DRAM interface (SDRAM) introduced synchronous operation. Data were sampled on the rising edge of the clock, and frequency of operation was in the range 66 to 133 MHz. Part of the speed improvement was due to the lower capacitive load of CIN = 2.5 pF. SDRAM I/O is still used today, although faster interfaces such as DDR I and II are available. DDR uses both the rising and falling edges of the clock to transport data. Hence, even with the same clock frequency as SDRAM, we can double the data rate. For example, a DDR200 provides the following transfer rate:

$$\text{data rate} = (\text{bus clock frequency})(2 \text{ for dual data rate})(8 \text{ bytes per cycle})$$
$$= (100 \text{ MHz})(2)(8 \text{ bytes}) = 1600 \text{ MB/s}$$

Note that in the DDR200 designation, the three numbers represent the effective clock rate of the memory. The effective clock rate is given by the actual clock frequency used (100 MHz in the example above), multiplied by 2. DDR I technology has been specified to work at a supply voltage of 2.5 V while DDR II of 1.8 V. With the high-data-rate requirements, DDR uses an I/O interface called stub series terminated logic.

Further improvements in speed are obtained in DDR II by having the clock run at double the rate of the data. Also, the output data are aligned with the clock signal using a delay locked loop (DLL). A frequency of operation of 333 to 400 MHz is possible using 1.8-V terminated signaling with a low input capacitance of CIN = 1.0 pF and a programmable on-die termination (ODT) to minimize signal integrity problems. A comparison of SDRAM, DDR I, and DDR II is presented in Figure 7.23. It is interesting to note that as memory I/O schemes have evolved, the power supply has decreased (Figure 7.24) and the memory bandwidth has increased (Figure 7.25). These trends are typical for VLSI and partially reflect improvements in underlying transistor technology and, in part, innovations in circuit design techniques over the years.

In closing this memory I/O discussion, it is worth pointing out that DDR III, quadruple date rate, and Rambus I/O schemes promise further improvements in memory I/O performance. In fact, data rates as high as 10 Gb/s per pin have already been demonstrated in the lab using sophisticated DLL/PLL techniques, as shown in Figure 7.26.

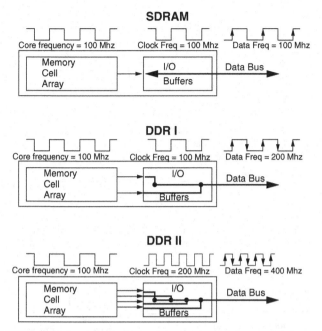

FIGURE 7.23   Comparison of SDRAM, DDR I, and DDR II.

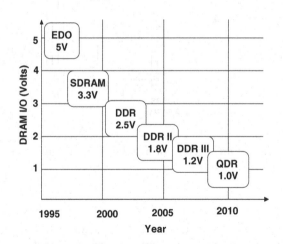

FIGURE 7.24   Evolution of the DRAM I/O power supply.

7.3.5   Microprocessor I/O Interfaces

A microprocessor interface serves as an interface to the higher levels of the OSI reference model. It is used to configure the networking device as well as to collect status information from the data received through the external

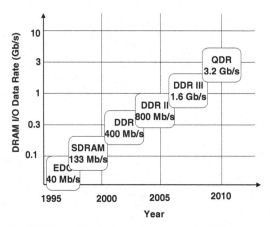

FIGURE 7.25   Evolution of the DRAM I/O data rate.

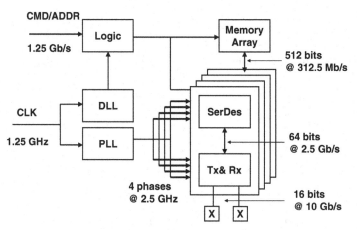

FIGURE 7.26   Internal architecture of a 10-Gb/s DRAM I/O scheme. (From Kim et al., 2005.)

interfaces. The choice of the processor used in various networking applications dictates the support of a variety of microprocessor interfaces designed in a networking chip. Some examples of the microprocessors that are used as an interface between software and networking devices are the Motorola MPC860, Motorola MC68302, and Intel processors with Ebus interface. A typical external microprocessor bus includes address, data and control signals. The control signals are responsible for indicating read or write operation, chip select, data burst, and data or address indicators when data and address are sharing the bus. The microprocessor interfaces can be asynchronous or synchronous, both solutions having advantages and disadvantages.

In the larger and more complicated networking devices, there are a very large number of configuration settings that can be controlled by a microprocessor. These settings are stored internally in the configuration registers. Similarly, the status of the data stream extracted by the networking device is stored in the internal status registers so that it can be read by the microprocessor when needed. The availability of the status data is usually communicated via interrupt control circuitry. Both configuration and status registers can be implemented as latches, flip-flops, or internal RAM bits. These registers are usually associated with the operation of individual blocks. Each register is given a specific address that is used in the same way as the memory address in RAMs. That is why registers within a chip are often referred to as memory maps. To access configuration and status registers, a microprocessor issues the underlying address within the memory map. Each chip specification describes this memory map in detail, in much the same way that it describes the external pin description.

The microprocessor interface of a networking device serves as a bridge between external microprocessor and internal microcontrollers of individual blocks. This is done for design portability, design reuse, and better design partitioning, which leads to a faster design cycle. A typical configuration of a microprocessor interface is shown in Figure 7.27. In the high-aggregation networking devices where the amount of information transferred to the higher OSI level is very high, there may not be enough bandwidth to communicate it through a simple microprocessor interface. The solution to this bottleneck is to move some of the software computations to the networking chip by including an embedded processor. The microprocessor interface is then moved to the internal part of the chip, allowing for faster communication between embedded processor and internal registers. At the same time, the external software interface is based on the instruction set understood by the embedded processor used.

7.4 EXAMPLES OF CHIP ARCHITECTURES

7.4.1 Time-Slice Architecture

In most applications the data passing through the VLSI digital networking devices is time-division multiplexed. The time multiplexing means that the data contain channels of information distributed uniformly in time. These channels of data are usually processed separately. For typical networking IC architectures this means that there are pieces of logic designed for one channel that are replicated a number of times in the chip and which perform the same function on all channels in parallel. A typical example of such

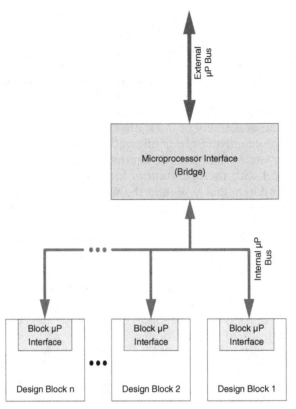

FIGURE 7.27 Block diagram of a typical microprocessor interface internal implementation.

architecture is the processing of the STS-12 SONET frame in the framer device where the time-division-multiplexed individual STS-1 subframes are processed independently in 12 parallel blocks (Figure 7.28). The parallelism of such data processing has a number of advantages. First, it is simple to implement. Second, one individual parallel stream can be processed at much lower clock speed than the original data stream. For example, an STS-12 SONET framer running at a bit rate of 622.08 Mb/s must have a clock of 77.76 MHz to process STS-12 octets of data. On the other hand, if the STS-12 data are demultiplexed into 12 STS-1 channels, the clock handling each channel is 6.48 MHz. This makes the design easier with a smaller switching power density and a more uniform current flow.

There is, however, another architecture that more naturally fits the nature of time-division-multiplexed data; the general idea behind this architecture, called *time-slice architecture*, is to reuse the combinatorial part of the logic for all the channels and to use channel addressable memory to store the states needed to calculate the next state for each channel. Consequently, the states

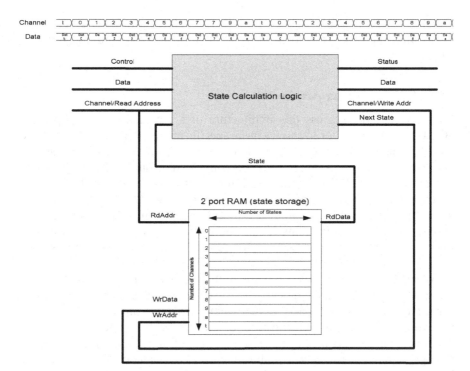

FIGURE 7.28   Generic time-slice architecture.

are stored again as if the set of flip-flops were used. This architecture works with any type of channelized data; however, one needs to evaluate its advantages versus the parallel processing architecture described above before making the final decision.

Generic time-slice architecture is presented in Figure 7.28. The time-division-multiplexed receive data present at the input of the time-sliced design can come at the predetermined order defined by a particular standard (as in the case of SONET/SDH) or can be received in any order. When the order at which data belonging to a particular channel is not known, the channel indicator input is necessary to determine the association of the data with a particular channel, as indicated in Figure 7.28. In addition to the data path being processed by the state calculation logic, the configuration control bus controls the desired behavior of the logic on a per channel basis. These control signals are inserted to the state bit words stored in the two-port RAM at specific locations and act as configuration register bits in the parallel architecture. The control signals can be provided directly by the microprocessor interface or by the control bus of some other block within the chip. Similarly, the status bits of individual channels can be presented to the outside world as sets of register status bits, or they can control the next block in the data path.

Some of the possible uses of time-sliced architecture include ATM cell delineation and processing logic, DS1/E1, DS3 framers, M13 multiplexers, and STS framers.

7.4.2 SONET Framer Architecture

In this section we discuss the architecture of a typical IC used to process SONET frames in order to show a more specific implementation of integrated circuits in data networking. A device like this, called a *SONET framer*, is shown in Figure 7.29. On one side the framer is connected through a SERDES device into an optical module using the SFI interface. This side is called the *line side*, as ultimately the signal is pushed into an optical fiber. On the other side, the framer is connected to a higher-layer device link, a link layer device, or a mapper. In some applications it may be connected directly to a switching fabric. As more frame processing is done at this side, it is called the *system side*.

A SONET framer IC has two paths, in the receiving and transmitting directions, which are largely independent of each other. The only dependencies come from various loop-back modes, where the data can be looped from a receiving side to a transmitting side, or vice versa, for various testing and diagnostic purposes. In a normal mode of operation the data received from the optical module (the lower path in Figure 7.28) is received by the SFI I/O block and after SONET processing is transmitted at the SPI interface I/O. Similarly, the data received from the system side (the upper path in Figure 7.28) is received by the SPI I/O block and after SONET processing is transmitted at the SFI I/O interface.

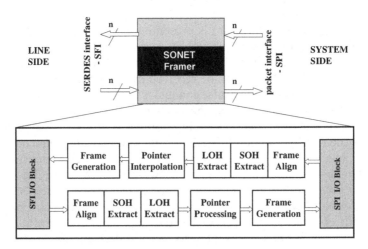

FIGURE 7.29   SONET framer chip.

SONET processing implies full analysis of all overhead bytes in the section, line, and path overhead. When the data are received, the first operations of the framer are to find A1 and A2 bytes and to find the position of the frame. Once the start of the frame is known, the section and line overheads can be extracted. Bytes H1 and H2 in the line overhead indicate the position of the path overhead. The operation of finding the position of the synchronous payload envelope is typically referred to as *pointer processing* in a downstream direction or *pointer interpolation* in an upstream direction.

Typical IC implementation supports SONET/SDH alarm detection and insertion functionality as well as bit-interleaved parity processing. Section, line, and path overhead are extracted and can be sent to the external control processor. Similarly, section, line, and path overhead can be supplied by the external processor and used to override the internal values. These features enable additional flexibility in system implementation where the external processor can implement additional, proprietary functions.

One can build various pieces of SONET networking equipment by using SONET framers and other ICs. As an example, schematic architecture of the SONET add–drop multiplexer (ADM) is shown in Figure 7.30; the individual line card was shown in Figure 7.13. Needless to say the drawings show only the key components of the system. For clarity, various additional pieces, such as power supplies, switches, reference clocks, and E²PROM memories, are not shown. The ADM consists of various line cards for working and protecting lines that connect to pairs of optical fibers. In Figure 7.30 the WEST and EAST links both contain two pairs; in addition, various tributary cards represent lower-rate data streams used to insert or drop SONET traffic into this WEST–EAST connection. The heart of this system is the cross-connecting platform,

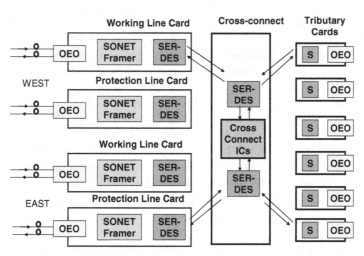

FIGURE 7.30 Schematic architecture for a SONET add/drop multiplexer.

where traffic from various line cards is cross-connected to appropriate outputs as required by information contained in the SONET frame overhead. SONET networking gear is described in more detail in Chapters 5 and 13. Here we just wanted to highlight how SONET integrated circuits might fit into a larger system.

7.4.3 Network Processor Architecture

As mentioned earlier, a network processor is a networking IC that is programmable through software. It offers higher flexibility compared to the hardwired ASIC, at the cost of slower processing speed. A network processor performs packet header operations, such as packet parsing, modification, and forwarding, between the physical layer interface and the switching fabric. The major advantage of a network processor is that it features one common hardware platform that can be utilized in different networking cards and various networking boxes. The major drawback of network processor solutions is difficulty keeping up with the line rate, as the line rate keeps increasing.

To illustrate network processor architecture, let us have a look at one of the early Intel processors, the IXP1200.[†] The Intel IXP1200 network processor is a highly integrated, hybrid data processor that delivers high-performance parallel processing power and flexibility to a wide variety of networking, communications, and other data-intensive products. The IXP1200 is designed specifically as a data control element for applications that require access to a fast memory subsystem, a fast interface to I/O devices such as network MAC devices, and processing power to perform efficient manipulation on bits, bytes, words, and longword data. As illustrated in Figure 7.31, IXP1200 combines the StrongARM processor with six independent 32-bit RISC data engines with hardware multithread support that, when combined, provide over 1 gigaoperation per second. The microengines contain the processing power to perform tasks typically reserved for high-speed ASICs. In LAN switching applications, the six microengines are capable of packet forwarding of over 3 million Ethernet packets per second at layer 3. The StrongARM processor can then be used for more complex tasks, such as address learning, building and maintaining forwarding tables, and managing the network.

The chip contains on-chip FIFO with 16 entries of 64 bytes each. It contains a 32-bit-wide PCI interface that communicates with other PCI devices. The IXP1200 interfaces to a maximum of 256 MB of SDRAM over a 64-bit data bus. A separate 32-bit SRAM bus supports up to 8 MB of SRAM and 8 MB of read-only memory (ROM). The SRAM bus also supports memory-mapped I/O devices within a 2-MB memory space. The 64-bit data bus supports the attachment of MACs, framers, or other custom logic devices, and an additional IXP1200.

[†]More information about this network processor and more advanced Intel chips can be found at the following Web site: http://www.intel.com/design/network/products/npfamily/index.htm.

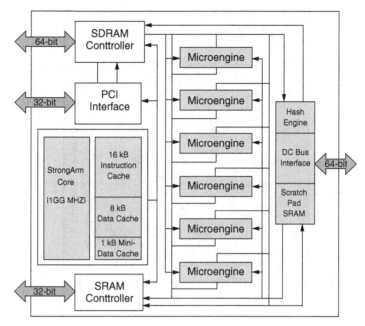

FIGURE 7.31   IXP1200 network processor. (Courtesy of Intel Inc.)

Implemented in 0.28-μm CMOS processes, the chip contains over 6 million transistors, dissipates about 5 W, and is packaged in a 432-pin ball grid array. The silicon die photo is shown in Figure 7.32. As can be seen from this description, even this simple IC is quite complex, although it delivers only 1-Gb/s data-throughput capabilities. One can imagine the complexities involved in designing a network processor for OC-192 applications that would require a data throughput which is higher by two orders of magnitude.

7.5 VLSI DESIGN METHODOLOGY

The process of VLSI chip design is a lengthy systematic process that consists of many tasks performed by numerous specialized engineers. A modern IC might require 30 to 50 engineers working together for a period of one to two years to complete the design. At the end of this process a chip containing several million transistors will have been created. One can easily imagine the complexities involved and requirements to partition the design work.

The partitioning of tasks and distribution of them among design engineers seem to become more important as the level of integration marches along Moore's law. Historically, the design process can be divided into two main groups of tasks: front-end design and back-end design. The front-end design consists of design concept and architecture, design entry, and functional

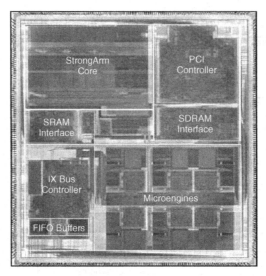

FIGURE 7.32   IXP1200 silicon die. (Courtesy of Intel Inc.)

verification. Back-end design takes on the data created by a front-end design process and creates the physical topology of an integrated circuit using rules for the particular manufacturing technology available.

VLSI design methodology is highly dependent on the technology used in the process of IC fabrication. The constant advance in fabrication process technology introduces new problems that have to be addressed by designers through the means of computer-aided design (CAD) automation. These problems are associated with power consumption and power distribution, signal integrity, accuracy of timing analysis, testability issues, design verification, and so on. The use of specialized design automation tools in solving the problems described above calls for specific sets of design rules that a designer has to be aware of while designing. Together with CAD tools, these design rules are constantly updated with the advances in IC process technology.

Design methodology depends on the fabrication path chosen. Typical examples of design methodologies classified based on fabrication path are custom IC and standard cell and gate array methodologies. Custom cell designs are used in leading-edge parts requiring special attributes, such as very high speed, low power, or small chip area. These designs are made of custom layout cells placed manually in the circuit topology. Once the custom design layout is finished, it can be used as a hard macro cell connecting to the rest of the chip's topology. This methodology is often used in implementing parts of microprocessors and usually requires knowledge of both front- and back-end design tasks.

Standard cell design is much more common than custom design. It relies on the library of standard cells that perform basic logic operations such as *and, or*, and *exclusive or*. These cells have a common physical topology height so that they can be arranged arbitrarily in rows of cells. After all the cells are placed, they can be connected through a network of wires (nets) using a process known as *routing*. Both placement and routing are based on a front-end design-produced *netlist*, a text representation of the design schematic using standard cells as basic components.

Finally, a third approach, gate array design methodology, is very similar to standard cells except the cells are prearranged and prefabricated on the wafer. Once the front-end netlist is generated, it is used to determine association of the prearranged cells with the gates used in the netlist (this process is still called *placement*). Subsequently, the nets are routed and the routed nets (metal layer wires) are superimposed on the prefabricated waver. This process is faster and less expensive than the standard cell design process but is not as efficient. As a result, ICs designed using gate array methodology are bigger, slower, and consume more power.

Based on the differences between fabrication paths described above, it is evident that the corresponding design methodologies must differ substantially as well. Different methodologies and their fabrication paths can have different economic impacts. For example, standard cell IC methodology yields smaller designs, which, in turn, can be packaged in smaller, usually cheaper packages. In high-volume production the low unit cost can offset high initial design-related costs.

In the remainder of this section we discuss a digital IC design process using a standard, cell-based design flow. The IC design process typically consists of the following stages:

1. *Design specification.* During this stage the function of an IC is determined. All architectural trade-offs are explored to determine what algorithms to use, how to partition hardware and software, what interfaces to use so that a device communicates most efficiently with the rest of the world, and the optimum CPU and memory architecture.

2. *Functional design and RTL coding.* Functional design is a process of translating very abstract design description, as written in the specification document, into a specific design entry format that is understood by CAD tools. Typically, Verilog or VHDL hardware description language is used for that purpose and referred to as RTL coding.

3. *High-level verification.* At this stage the RTL code is verified against the design intent by exhaustive simulations. This is accomplished by writing a behavioral simulation test bench within which the RTL code of the design is placed and verified. When verification is complete, the chip is ready for synthesis and physical implementation.

4. *Design synthesis.* Design synthesis comprises design compilation and mapping to a particular technology and subsequently optimizing the result so that it meets the timing, area, and power constraints imposed

by the design specifications. Usually, test circuitry is inserted during that stage. A gate-level netlist is produced as the result of that design stage.

5. *Physical design and verification.* At this stage the gate-level netlist is transferred into a physical representation of the chip. The process is usually iterative and involves cell placement and detailed metal routing. Once the design is placed and routed, the resulting physical layout needs to be verified for correctness against the transistor-level netlist. The process is known as *layout vs. schematic* (LVS) *verification.* In addition, the physical layout needs to be verified against the design rules required by wafer manufacturing. This process is called *design rule checking* (DRC). Chip-level LVS and DRC are final steps before sending the design in the form of GDSII file for mask making and subsequent manufacturing.

A flowchart showing the steps described above is often used to capture the design methodology adopted by a design team. The chart, frequently referred to as a *design flow*, shows the relationship between various tasks so that it is easier to coordinate the work of various members of the design team. A typical simplified design flow represented graphically is shown in Figure 7.33. Note the recursive nature of the design flow. Each of the design tasks includes an initial stage followed by the verification and, consequently, modification of the

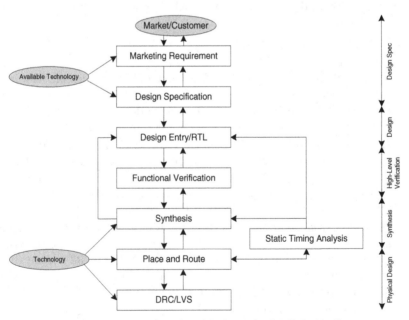

FIGURE 7.33   Typical simplified standard cell design flow.

original task. The design efficiency is highly dependent on the number of iterations performed within the design task or a set of tasks. With an assumption of the recursive nature of the design cycle, the design flows are planned to adopt concurrency of the design process. In the design flowchart, the tasks are often linked to appropriate CAD tools that are used in the design process. This is to communicate promptly to the design team the tools to be used in the design process.

7.5.1 Design Specification

A design of networking chips does not differ much from the design of many other types of ICs used in a variety of applications. The first step is always to create a design specification that determines the function to be performed by a device. The specification is based on the system requirements. The system, to a large extent, determines the partition of the tasks performed by hardware and software. Within the hardware part of the system, one can partition a set of functions to be performed into one or more ICs. The last partition depends heavily on the technological limitations as well as economic factors. Architecting the design from the specifications means that one has to partition the design into smaller blocks of logic that can be handed to individual designers. With the advent of design automation, a chip architect has to evaluate possible design architectures using CAD tools. Once the architecture of a chip has been determined, a set of specifications has to be written to cover block design issues.

7.5.2 Functional Design and RTL Coding

Functional design is the process of translating very abstract design descriptions, as written in the specification document, into a specific design entry format that is understood by CAD tools. It can be further transformed under the guidance of a designer into a circuit topology that at the end is used in an IC fabrication process. Various design entry formats are in use today, but the majority of digital VLSI circuits are designed using a register transfer language (RTL).

There are two mainstream RTL choices (with some variations of each): VHDL and Verilog. VHDL stands for very high speed integrated circuit (VHSIC) hardware description language (VHDL), and its roots lie in work done by the U.S. military in 1980. The newer RTL choice, Verilog, was originally developed by Gateway Design Automation Inc. in 1984. After years of evolution and changes in ownership, IEEE standardized both VHDL and Verilog and now they are known as IEEE Standards 1076 and 1364, respectively. For years, despite arguments by Verilog and VHDL proponents "proving" the superiority of one language to the other, both languages are equally popular in the digital VLSI design community. Examples of simple

VHDL	VERILOG

```
bit_counter:
PROCESS(clk, reset)
BEGIN
  IF(reset = '0') THEN
    count <= 0;
  ELSIF (clk'EVENT AND clk = '1') THEN
    IF (count = 0) THEN
      count <= dat_width-1;
    ELSE
      count <= count - 1;
    END IF;
  END IF;
END PROCESS bit_counter;
```

```
always @(posedge clk or posedge reset)
begin
  if (reset)
    count <= 0;
  else
    if (count == 0)
      count <= dat_width-1;
    else
      count <= count - 1;
end
```

FIGURE 7.34 Examples of binary counter RTL representations in VHDL and Verilog.

synthesizable RTL code snippets written in VHDL and Verilog representing an implementation of a binary counter are presented in Figure 7.34.[†]

Most recently, in an effort to further increase the productivity of digital designers, new language extensions were introduced to Verilog. As a result, a new language called System Verilog has emerged. System Verilog is Verilog backward compatible. In 2005 System Verilog became IEEE Standard 1800–2005. Since the IEEE standard ratification, many CAD tool vendors adopted support for System Verilog, giving a designer the ability to design and verify digital ICs using less verbose and more efficient language.

A digital circuit is a combination of logic gates, some of which are complex combinations of basic logic functions such as *and, or,* and *not* (combinational logic) as well as memory elements such as flip-flops or latches (sequential logic) which hold states of logic calculated by logic gates. Sequential logic is timed by a clock signal that alternates between states 0 and 1. The time between clock transitions is the interval dedicated for calculating states in combinational logic so that it can be stored in the sequential elements. That is why delay through the combinational logic and interconnecting nets determines the maximum frequency of the clock the circuit operates in.

The absolute value of delays through the combinational gates depends on many factors that are technology dependent. A front-end designer can, however, control the delay by introducing more sequential elements and can thereby reduce the amount of combinational logic between flip-flops or latches. The decision on how to break the combinational logic with sequential elements has to be reached early in a design cycle. At the end of the RTL coding phase the design architecture has to be evaluated against the timing requirements established in the specification document.

When an RTL code is written, a designer uses functional verification tools that include a syntax checking tool and a simulator combined with a waveform

[†]Note that the VHDL and Verilog examples do not represent complete RTL implementation of a binary counter, as they do not include declaration of ports, signals, and parameters.

viewer. These tools perform basic checks for desired functionality and enable a designer to add RTL code incrementally with a certain degree of confidence. For this to happen, a designer has to develop a basic simulation environment that consists of a stimulus generator and a basic result checker in the form of a simulation testbench. Using this environment, a designer can inspect signal waveforms within the design to make sure that the stimulus causes the desired response. RTL code that becomes part of the design functionality is written using a synthesizable subset of a given language (VHDL or Verilog), whereas testbench code used to verify design functionality is written using constructs of the language often referred to as behavioral constructs.

Apart from basic RTL design verification, a synthesis exploration phase is often required to make sure that the design can meet the electrical and physical constraints imposed by the specification. A typical RTL design process is iterative, and usually a number of iterations associated with the correction of functional behavior as well as electrical characteristics of the design are quite large.

7.5.3 Functional Verification

There are many approaches to functional verification that are used in the semiconductor industry today. Some involve hardware–software cosimulation; others involve hardware verification and platform verification approaches. We concentrate on the traditional testbench approach where the design under verification is verified within the simulation code responsible to produce input stimulus and observe the resulting response. Testbenches are traditionally written using the same hardware description language that was used to write the RTL code.

With an increasing design complexity, functional verification is gaining the reputation of being the most complex task in the digital design process. It is important that an engineer other than the original circuit designer perform the design verification. This is to ensure that the interpretation of the design specification that the RTL designer relied on is the same as the spec interpretation of the verification engineer. In complex networking chip design projects, the number of verification engineers usually exceeds the number of RTL design engineers.

The typical testbench architecture in a data networking device consists of bus functional models (BFMs) which are responsible for creating stimulus, clock generators, and reference models. These BFMs may include data generators and data analyzers with appropriate bus protocol interfaces. They also include microprocessor interface models so that the testbench can control the design under verification using the same networking protocol that will be used when the device is fabricated.

To be able to verify a design successfully with a certain amount of confidence, special simulation tools that are capable of reporting verification coverage are needed. Verification coverage may be represented in a variety of ways, including toggle coverage, line coverage, finite state machine coverage, statement coverage, and assertion coverage. By analyzing the results of the verification coverage numbers, one can conclude whether or not a particular design has reached sufficient verification coverage.

7.5.4 Design Synthesis

Design synthesis comprises design compilation and mapping to a particular technology cells specified in the technology library. The technology library describes the functions of each cell in the library so that it can be matched with compiled logical functional primitives. It also describes the electrical and physical parameters of each cell, such as load capacitance, intrinsic delays, power dissipation, and area. Based on that information, a synthesis tool has to calculate the delays through the wires connecting logical cells by taking into account a particular wire load model defined based on the area of the synthesized circuit. All of these parameters play a role in cell selection and substitution during design optimization. During design synthesis the designer has to place a set of design constraints such as clock frequency, input and output delays, input driving strength, output load, area, and power.

The two most common approaches to design synthesis are the bottom-top and top-bottom methodologies. The *bottom-top methodology* refers to synthesizing design in a particular order starting at the bottom of the design hierarchy and ending at the top level. This method was historically prescribed to large designs, due to the capacity limitations of the synthesis tools. Although synthesis capacity limitations are largely alleviated in today's synthesis technologies, many design houses maintain this methodology. *Top-bottom methodology* refers to synthesizing the entire design at once from the top level. This method is much simpler, as the design constraints are applied at the top level and propagated down the hierarchy by the synthesis tool.

As the technology geometries shrink, the synthesis tasks become more intricate and the synthesis methodologies evolve accordingly. One example of this evolution is the incorporation of the elements of physical design into synthesis flow. This is done in response to the challenge of representing wire delays in the deep-submicron technologies. Another example is that if automatic test structure insertions such as internal scan and build-in self-tests for both embedded memories and logic.

7.5.5 Physical Design and Verification

Physical-level design involves the process of taking the gate- or transistor-level netlist that represents the design schematic and producing a physical layout that will be used for IC manufacturing. *IC layout*, also known as *mask layout*, is a physical representation of the chip in terms of planar geometric shapes that correspond to the patterns of individual transistors and metal interconnect lines. Modern IC layouts are performed with the aid of IC layout CAD software in a semiautomated fashion.

The first phase of the physical layout design is floor planning. Based on the estimated sizes of all blocks and known chip architecture, the positions of individual blocks are distributed in a manner that provides the most effective signal distribution. The power distribution network and clock trees are planned at this stage. Floor planning is followed by an actual placement, where the precise positioning of blocks and cells takes place. Placement is performed iteratively with routing where the interconnections between blocks and individual cells are conducted. The final layout needs to be extracted for parasitic elements (distributed resistances and capacitance) that might affect delays between logical gates. Using extracted parasitic components, postlayout simulation and static timing analysis have to be performed to verify that the designed layout still complies with the chip specifications.

At the last stage the layout must pass a series of checks in a process known as *layout verification*. The two most common checks in the verification process are design rule checking and layout vs. schematic. When all verification is complete, the data are translated into an industry standard format called GDSII and sent to a semiconductor foundry. The process of sending these data to the foundry is called a *tapeout*, due to the fact that the data used to be shipped out on a magnetic tape. Today, the process is done electronically, but the old term remains. The foundry converts the data into another format and uses them to generate the masks used in the process of chip manufacturing.

KEY POINTS

Networking ICs:

- Networking ICs can be divided based on their functionality into layer 1, layer 2, and layer 3 devices using an ISO model as a reference.
- Layer 1 devices perform modulation and amplification of electrical signals for transmission through a physical medium (an optical fiber, a twisted copper pair, or a coaxial cable). Similarly, in the receiving direction: demodulation and equalization for reception of electrical signals. These devices are typically referred to as physical medium devices and physical layer devices and are considered to be layer 1 devices.
- Layer 2 devices format data into frames or cells using predefined protocols (ATM, Ethernet, fiber channel, SONET/SDH). These devices are typically referred to as framers or mappers.
- Layer 3 devices perform data packet processing. These processing functions include protocol conversion, packet forwarding, policing, lookup, classification, encryption, and traffic management. These devices are

typically referred to as network processors, classification engines, or data link devices. Layer 3 ICs perform in hardware networking functions assigned in the OSI model. Examples include IP routing, ATM layer policing, or fiber channel interworking with SCSI protocol in storage area networks.

- Layer 1 and 2 devices are typically implemented as ASICs. Layer 3 devices can be implemented as ASICs or as network processors, the latter providing flexibility at the cost of complex firmware. FPGAs can also be used for layer 3 processing if the required gate count is low.

I/O interfaces:

- The networking IC has data interfaces, microprocessor interfaces, and memory interfaces. The data interfaces are divided into a line side, closer to the optical fiber transmission, and a system side, close to the switching fabric core.
- A microprocessor interface is used to send control information. It includes address, data, and control signals. The control signals are responsible for indicating read or write operation, chip select, data burst, and a data or address indicator when data and address are sharing the bus.
- To enable interworking between various networking ICs, international standards are required that specify electrical conditions at the data I/O interfaces.
- The Optical Internetworking Forum (OIF) has defined two electrical interfaces within SONET-based communication systems: SFI and SPI. The OIF has published numerous implementation agreements for the SPI and SFI interfaces to which architects of networking ICs must adhere.
- The SERDES (serializer/deserializer) framer interface (SFI) defines an electrical interface between a SONET framer and the high-speed SONET SERDES logic.
- SERDES serializes (parallel-to-serial) and deserializes (serial-to-parallel) data.
- In some hardware implementations the SERDES device is integrated with the framer, effectively eliminating the SFI interface. In other implementations the SERDES device is integrated within an optical module, in which case the SFI interface becomes internal to the OEO module.
- The system packet interface (SPI) is the electrical interface between the physical and link layers in a SONET system. It separates the synchronous physical layer from the asynchronous packet-based processing performed by the higher layers of the OSI reference model. It is designed for the efficient transfer of both variable-sized packets (e.g., Ethernet) and fixed-sized cell data (e.g., ATM).

- SFI and SPI interfaces are defined for various speeds: 2.5, 10, and 40 Gb/s.

IC design process:

- The design of ICs is a complex process that comprises the following stages: design specification, functional design, high-level verification, design synthesis, and physical design with its corresponding verification.
- Design specification captures IC functionality using high-level computer language. All architectural trade-offs are being explored to determine what algorithms to use, how to partition hardware and software, and the optimum central processing unit and memory architecture. The RTL (register transfer language) description is the final result of this design stage.
- Functional design is the process of translating very abstract design description, as written in the RTL specification document, into a specific design entry format that is understood by CAD tools. Typically, Verilog or VHDL hardware description language is used for RTL coding.
- High-level verification is used to verify the VHDL/Verilog code against the design intent.
- Design synthesis comprises design compilation and mapping to a particular technology. A gate-level netlist is being produced as the result of that design stage.
- Physical design takes the gate-level netlist and transfers it into a physical representation of the chip. At the end of this process, the final layout, captured as a gdsii file, is complete and is used later for mask making.

Acknowledgments

We would like to acknowledge major contributions of Jerzy Świć, of Silicomotive Solutions, to this chapter.

REFERENCES

Intel network process Web site: http://www.intel.com/design/network/products/npfamily/index.htm.

Kim, K., et al., A 20 GB/s 256 Mb DRAM with an inductorless quadrature PLL and a cascaded pre-emphasis transmitter, presented at the IEEE International Solid-State Circuits Conference (ISSCC), San Francisco, CA, 2005.

Liu, S., J. Kramer, G. Indiveri, T. Delbrück, and R. Douglas, *Analog VLSI: Circuits and Principles*, MIT Press, Cambridge, MA, 2002.

OIF Forum Web site: http://www.oiforum.com/.

Piguet, C., Ed., *Low-Power Electronics Design*, CRC Press, Boca Raton, FL, 2005.

Rabaey, J. M., A. Chandrakasan, and B. Nikolic, *Digital Integrated Circuits: A Design Perspective*, Prentice Hall, Upper Saddle River, NJ, 2003.

Sharma, A., *Semiconductor Memories: Technology, Testing and Reliability*, IEEE Press, Piscataway, NJ, 1997.

8

CIRCUITS FOR OPTICAL-TO-ELECTRICAL CONVERSION

8.1 INTRODUCTION

Although light seems to be ideal for sending signals in the core of a network, electrical signals are obviously used on the network's edges, as all terminal

Network Infrastructure and Architecture: Designing High-Availability Networks,
By Krzysztof Iniewski, Carl McCrosky, and Daniel Minoli
Copyright © 2008 John Wiley & Sons, Inc.

equipment, be it PCs, cell phones, or large servers, utilizes electronics. The fact that the edge of the network is electrical and the core is optical implies that somewhere in the network a conversion process from optical to electrical (O-E) needs to take place. So, how is O-E conversion done? The goal of this chapter is to show you how electrical signals are converted to optical siguals, and conversely, at the end of the optical transmission, how they converted back to the electrical domain. Since we have already discussed light sources and detectors in Chapter 2, it is probably quite intuitively clear how this task can be accomplished. We will, however, look with more detail on what electrical components are required to perform electrical-to-optical (E-O) conversions, and vice versa.

We begin with a generic description of the optical transceiver. An optical transceiver is a piece of networking equipment that contains both the optical and electronic components needed to perform O-E and E-O functions. We discuss how optical transceivers are built, what components they contain, and what types of interface specifications are needed to ensure proper transmission parameters in both signal domains. We introduce some basic circuits that are used to amplify distorted signals, generate and recover high-speed clocks, serialize parallel data streams, and deserialize serial streams. Finally, we describe some signal impairments and ways to minimize them through an innovative type of circuitry called preemphasis and equalization. The circuits developed for these applications are also used in chip-to-chip communication links called high-speed serial links.

8.2 OPTICAL TO ELECTRICAL-TO-OPTICAL CONVERSION

8.2.1 Principle of Operation

The system architecture of a typical Gb/s optical communication transceiver IC, shown in Figure 8.1, is composed of a number of optical and electrical components. Optical components consist of a continous-wave laser, an optical modulator, and a photodiode. Eletrical components consist of various integrated circuits (ICs) used in transmitting and receiving directions. Transimpedance amplifiers (TIAs), limiting amplifiers, demultiplexers (demuxes), and clock data recoveries are examples of receiving path ICs. Clock multiplications (CMUs), multiplexers (muxes), and modulation drivers are examples of transmitting path ICs.

A photodiode converts the incoming optical pulses from a Gb/s fiber channel into current pulses that are later transformed into a signal voltage by a TIA. The voltage signals are amplified by a limiting amplifier to a sufficient level for reliable operation of the CDR circuit. The demux (e.g., 1:16) separates the regenerated Gb/s data stream from CDR into multiple, lower-speed channels which are processed further to comply with layer 1 (Ethernet or SONET/SDH) protocol standards. These data can be multiplied by mux (e.g.,

16:1) and amplified by a laser driver to be further generated by laser-to-optical pulses that can be transmitted by fiber optics. In the remainder of this chapter we discuss how these circuits are built and operate in an optical transceiver environment.

8.2.2 Optical Transceiver Architectures

A practical implementation of the optical transceiver is shown in Figure 8.2. Optical components are typically grouped and packaged in two subsystem components: optical Rx for the receiving side, and optical Tx for the transmitting side. Electrical components can be packaged in various forms, depending on the degree of VLSI integration. In a typical implementation, shown in Figure 8.2, three ICs handle electronic functions: the laser driver, the SERDES, and the control chip.

The SERDES chip, shown in Figure 8.3, recovers clock information from the data stream using phase-locked-loop or digital phase-picker circuitry. The need for clock extraction arises from the fact that only data are sent along

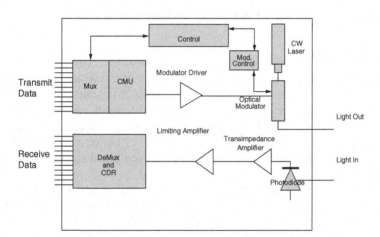

FIGURE 8.1 Optical transceiver.

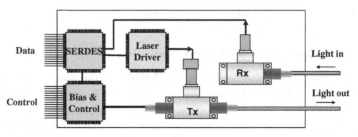

FIGURE 8.2 Practical implementation of an optical transceiver. Rx indicates receive and Tx transmit optical subassembly.

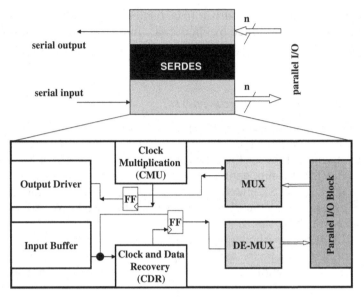

FIGURE 8.3   SERDES integrated circuit.

optical fiber, and therefore the clocking information needs to be "retrieved" or extracted from the data stream. The process of clock and data cleanup is called clock and data recovery (CDR). In the transmitting direction, the SERDES IC synthesizes a clock from a given reference and uses it to retime and multiplex the transmitted data. The synthesis, or clock multiplication, is carried out by clock multiplication unit (CMU) PLL circuitry. The data are then sent to the laser driver, which modulates the optical signal coming out of the laser.

Optical transceivers are sometimes called *transponders*. Although there is some confusion in the naming of these devices in the marketplace, we will use the following terminology: The *optical transponder* is an optical module that contain a SERDES device. The *optical transceiver* is an optical module without a SERDES device. This naming convention is illustrated in Figure 8.4. In some cases it is possible to use transceivers directly on a printed circuit board, in others, it is easier to deal with transponders.

8.2.3   Integrated Circuits for Optical Transceivers

One of the complexities and challenges of building the electronic devices in O-E-O converters is that TIAs are the most sensitive to noise. As a result, they resisted CMOS chip integration for a long time and were frequently manufactured in more exotic bipolar technologies. TIAs are therefore sometimes integrated at the package level with avalanche/PIN photodiodes. However, in recent years a performance of deep-submicron CMOS processes has become

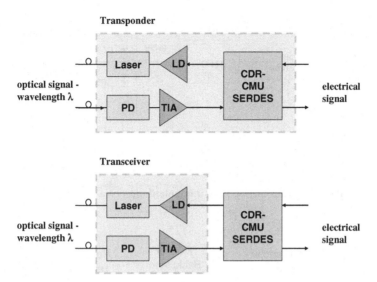

FIGURE 8.4 Optical modules with (transponder) and without (transceiver) a SERDES chip.

sufficient to manufacture high-speed (up to 10 Gb/s) CMOS TIAs, allowing for integration with SERDES devices.

SERDES IC are used to serialize a stream of parallel data into a serial data stream. SERDES and CDR IC implementations are critical to the performance of most networking systems, including storage subsystems. A typical SERDES device, shown in Figure 8.3, consists of the following circuit blocks:

- Input buffer and output driver to handle high-speed serial I/O signals on the line side of the chip
- A CMU to synthesize a high-frequency clock on the chip from high-quality, low-frequency references
- CDR circuitry to recover a high-speed clock from input data streams and to filter accumulated jitter from the data stream
- Muxes and demuxes to convert serial data streams to parallel, and vice versa
- Parallel I/O block to handle a low-speed system interface

In the transmitting direction, SERDES synthesizes a clock from a given reference and uses it to retime and multiplex the transmittingted data. The synthesis, or clock multiplication, is carried out by CMU PLL circuitry. A typical implementation of a transmitting path for high-speed SERDES is shown in Figure 8.5. The data are processed from a lower-speed parallel interface to a high-speed serial line interface operating at wire speed. The parallel

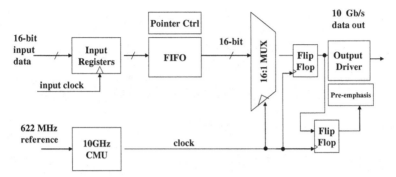

FIGURE 8.5 Transmitting path implementation of a 10-Gb/s SERDES.

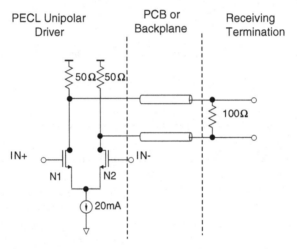

FIGURE 8.6 PECL output driver implementation.

input data signal is received by parallel input registers and send to a FIFO buffer. FIFO is a first-in first-out delay buffer that accommodates timing differences between the receiving and transmitting sides. The FIFO operation is governed by movement of a pointer control that monitors the state of the FIFO. The parallel data from the FIFO are serialized by the multiplier block. A line rate clock required for serialization is provided by the CMU. The serial data from the mux is sent to the output driver. Finally, optional preemphasis is implemented by subtracting bit-delayed data from the signal sent out. We explain preemphasis in Section 8.6.

Output driver design typically uses an ac capacitive coupling scheme differentially terminated with 100Ω at the receiving end, as shown in Figure 8.6.

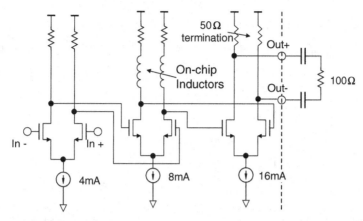

FIGURE 8.7   Output driver implementation of a 10-Gb/s SERDES.

The differential scheme is more robust from the power supply noise perspective. Pseudo-emitter-coupled logic (PECL) is typically used as a signaling scheme for high-performance applications. As PECL is a power-hungry technology, low-voltage differential signaling is frequently selected today instead. For proper impedance matching, the output buffer implements its own on-chip 50-Ω resistive termination for each branch of the final output stage. For high-speed 10-Gb/s implementations, on-chip inductors have to be used to improve signal bandwidth, as shown in Figure 8.7. At the chip boundary, electrostatic discharge and latch-up constraints have to be considered carefully.

The SERDES IC recovers the clock information from the data stream using phase-locked loop or digital phase-picker circuitry. The need from clock extraction arises from the fact that only data are sent along chip-to-chip interfaces, so the clocking information needs to be "retrieved" or extracted from the data stream. Similar considerations apply to a backplane—it is important to minimize the number of traces. CDR is the process of clock extraction from the data and subsequent resampling of data.

A typical receiving path implementation is shown in Figure 8.8. The data received are amplified and equalized before being sent to a slicer that decides whether the data is a 1 or a 0. Amplification is required here, as high-speed electrical signals are attenuated very quickly, so any data transfer between chips or from optics involves substantial signal loss. Equalization is an optional process that boosts high-frequency components of the data stream received. Equalization is discussed in more detail in Section 8.6.

The data from the slicer is sent to the CDR unit. Using one of the PLL techniques, the CDR is capable of recovering the clock signal from the serial data stream. The high-speed recovered clock is used to retime and clean up the data using a high-speed flip-flop. After recovery, the data are sent to the

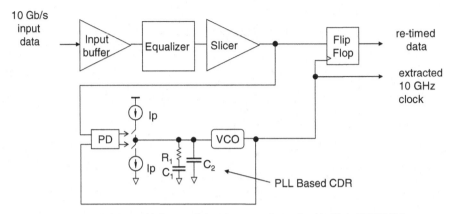

FIGURE 8.8   Receiving path implementation of a 10-Gb/s SERDES.

demux block, which demultiplexes serial signals into an *n*-bit-wide parallel stream. Finally, a parallel output block sends the data out of the chip.

From a high-speed analog design perspective, a SERDES transceiver is quite a demanding device to build. The transceiver needs to be compliant with the rigorous Bellcore SONET/SDH jitter specification. In the first generation these high-speed devices were typically manufactured in an exotic process such as one based on gallium arsenide (GaAs) or silicon germanium (SiGe). Since neither GaAs nor SiGe offers integration possibilities, and both processes have lower yields than CMOS and are power hungry, they have been displaced in the second generation by CMOS parts. Presently, 0.13-µm CMOS processing power is sufficient to design and manufacture 10-Gb/s-compliant transceivers that consume less than half the power of their SiGe counterparts.

8.3   SIGNAL AMPLIFICATION

8.3.1   Trans-Impedance Amplifier

The current generated by photodetectors is typically of insufficient magnitude to drive decision-making and clock recovery circuits. For example, after 50 km of fiber with a 0.2-dB/km signal loss, a 1-mW laser will produce 100 µA of photocurrent if the photodetector's responsivity is 1 A/W. As mentioned before, a TIA is commonly used to convert and amplify photocurrent into a voltage signal of a few hundred millivolts. In the example above, a TIA with a 2-kΩ trans-impedance will produce an output of 200 mV. A frequently used circuit topology is shunt-feedback TIA, shown in Figure 8.9. The photodiode, shown with junction capacitance C_D, feeds the photocurrent into an inverting amplifier with a gain of G. With a high input impedance to the inverting amplifier, most of the photocurrent flows through the feedback resistor R_f, making the

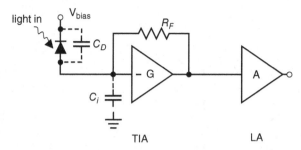

FIGURE 8.9 Photoreceiver: a photodiode connected to a shunt-feedback TIA, followed by a limiting amplifier.

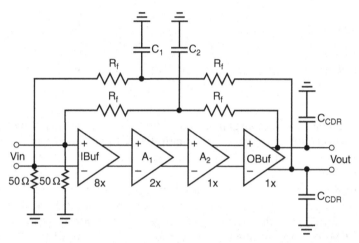

FIGURE 8.10 Limiting amplifier. (From Wang et al., 2005, with permission © IEEE 2005.)

trans-impedance gain at low frequency approximately equal to R_f. A limiting amplifier can also follow the TIA to further increase the gain. The 3-dB bandwidth of the TIA is ultimately limited by the feedback resistance, the capacitances C_D and C_i, and the bandwidth of the inverting amplifier. Noise produced by the TIA is commonly specified as an equivalent noise current at the input to the TIA, with values typically between 1 and 2 μA for 10-Gb/s TIAs.

8.3.2 Limited Amplifier

A TIA is frequently followed by a limiting amplifier (LA) that provides additional signal gain. In CMOS LA design, wide bandwidth and sufficient gain are the main concerns. An example of a simple architecture of a fully differential LA is shown in Figure 8.10. It consists of an input buffer stage for 50-Ω

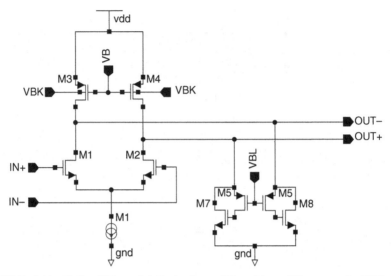

FIGURE 8.11 Fully differential limited amplifier. (From Wang et al., 2005, with permission © IEEE 2005.)

input matching, two similar gain cells (A_1 and A_2) to provide enough voltage gain, an output buffer to drive the on-chip CDR load, and a pair of feedback networks for dc offset compensation due to process, temperature, and voltage supply variations.

Figure 8.11 shows the circuitry used in input buffers. M_1 and M_2 act as a common-source differential pair, and M_3 and M_4 are PMOS active loads bulk-biased to V_{BK} rather than V_{DD} in order to decrease the threshold voltages of PMOS and to allow for low-power-supply operations. The output stage uses active inductors to increase signal bandwidth by canceling the parasitic capacitance at the output nodes. The M_5 and M_6 transistors of active inductors are biased by V_{BL} in the triode region in order to tune the inductance value.

8.3.3 Laser Driver

A laser driver has very specific requirements which are somewhat different from those of other PMD components discussed so far. It has to operate at a high date rate but must deliver the high-voltage levels required by the laser as well. For conventional Mach–Zehnder or electroabsorption structures to achieve a good extinction ratio, the laser driver needs to provide a signal with a large (at least 3V) voltage swing, which is difficult to do in a CMOS chip with a low power supply. For these reasons, laser drivers are implemented in non-CMOS technologies such as GaAs or InP that can meet large voltage-swing requirements for high-speed signals. As a result of different manufacturing technologies, laser driver chips currently cannot be integrated with the rest of the CMOS transceiver circuitry.

8.4 PHASE-LOCKED LOOP

8.4.1 Phase-Locked-Loop Architecture

A phase-locked loop (PLL) is a specialized circuit that synchronizes an output signal with an input reference signal. PLLs find use in a wide range of applications, including clock skew suppressions, timing jitter reductions, high-frequency clock syntheses, and clock/data recoveries. In this section we briefly review a basic principle of PLL operation. A PLL is a feedback system that operates on the excess phase of nominally periodic signals. A schematic PLL architecture is shown in Figure 8.12. The phase detector serves as an "error amplifier" in the feedback loop, thereby minimizing the phase difference between the reference clock and the divide-down clock. The voltage-controlled oscillator (VCO) is an oscillator whose oscillation frequency is controlled by a voltage. The loop filter provides a memory for the loop, preventing the VCO from rapidly changing the output frequency.

For a typical application, shown in Figure 8.12, where the divider divides by an integer number n, the relationship between the output frequency f_{OUT} and the input frequency f_{REF} is very straightforward: their ratio is given by the n number. There are more complex PLL circuits, called *fractional PLLs*, that do not have that limitation and n can be made to be any real number, including, in particular, fractional numbers.

8.4.2 Voltage-Controlled Oscillator

The VCO is the heart of the PLL body. It "beats" in a regular rhythm, with the "beating" frequency regulated by a control voltage. The on-chip VCO can generate high-quality periodic signals (clocks) that can be used by the CMU or CDR circuits. Depending on its design and the semiconductor process used for fabrication, VCO frequency can be as high as 10 GHz, with circuits approaching 100 GHz appearing on the research horizon.

VCO oscillators are negative-feedback circuits. To achieve oscillation, the *Barkhausen criteria* have to be satisfied, stating that the loop gain has to be

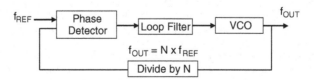

FIGURE 8.12 Phase-locked-loop architecture.

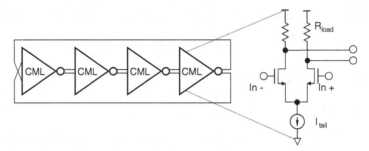

FIGURE 8.13   Ring oscillator.

larger than 1 and the phase shift equal to 180 degrees at the frequency of oscillation. Making sure that the on-chip VCO starts and oscillates under all possible temperature and power supply conditions is not trivial.[†]

VCOs have the following requirements: phase stability, a large frequency tuning range, linearity of frequency vs. control voltage characteristics, and low phase noise/timing jitter properties. As with most modern circuits, they should also dissipate low power, as VCOs/PLLs end up being used in battery-operated hardware. The requirement for low power is typically a trade-off with low-jitter, high-performance requirement.

There are three types of VCO: ring oscillators, LC oscillators, and relaxation oscillators. *Ring oscillators* can be as simple as a chain of CMOS inverters connected in a ring.[‡] A more practical solution consisting of four differential CML (coupled emitter logic) stages is shown in Figure 8.13. In this circuit the frequency of oscillation is determined by a signal delay of the CML stage that can be tuned by adjusting the load resistance R_{load} or source current I_{tail}.

LC oscillators utilize the fact that capacitors and inductors have opposite impedance imaginary values so that with sufficiently high impedance (L), one can compensate for the capacitance (C), leading to sustained oscillations. Conceptually simple LC oscillators can be difficult to implement, as on-chip inductors have low values of inductance and high resistive, and capacitive losses. However, they provide the best quality of generated signals and have lower timing jitter/phase noise than the ring oscillators.

An example of the LC-based VCO structure is shown in Figure 8.14. The cross-coupled transistor pair effectively provides a negative resistance by looking into the drains of the transistors while the inductor sets the oscillation frequency together with the capacitance. The capacitance is tuned to change the frequency of oscillations as desired. An example of silicon implementation

[†]In practice, it is not uncommon to see on-chip oscillators that do not oscillate under certain conditions. Conversely, there are some high-frequency circuits that are not supposed to oscillate, (e.g., amplifiers), which occasionally show signs of oscillations under some conditions. Murphy's law shows its presence in high-frequency oscillations!

[‡]When designed properly, a chain of inverters can oscillate but cannot be tuned.

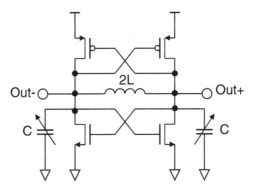

FIGURE 8.14   LC VCO oscillator.

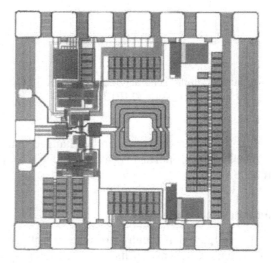

FIGURE 8.15   Die photo of the LC oscillator. (From Magierowski et al., 2004, with permission © IEEE 2004.)

is shown in Figure 8.15. As easily seen, the inductor occupies a large silicon area. This observation emphasizes the point we made earlier: that designing a noticeable on-chip inductance is a challenge in CMOS technology. The tuning characteristics, shown in Figure 8.16, indicate that this particular VCO can be tuned around the target frequency of 2.488 GHz (OC-48 rate) from about 2 to 2.9 GHz. This tuning range of 38% is typical for modern CMOS circuits.

The third class of VCOs are the *relaxation oscillators*. Their principle of operation is to charge and subsequently discharge a reference capacitor through use of a control current. By adjusting the value of the control current, the frequency of oscillation can be changed. Relaxation oscillators can have a

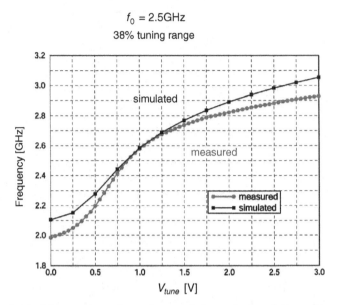

FIGURE 8.16 Measured tuning characteristics of the LC oscillator. (From Magierowski et al., 2005, with permission © IEEE 2005.)

large tuning range as the control current can be changed in a wide range but suffer from poor phase noise properties. As a result, they are typically not used in modern microelectronics.

8.4.3 Phase and Frequency Detectors

A *phase detector* is a circuit that ideally produces a voltage that is linearly proportional to the phase difference of its two (periodic) inputs, as shown in Figure 8.17. Similarly, a *frequency detector* is a circuit that produces a voltage that is linearly proportional to the frequency difference of its inputs. Frequently, circuits are sensitive to both phase and frequency and are called *phase–frequency detectors* (PFDs).

The various phase and frequency detectors presented in the literature rely on XOR gates, S-R flip-flops, or double-balanced mixers. We mention only

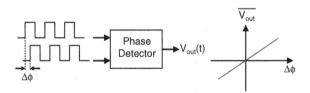

FIGURE 8.17 Concept of a phase detector.

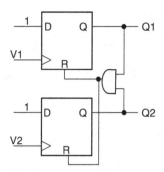

FIGURE 8.18   Simple implementation of a PFD detector.

briefly the operation of the most popular circuits used in CMU and CDR applications. Consider the simple detector shown in Figure 8.18. It consists of two edge-triggered D flip-flops, each having one of its inputs connected to its logical high. A combination of the two output signals is used to reset both flops. A simple analysis of input waveforms shows that the output voltage is proportional to a phase difference between two periodic input signals, V_1 and V_2. The circuit is also sensitive to a frequency difference—hence the name PFD. Frequency sensitivity is a characteristic that is desirable in CMU applications.

Frequently, as in CDR applications, it is desirable to have a circuit that is only sensitive to a phase difference. Figures 8.19 and 8.20 illustrate two circuits that possess this property. The first (Figure 8.19) is a *linear-phase detector* (also called a *Hogge-type detector*), in which the output voltage is linearly dependent on the input phase difference. The second is a *binary-phase detector* (also called an *Alexander* or *bang-bang detector*), which produces, at its output, only digital information indicating whether the clock is running late or early. The linear detector leads to "smoother" CDR operation, but its implementation at very high frequencies might not be possible. In these cases the binary detector that is used is the one that provides a simple, digital notification as to whether the clock is late or early. The binary detector results in more abrupt CDR operations, resulting in some degradation in PLL jitter performance.

8.5   CLOCK SYNTHESIS AND RECOVERY

8.5.1   Clock Synthesis

Clock synthesis, or clock multiplication, relies directly on PLL operation. Figure 8.12 illustrated a basic clock synthesis concept: that using a high-purity reference clock PLL can generate a high-frequency clock that is an integer multiple of the reference frequency. In practice, due to some design and stability issues, a somewhat different circuit is used as shown in Figure 8.21. This

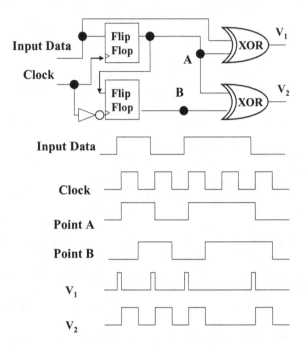

FIGURE 8.19 Linear-phase detector.

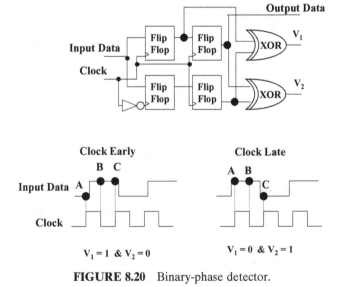

FIGURE 8.20 Binary-phase detector.

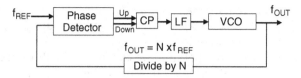

FIGURE 8.21 Charge-pump-based PLL for clock synthesis.

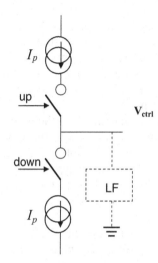

FIGURE 8.22 Charge pump.

circuit contains one additional component, a *charge pump*—hence the name *charge-pump-based PLL*.

Charge-pump circuitry is shown in Figure 8.22. The circuit contains two inputs, up and down, that control whether the charge pump current I_p is to charge the loop filter (up operation) or discharge it (down operation). By increasing or decreasing the control voltage across the loop filter, VCO operations can be sped up or slowed down. Like any other PLL, charge-pump PLL is a feedback system. It has many interesting properties that require complex mathematics to explain: its stability criteria, lock-in dynamics, and jitter properties. Next, we examine briefly basic PLL properties important in practical clock synthesis units.

The PLL can operate in various modes. The first phase is an acquisition phase. The acquisition procedure for a PLL is to become locked when initially unlocked. The two steps involved are frequency acquisition (frequency pull-in) and phase acquisition (phase lock-in). Due to PFD operations, first frequency pull-in is achieved, then phase lock-in. Once locked, the PLL operates in a tracking phase. During tracking the output phase tracks the input phase. In lock, the VCO's frequency equals the input frequency and the VCO's phase and input phase have a constant difference. Similar modes of operation can be defined when PLL is losing its lock.

Based on these modes of operation we can define the PLL range of operation terms: hold range, pull-in range, pull-out range, and lock range. The *hold range* is defined as the frequency range in which a PLL can maintain phase tracking. The *pull-in range* is the range within which a PLL will always become locked through the acquisition process. The *pull-out range* is the value of a frequency step applied to the reference frequency that causes the PLL to unlock. Finally, the *lock range* is the frequency range in which a PLL locks a single-beat note between the reference frequency and the output frequency.

Achieving PLL lock is clearly a very important characteristics of the PLL. For fast operation, which is required in a data networking environment, the PLL needs to lock very quickly. The *acquisition time* is the amount of time the PLL takes to converge within a certain phase error of the input signal. The acquisition time is inversely proportional to the loop bandwidth. However, the acquisition range is directly proportional to the loop bandwidth of the PLL. As a result, selection of the PLL bandwidth is a difficult compromise. One can have a PLL that acquires lock very quickly, but only if the starting point in the frequency domain is already very close to the target. Alternatively, one can have a PLL that locks in a wide range of starting frequencies but whose acquisition process will take a long time.

What are the important parameters for PLL? In addition to the acquisition time and range already discussed, PLL phase noise and jitter are frequently critical in data networking. Phase noise in the frequency domain translates into jitter in the time domain. Phase noise is measured in dBc/Hz at a given $\Delta\omega$ deviation from a central frequency of operation ω_0 (Figure 8.23). Phase noise generated in a PLL comes from the VCO oscillator and is high-pass filtered through the loop. This noise is minimized as the PLL bandwidth increases.

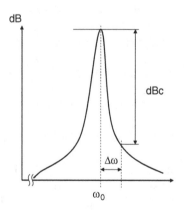

FIGURE 8.23   Phase noise characteristics.

Phase noise injected into a PLL (through the input or power supplies) is low-pass filtered. This noise component decreases as the PLL bandwidth decreases.

Clearly, selection of the PLL bandwidth needs to be considered carefully during design. In CMOS design practice, one increases the PLL bandwidth as far as possible, resulting in a reduction in the acquisition time, an increase in the acquisition range, and a reduction in the VCO phase noise contribution. This PLL bandwidth increase is limited by the increased contribution of external sources of jitter to output.

One final important consideration for PLL is its stability. Recall that PLL is a negative-feedback system. Like any feedback system, depending on phase margins, the system can be stable or oscillate.[†] The occurrence of stability problems depends strongly on details of the loop filter implementation. Depending on the number of poles and zeros in the PLL transfer function stability criteria can be formulated and solved mathematically. In practical CMOS PLL design a phase margin of at least $60°$ is required to ensure that PLL is stable under all operating conditions.

8.5.2 Clock and Data Recovery

CDR is an important function in optical networking, as it enables the extraction of a signal clock out of a data stream sent through an optical fiber. Without this functionality one would have to send data and clock signals on separate optical fibers, effectively doubling the required number of fiber strands. In addition to minimizing the number of fiber strands, the clock extracted is used to clean the received data from all distortions that have accumulated during signal transmission, as shown in Figure 8.24.

One popular CDR technique relies on PLL operations (Figure 8.25). In that circuit the phase detector compares phases of its internal VCO clock and the

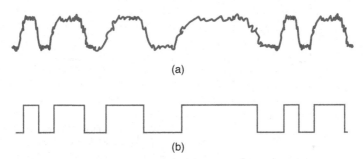

(a)

(b)

FIGURE 8.24 Principle of CDR operation: (a) distorted received signal; (b) retimed signal by CDR.

[†]In the context of a PLL operation, the term *oscillate* is meant here as a system. The PLL system needs to be stable, but its internal VCO will obviously oscillate.

jittery data stream received. If there is a finite phase difference between these signals, the VCO will speed up or slow down to have the phase difference disappear (in an ideal case). Note that for proper operation of CDR, the internal VCO has to oscillate at the correct frequency prior to clock extraction from the serial data stream. This functionality is achieved through some additional circuitry not shown in Figure 8.25, called *training loops*.

A CMOS implementation of CDR functionality was shown in Figure 8.8. An alternative architecture for CDR is shown in Figure 8.26. In that solution, PLL generates a number of equally spaced timing samples that are compared against the input jittery signals. Again, depending on whether the phase difference is negative or positive, the internal VCO is forced to oscillate faster or slower until that phase difference disappears. The main difference between the architecture from Figure 8.8 and Figure 8.26 is that the first uses completely analog circuitry whereas the second uses a significant amount of digital circuitry.

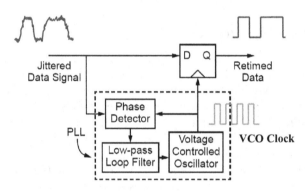

FIGURE 8.25 CDR block diagram.

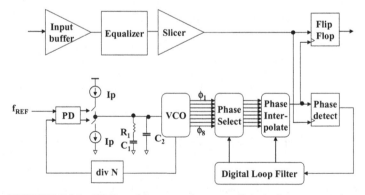

FIGURE 8.26 CDR architecture that uses digital phase processing.

8.5.3 Jitter Requirements

We have already mentioned timing jitter a few times in this book. Now let us take a closer look at how jitter is defined for data networking applications. A communication link can be modeled with three elements: a transmitter, a channel that propagates the signal, and a receiver. In an ideal system, the edges of an electrical or optical signal will always occur at integer multiples of the signal period. In a real system, the edges of a digital signal will occur in a distribution around the center point, which is the average period of the digital signal. *Jitter* is defined as the variation in the edge placement of a digital signal. Three jitter components are usually specified: *Jitter generation* is the amount of jitter created by a device assuming that the device's reference clock is jitter-free; *jitter tolerance* is the maximum amount of jitter that a device can withstand while still being able to receive data reliably; and *jitter transfer* is a measure of the amount of jitter transferred from the receiving side of a device to the transmitting side.

Jitter requirements for high-speed signaling standards vary widely. Typically, they are specified in terms of the following jitter components: *Deterministic jitter* is jitter generated by either insufficient channel bandwidth, leading to intersymbol interference, or by duty-cycle distortion, which leads to timing errors in data clocking; *random jitter* is usually assumed to have a Gaussian distribution and is generated by physical noise, such as thermal noise; *sinusoidal jitter* is used to test the jitter tolerance of a receiver across a range of jitter frequencies and is not really a jitter type that would be encountered in a deployed system. Sinusoidal jitter is injected artificially into the receiving side of a circuit to measure the performance of the receiver in the presence of the user-defined sinusoidal noise source. With this sinusoidal jitter technique, the receiver's jitter tolerance vs. frequency can be measured.

Networking standard bodies have introduced a number of very detailed jitter specifications to ensure that all electrical and optical components can interoperate in real deployment conditions. These standards regulate how much jitter networking equipment can generate (jitter generation) and how much it should be able to accept (jitter tolerance). They also define a ratio of output to input jitter called *jitter transfer*.

Examples of jitter specification for data networking standards are shown in Figure 8.27 and 8.28. Figure 8.27 shows the jitter tolerance specification for SONET/SDH links. To be compliant with the specification, the actual IC (its CDR unit) has to be able to accept at least the specified minimum amount of sinusoidal jitter injected at a certain frequency. As the characteristics plotted indicate, at low frequencies CDR has to be able to deal with a jitter of several unit integrals (UIs). However, as the UI value determined by a period of injected jittery signal is small, this requirement is usually easy to meet. The challenge is on the high-frequency side. CDR has to be able to accommodate at least 0.15 UI of high-frequency jitter. Other communication standards (e.g., Ethernet) can be even more demanding than SONET/SDH in

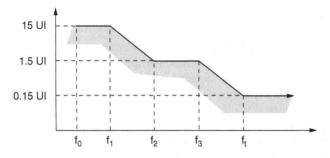

Data Rate		f_0 [Hz]	f_1 [Hz]	f_2 [Hz]	f_3 [kHz]	f_t [kHz]
OC-3	155 Mb	10	30	300	6.5	65
OC-12	622 Mb	10	30	300	25	250
OC-48	2.488 Gb	10	600	6000	100	1000
OC-192	10 Gb	10	2400	24000	400	4000

FIGURE 8.27 Jitter tolerance specification for SONET/SDH standard.

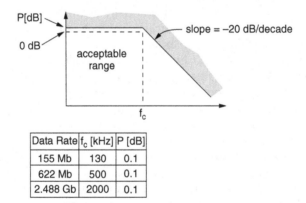

Data Rate	f_c [kHz]	P [dB]
155 Mb	130	0.1
622 Mb	500	0.1
2.488 Gb	2000	0.1

FIGURE 8.28 Jitter transfer specification for SONET/SDH.

this regard.[†] Figure 8.28 shows jitter transfer specifications for SONET/SDH. To be compliant, the IC has to provide less than $0.1\,dB^{†}$ jitter peaking up to a certain corner frequency. The idea behind this specification is to prevent the accumulation of jitter when forming repeater links.

[†]Ethernet requires a high-frequency jitter tolerance of 0.6 UI. However, SONET/SDH has much more stringent requirements for jitter generation than those for Ethernet, requiring jitter generation to be less than 0.1 UI.
[‡]Since jitter transfer is a ratio between two jitter numbers, it is effectively a unitless quantity, typically expressed in decibels.

Multiple approaches to meet the jitter requirements can be taken. A typical solution involves using a high-quality crystal as a reference to a CMU PLL to generate the transmitting clock required to send the data, as discussed previously in this chapter. Alternatively, the clock recovered from the data received can also be used, although its jitter properties are inferior to that of a good-quality crystal oscillator. Dealing with jitter requires a great deal of engineering effort and is one of the major challenges in managing the operation of data networks.

8.6 PREEMPHASIS AND EQUALIZATION

8.6.1 High-Speed Signal Impairments

Although the SERDES circuits presented in previous sections represent the typical design for numerous applications, additional considerations have to be taken into account when driving high-speed signals over long distances. What "high speed" and "long distance" are is obviously relative, but suffice it to say that transmittingting 10-GHz signals over PCB traces a few centimeters long is very challenging. In addition, a special consideration is needed when driving signals over the backplanes, which are the communication backbones of networking hardware equipment such as switches and routers. As we will see shortly, these backplanes contain additional PCB vias and connectors that introduce additional signal impairments. A typical backplane (Figure 8.29) consists of long PCB traces connected to the corresponding line cards by two connectors. An example of system architecture using a backplane is shown in Figure 8.30. In Chapter 12 we discuss how switches and routers are built using backplane connectivity.

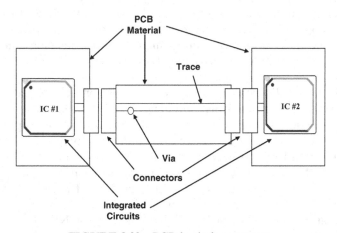

FIGURE 8.29 PCB backplane system.

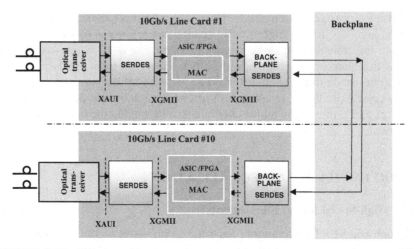

FIGURE 8.30 Switch architecture that uses a backplane to connects several line cards.

To define a system architecture that meets the challenges of data transmission over high-loss, low signal-to-noise ratio environments, it is important to identify and simulate system impairments so that they can be addressed properly. The major contributors to signal deterioration in most backplane or cable transmission environments are channel loss, crosstalk, and reflections.

Channel Loss   Channel loss is a phenomenon where the limited bandwidth of a given channel attenuates the high-speed components of a data signal. This attenuation effect creates *intersymbol interference*, in which the signal sent is smeared and thus has an impact on subsequent bits that are sent. For the operating frequency of a 10-Gb/s data signal, where the UI value is 100 ps, the minimum cyclic transition T_{ct} is twice the UI value (200 ps), and the F_{ct} frequency is 5 GHz (1/200 ps). At 5 GHz, and with a typical backplane system, the loss experienced by the signal is approximately −30 dB, meaning that only 3% of the signal transmittingted is reaching the receiver.

Crosstalk   Crosstalk occurs in any system with two or more conductors. Each wire segment acts individually as an inductor and capacitor and as an antenna. Together, they act as coupled antennas, due to their mutual coupling, which can be expressed in terms of capacitive and inductive components. To achieve higher serial bit rates, it is necessary to increase the rate of transition of the signal. This is detrimental to the crosstalk problem because coupling increases proportionally to the rate of change of the aggressor. The frequency-domain transfer function of near end crosstalk (NEXT) and far end crosstalk (FEXT) indicate that a major contributor to crosstalk noise is NEXT and that the impact increases with frequency.

Signal Reflections Impedance discontinuities in the channel path cause signal reflections, which cause significant deterioration of the SNR at the receiver. Any time that there is a discontinuity in a signal's propagation path, a certain amount of energy is reflected back toward the source. In the context of a receiver operation, this means that energy from past bits is being received at the same time as new bits are arriving.

8.6.2 Preemphasis

Preemphasis, or *equalization*, means essentially that high-frequency components are boosted so as to compensate for channel loss. The term *preemphasis* is typically applied to the transmittingting terminal, whereas *equalization* is used for the receiving end. A concept of preemphasis implementation is shown in Figure 8.31. For each high (logical 1) and low (logical 0) there are two voltage levels. Each level can have two different voltage values, one for a steady state (when the preceding bit is the same as the current bit), and the other for a transition state (when the preceding bit is different than the current one). The transition state-voltage is higher than the steady-state voltage. In this way the high-frequency content in the data is amplified (or emphasized) more than the low-frequency portion of the signal. The preemphasis is implemented by subtracting a scaled version of the delayed bit from the bit that is currently being transmittingted.

8.6.3 Equalization

Equalization essentially means that high-frequency components are boosted in such a way that the channel loss is compensated completely for, as shown

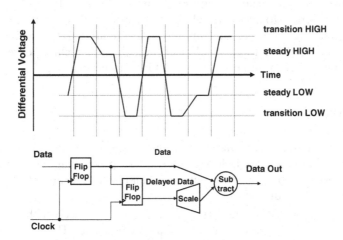

FIGURE 8.31 Signal preemphasis concept.

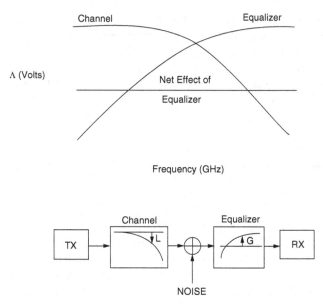

FIGURE 8.32   Signal equalization concept.

in Figure 8.32. Equalization is applied at the receiving end of the channel. Typically, IC implementation of the equalization is more complex than that of preemphasis. However, equalization is more flexible than preemphasis, as it does not require any knowledge of the channel characteristics.

Equalization is a relatively simple concept, but there are many issues involved in its implementation. Consider first channel compensation loss. Channel loss may easily vary from −10 dB down to −30 dB, which means that the level of high-frequency boost must be very different in the two cases. The second problem is noise amplification. When there must be heavy equalization, it is very dangerous to apply straight gain because the SNR before the gain is very low, and as a result, amplification of noise relative to the signal will also be very high. Despite these challenges, equalization techniques have been applied universally in numerous ICs used in data networking.

KEY POINTS

- Optical signals are suitable for transmission over long distances. Electronic signals are suitable for information processing. As a result, data networks perform optical-to-electrical and back to optical (O-E-O) operations very frequently.
- O-E-O conversion is performed by an optical transceiver that contains a laser driver, photodiode, and associated electronics. The electronics consists of various amplifiers, drivers, and SERDES integrated circuits.

- A SERDES device, which serializes and deserializes parallel data into a serial stream, and vice versa, is an important component of the overall hardware system, as it contains clock and data recovery (CDR) circuitry. SERDES and CDR IC implementations are critical to the performance of most networking IC.

- In the transmitting direction, SERDES synthesizes a clock from a given reference and uses it to retime and multiplex the transmittingted data. The synthesis, or clock multiplication, is carried out by clock multiplication (CMU) PLL circuitry. CMU is used to synthesize a high-frequency clock on the chip from a high-quality, low-frequency reference.

- In the receiving direction, SERDES recovers the clock information from the data stream using phase-lock-looped circuitry. The process of clock extraction from the data and subsequent data resampling is called clock and data recovery.

- A FIFO is a first-in first-out delay buffer that accommodates timing differences between the receiving and transmitting sides. The FIFO operation is governed by movement of a pointer control that monitors the state of the FIFO.

- A multiplexer converts serial data stream to parallel, and vice versa. A demultiplexer performs the opposite operation.

- A phase-locked loop (PLL) is a feedback system that operates on a signal phase. Typical PLL circuitry contains a voltage-controlled oscillator, a charge pump, a phase and frequency detector, and a divider. PLLs are used to synthesize or recover high-speed clocks.

- High-speed signals experience various impairments, including frequency-dependent signal attenuation, crosstalk, and signal reflections caused by impedance discontinuities.

- Frequency-dependent signal loss is a phenomenon whereby limited bandwidth of a given channel attenuates high-speed components of the data signal more than low-frequency components. This phenomena leads to the intersymbol interference effect, where the given bit affects subsequent bits that are sent.

- Crosstalk occurs in any system with two or more conductors, as each wire segment acts individually as an inductor, capacitor, and an antenna. Crosstalk leads to signal corruption by operation of its neighbor.

- Reflections are caused by impedance discontinuities and lead to degradation of signal quality.

- Preemphasis and equalization are processes of enhancing the high-frequency components of a signal. Preemphasis is used in the transmitting direction (therefore sometimes referred to as transmitting equalization), and equalization is used in the receiving direction.

- The process of equalization helps in more clearly "seeing" distorted and jittery input signal. After equalization, the serial data are sliced to determine what is 0 and what is 1 in the data stream. Equalization implementation in silicon chips is quite complex. However, preemphasis implementation can be quite simple: for example, by relying on subtracting bit-delayed data from the signal sent out.

REFERENCES

Choudhary, V., and K. Iniewski, Phase-locked loop based integer-N RF synthesizer, in *Wireless Technologies: Circuits, Systems and Devices*, K. Iniewski, Ed., CRC Press, Boca Raton, FL, 2007.

Iniewski, K., V. Axelrad, A. Shibkov, A. Balasinski, S. Magierowski, R. Dlugosz, and A. Dabrowski, 3.125 Gb/s power efficient line driver with 2-level pre-emphasis and 2 kV HBM ESD protection, *Proceedings of the IEEE International Conference on Circuits and Systems (ISCAS)*, Kobe, Japan, pp. 64–67, May 2005.

Iniewski, K., S. Voinigescu, and M. Syrzycki, Process and device requirements for mixed-signal integrated circuits in broadband networking, *Journal of Telecommunications and Information Technology*, pp. 88–96, January 2004.

Johns, D., and K. Martin, *Analog Integrated Circuit Design*, Wiley, Hoboken, NJ, 1997.

Laude, J. P., *DWDM Fundamentals, Components, and Applications*, Artech House, Norwood, MA, 2002.

Magierowski, S., K. Iniewski, and S. Zukotynski, A wideband LC-VCO with enhanced PSRR for SOC applications, *ISCAS*, pp. 173–176, 2004.

Ramaswami, R., and K. N. Sivarajan, *Optical Networks: A Practical Perspective*, Academic Press, San Diego, CA, 1998.

Razhavi, B., *Design of Analog CMOS Integrated Circuits*, McGraw-Hill, New York, 2001.

Sameni, P., C. Siu, K. Iniewski, M. Hamour, S. Mirabbasi, H. Djahanshahi, and J. Chana, Modeling of MOS varactors for 5–6 GHz LC VCO, *EURASIP Journal on Wireless Communications and Networking*, vol. 2006.

Wang, Y., and K. Iniewski, A 2.3 GHz CMOS transimpedance preamplifier for optical communication, *IWSOC*, pp. 243–246, 2005.

9

PHYSICAL CIRCUIT SWITCHING

9.1   INTRODUCTION

This chapter has two primary goals: (1) to divide the world of switching architectures into three classes: switches for physical circuits, switches for time-

Network Infrastructure and Architecture: Designing High-Availability Networks,
By Krzysztof Iniewski, Carl McCrosky, and Daniel Minoli
Copyright © 2008 John Wiley & Sons, Inc.

division-multiplexed signals, and switches for cells and packets, and (2) to describe in some detail the architectures of the first class of switches, the physical circuit switches. We shall see that physical circuit switches fall into two categories: (1) single-stage switches and (2) multistage switches. The conceptual simplicity and beauty of single-stage crossbar switches are described along with the reasons that we cannot always use these simple switches. Then multistage switches are explored as an alternative to single-stage switches. Although the research literature presents many interesting mulitstage switches, and some of these proposals have been used in products, we restrict ourselves to the fabrics most commonly used for physical circuit switching: Clos networks. In the chapter we study the blocking and rearranging complexities that arise with Clos switching solutions.

With the goal of eventually building simple but reasonably accurate VLSI cost models for all of the switches presented in this portion of the book, we begin in this chapter with the development of a cost model for single and multistage crossbar switches. In addition, we look briefly at micro electrical mechanical system (MEMS) switches for optical signals, protection switching issues and their impact on multistage fabrics, and a family of commercial crosspoint switches.

9.2 SWITCHING AND WHY IT IS IMPORTANT

Wherever a communications system brings multiple communications links together, and each of the ingress links carries one or more signals that must be sent out any of the egress links at one time or another, the functional unit that allows the multiple inputs to be varyingly connected to the multiple outputs is called a *switch*. The act of making these varying connections is called *switching*. The ingress and egress links are normally paired and used to connect *N* nodes. Figure 9.1 illustrates this abstract definition.

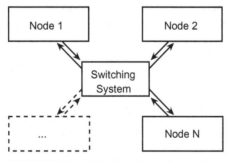

FIGURE 9.1 Abstract view of switching.

The preceding simple definition of switching captures the essence of the material of the next three chapters. The simplicity is beguiling and misleading. As we shall see, many interesting and challenging issues are involved in the design of appropriate switches. The study of switching is interesting and instructive on its own, but there is an additional advantage to looking carefully at switching systems and solutions. In many applications, from telecommunications to computer systems to storage systems, switches lie at the heart of the application. To understand the switching problem, it is necessary to understand the application thoroughly; similarly, to understand the application, it is necessary to understand the switching capabilities required. Thus, being closely related to the essential ideas and issues embedded in an application, the switching perspective is a good key with which to open one's understanding of an application.

Suppose that we require a network to connect N nodes in such a way that any node can communicate with any other node, but being lazy we do not want to have to read the remainder of the switching chapters in this book. Can we build a network to connect these N nodes without employing a switch of any sort? Can we avoid a lot of work? Suppose that each of the N nodes is given a point-to-point (unswitched) duplex link to each of the other $N - 1$ nodes, as illustrated in Figure 9.2. This system is completely connected in the sense that each node has a direct connection to every other node. For instance, to communicate with node 3, node 1 uses $link_{1,3}$. Node 1 assumes that node 3 is prepared to accept its signal on the 1-to-3 half of $link_{1,3}$ and knows that this channel is not shared in any way. Node 1 can send its communication to node 3 at any time. Before you put this book down relieved that you need not study these switching chapters, let's consider some of the properties and limitations of this solution:

1. For N nodes, the system requires $N(N - 1)$ links. These links require $N(N - 1)$ transmitters and $N(N - 1)$ receivers. Thus, the number of link terminations

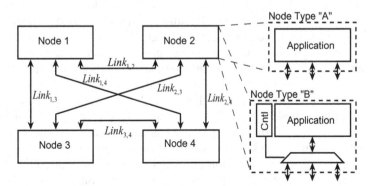

FIGURE 9.2   Completely connected network, with node types A and B.

is quadratic in the number of nodes. This cost is acceptable for small N but is absurd for large networks (e.g., where N is the number of computers on your campus). As we shall see, other solutions have much lower link termination costs than those of completely connected networks.

2. Should the system extend over an appreciable geographic extent (e.g., Europe), the total distance of communications channels becomes a significant cost. Where *length* is a function giving the length of each channel, the total cost is $\sum_{i=1}^{N-1}\sum_{j=i}^{N} length(link_{i,j})$. In general, this channel cost can be a stronger limit than the cost of terminations. There are applications with small N that remain prohibitive, due to excessive link costs. Again, we shall see that other solutions have much lower channel distance costs.

3. Figure 9.2 shows two alternatives for the internal structure of each of nodes A and B. With alternative A, the application embedded in the node must be constructed to emit and accept $N - 1$ times the bandwidth of a single link, because there is no restriction that prevents $N - 1$ other nodes from requiring work by node 1 at any time. Although this increases the capabilities of the overall system (by giving it $N - 1$ times the overall processing capacity), it is an expensive design. Also, if fewer than $N - 1$ other nodes are prepared to send signals to some particular node, a portion of that node's processing resources will be wasted.

4. The system illustrated in Figure 9.2 has $N = 4$ nodes. Each component node has been built to support $N - 1 = 3$ communications channels. Should the evolution of this system over time require $N' > N$ nodes, it is necessary to replace each node with a version that supports more communications channels. Replacement or field upgrade to $N' - 1$ is generally a disruptive and expensive alternative.

5. Alternative B in Figure 9.2 uses a multiplexer between the $N - 1$ links and the application. This allows the application implementation to handle only one duplex link of signal at a time (and hence be less expensive). Notice that the multiplexer is assumed to be a duplex device, to allow both directions of traffic to be selected (i.e., there are two devices, a unidirectional ingress multiplexer and a unidirectional egress demultiplexer, each handling one direction of the full duplex link). Additional control logic (Cntl) is required to select the active path through the multiplexer. This approach has the advantage that it is presumably easier to replace just the multiplexer component should we need to grow N to N'. However, this structure ensures that each node can use only one channel at a time. This means that only N of our $N(N - 1)$ links can be active at a time, and we are forced to idle the remaining $N(N - 2)$ links. With large N this is very wasteful. Alternative B has brought us to the idea of switching, albeit through the back door. The structure in Figure 9.2 with node alternative B can be viewed as a distributed switch, in which the N distributed parts of the switch (the duplex multiplexers and their controllers) are embedded with each application as a node. Distributed switching structures are discussed in more detail later.

Thus, we see that fully connected networks such as that shown in Figure 9.2 have serious limitations. To summarize: At least for nontrivial N and nontrivial geographic distributions, they have too many link terminations and too much channel distance; nodes must be built to some specific N, and this becomes a serious obsticle to growth; and they force overly expensive nodes and/or imbalances between node and network capabilities. Nevertheless, these fully connected networks are useful in some applications, for instance on the surface of a silicon die, where distances are small and only a few components need to be connected. Networks with high N or with large internode distances require some solution other than a fully connected network. Thus, we must conclude that it is necessary to look deeper into the issues of switching, to search for more appropriate network structures, so I will continue to write and you should continue to read.

Before we proceed to look at switching in more detail, let's consider the statement that departure from a completely connected mesh forces us to introduce a switch of some sort. Is this really the case? Consider any node A and its connections to the rest of a small network, consisting of nodes B, C, and D. If the network is not a full mesh, A cannot be connected directly to all of B, C, and D. Yet we assume that it is necessary for A to communicate with B, C, and/or D at varying times. There are two solutions: a distributed switching fabric embedded with the application nodes, or the introduction of a central switching unit.

Distributed switching fabrics work in the following manner: The set of application nodes (our A through D) are connected in the sense that it is possible to get from any node to any other node by making one or more hops across the node-to-node links. Information to be communicated is formed into individual packets. Logic is added to each node to recognize the destination of any packet. Then nodes originating packet traffic forward their packets on a link toward the intended destination node. Intermediate nodes along the intended path accept the packets, determine where they should be sent, and forward them. The destination node recognizes packets addressed to it and accepts them.

Distributed switching fabrics have received a fair amount of academic research but have failed to make much impact in the world of practical networks. One of the most important reasons for not using distributed switching fabrics or networks is that the communications capacity assigned to any particular node depends on its position within the network and the load generated by its various neighbors at varying distances through the network. The result is that either each node must have scalable communications resources, or all nodes must have some maximum communications resources. Although distributed switching networks have had little direct practical impact, you will recognize a strong similarity with IP packet routing strategies.

The generally accepted practical solution is to introduce some intermediate node(s) that allow A to pass signals varyingly to B, C, and/or D without being connected directly to all of B, C, and D. The standard obvious solution is to introduce one new switching node, S, to which A through D are connected, as

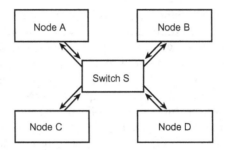

FIGURE 9.3   Switch as a distinct central element.

illustrated in Figure 9.3. S is a switch that allows any or all of the attached nodes to communicate with any of the attached nodes. Although this is the most straightforward switched network solution (i.e., to have one central switching element), we shall see later that there are many other solutions.

9.3   THREE TYPES OF SWITCHING

In Section 9.2 we discussed the need for switching components without much discussion of what was being switched or how the switching was accomplished. The discussion merely referred to the need to communicate signals from one node to another. As we learned earlier in the book, there are many different signals, or *protocols*. Interestingly, differing protocols generate quite different switching requirements and solutions. We begin this section with an examination of the effect of protocols on switching. We lump all switching into three groups based on the nature of the protocol being switched: switching of physical circuits, switching of time-division-multiplexed (TDM) signals, and switching of sequences of cells or packets. Of course, in real networks we shall find that a signal containing streams of packets may be switched as packets in one network element, as a single physical signal in another network element, and as a set of TDM signals in yet another network element.

Alert readers will realize that the switching of higher-layer protocols (e.g., the Internet protocol) is called *routing*. This terminology is correct; the router as a network element has a responsibility to find a useful route for each packet it is given, and is thus different from a switch. However, each router has an embedded switch that provides the varying connectivity required once a useful route has been determined. In this section of the text we do not consider the routing of higher-layer protocols but concentrate on switching issues.

9.3.1   Switching of Physical Circuits

The physical layer of a protocol is either an electrical or an optical signal. In some applications it is necessary to switch at this fundamental layer. Such switches typically have N ingress ports and N egress ports. Connections can be made from any ingress port to any one or more egress ports. In their pure form, such switches transmit the received analog signal to the egress links without even finding the clock inherent in the signal. However, in practice such switches commonly capture the ingress clock, capture each ingress bit, carry out some protocol logic, and emit a repeated but not literally identical signal with improved clock and transitions. Some switches that appear to be physical circuit switches also perform various protocol retiming functions, such as pointer processing in SONET/SDH.

At any time, a physical circuit switch will be asked to establish a certain set of connections, depending on the user's needs or some system's needs. The connection pattern requirements of a physical circuit switch of N ports can be expressed as a one-to-one function or as a one-to-many relation on the domain $\{1, \ldots, N\}$. Although there are a number of ways to specify functions and relations, the most convenient for this purpose is to generate a vector of N elements, with each element of the vector representing one of the egress ports of the switch. Each element is then filled with the ingress source that should be copied to that egress. For example, with $N = 4$, the vector [3, 1, 2, 4] specifies the following ingress-to-egress connections: $3 \rightarrow 1$ (i.e., port 3's input stream is switched or forwarded to become port 1's output), $1 \rightarrow 2$, $2 \rightarrow 3$, and $4 \rightarrow 4$. This simple concept will grow to become the "connection pages" of switches later in these chapters.

9.3.2 Switching of Time-Division-Multiplexed Signals

As we saw earlier in the book, some protocols divide the bandwidth provided by the underlying physical layer into separate time slots in some repeating cycle of allocated time slots. Useful switches for these TDM protocols are significantly more complex than circuit switches. If we have N attached nodes, each of which source and sink a TDM protocol with G time slots, the overall switching problem is to accept all NG incoming time slots, and to fill each of the NG outgoing time slots with any arbitrary choice of input time slots. In its simplest form, this switching problem is to implement any permutation of NG inputs to NG outputs. This permutation problem corresponds to a one-to-one mapping of input samples to output samples. There are $(NG)!$ such possible mappings. However, in practice we do not always want to limit ourselves to permutation mappings. A more common definition of the TDM switching problem is that any egress sample can be filled from any ingress sample without concern for whether the input sample has been used previously to fill some other egress sample. Thus, some input sample(s) could each be used to fill more than one egress sample. When this is the case, some other input sample(s) may be unused by any egress sample, or dropped. This problem has more possible mappings than the one-to-one permutation mapping: There are $(NG)^{NG}$ mappings. NG^{NG} is much greater than $(NG)!$ as NG grows.

TDM switching has three inherent components: capturing the ingress samples, holding the ingress samples in a memory, and communicating the ingress samples to the required outputs. Capturing the ingress sample means that logic in the receiver must discover the boundaries of bits, bytes, samples, and the repeating frame structure. This is necessary for the switch to sample the ingress grains properly. Memory is required because an ingress sample may appear in a time slot before it is assigned to be emitted on some egress port, so it is necessary that (in general) all ingress samples be stored in some memory. Captured ingress samples must be communicated from their input to their output in time to be emitted; this communication action may take place before or after the sample is captured in memory.

The most important application of TDM switching is in SONET/SDH. In these protocols, TDM switching applies at the broadband level (STS-1's or STM-0's), at the wideband level (virtual tributaries or TUs) and/or at the narrowband level (a single telephony connection, known as a *DS0*).

9.3.3 Switching of Cells and/or Packets

Many protocols divide the available bandwidth into cells of fixed length or packets of varying length. Recall that in TDM protocols each ingress sample position is switched to some fixed egress sample position in a long-term repeating pattern. Cell and packet protocols have no such semistatic repeating switching pattern. Instead, each cell or packet carries an address (or other routing instructions) that directs it to a particular output. The most critical difference from TDM protocols is that cell and packet protocols require that the switch route each cell or packet, depending on requirements found within the cell or packet. Thus, the fundamental switching mechanisms must be driven by the cells or packets themselves, not by a static control pattern derived from the control system. This difference is crucial: With cells and packets there is no assurance that the aggregate flow of cells or packets from the multiple inputs will not overwhelm some particular egress port(s) with more cells or packets than can be absorbed. Thus, traffic patterns are determined by the traffic itself, and there can easily be periods when the traffic pattern will overload "hotspot" egress ports. Such hotspots can exist for any duration, from roughly the time required to transmit a single cell or packet to any extended period of overload.

In Chapter 11, we shall see that this fundamental difference produces a much different approach to switch architecture than we will have seen in this chapter or in Chapter 10.

9.4 QUALITY OF SERVICE

In its broadest interpretation, *quality of service* (QoS) is a general catch phrase for a wide variety of measures that deal with the degree to which users of a communications system are pleased with the service they receive. This general concept is backed up by many different detailed concepts and measures, depending on the nature of the service being provided and the nature of the switching elements in the signal's path through a network. In this section we provide a general introduction to some of the component concepts behind the term QoS.

1. *Call rejection.* Consider a customer wishing to establish a new service or signal between one or more nodes. In general, it is possible that a switch or network will reject such a request due either to a clear lack of resources to support the request or to a judgment by traffic control portions of the network that adding the new service will compromise existing services and/or the new service. The frequency or probability of such denial events is an important QoS measure for some switches and networks. Call rejections can occur at all three types of service: denial of an attempt to establish a physical connection, denial of an attempt to establish a TDM connection, and denial of an attempt to establish a new cell or packet connection. These rejections can occur due to a number of distinct causes: (a) the lack of capacity on a network link, (b) the inability of a switch to connect an idle input signal to an otherwise idle output signal, (c) the lack of sufficient buffers to support a new cell or packet flow, or (d) the judgment by a traffic control component that adding a new cell or packet flow will cause QoS assurances for existing calls and/or the new call to be violated.

2. *Lost information.* User signals are always subject to some risk of loss. Even at the physical signal and TDM levels, it is possible for established connections to fail due to hardware or operational errors, and for customer information to be lost while the network switches to protection paths to restore the service. Lost information is more common in cell or packet services, where bursts of user traffic overwhelm links, buffers, or switches and must be dropped due to a lack of alternatives or due to an established policy for a particular class of service. Lost signals are a network issue but commonly originate in switching equipment, so must be considered in these chapters on switching.

3. *Latency.* The time a customer's signal takes to transit a network or a switch is often an important issue. The issue is obvious in telephony networks, in which latency is apparent to customers in the form of pauses in conversations, but many applications have important latency constraints. In general, networks and their switches must offer some quantifiable latency assurances. In physical circuit switches these latencies may be very small. In TDM switches, we shall see that it is necessary to delay signals by at least a TDM sample repeat group. Such switches attempt to minimize additional latencies. Cell and/or packet switches have more complex latency issues, as traffic may be delayed in buffers for varying periods, depending on the class (priority) of traffic and the current network and switch load conditions. In general, minimizing latency will be an important objective for all the types of switches that we examine.

4. *Jitter.* Jitter occurs at several layers. We can consider the jitter of clocks and bit edges in physical links. Protocol standards for signals at these layers (e.g., SONET/SDH or the physical layer of fiber channel) establish clear boundaries on signal jitter. However, we can see jitter phenomena at the cell and/or packet layer, where a stream of cells or packets may suffer movements in time that cause the cells or packets to cluster more closely together or spread farther apart in time as they progress through a network. Jitter at this level is difficult to manage at the switching level. More commonly, cell and/or packet jitter is addressed by traffic management components that serve as port interfaces on some of the more sophisticated cell and/or packet switches.

9.5 SPECIAL SERVICES

The basic service provided by a switch or a network is a unicast service, in which one transmitting node emits a single signal that is carried by the network or switch to a single receiving node. The normal service provided is a duplex pair of these unicast services, forming a full duplex path between the two nodes. However, three common forms of special services are commonly offered by networks and their switches:

1. *Broadcast.* Broadcast is a special service in which a switch with N duplex ports accepts a signal from one ingress and replicates that signal to the other $N - 1$ ports (or possibly to all N egresses). Broadcast can occur at the physical signal, the TDM, or the cell and packet levels. In addition, some protocols have special control symbols that require broadcast (e.g., serial RIO). Broadcast is generally straightforward to build in physical signal and TDM switches but is more challenging in cell and packet switches.

2. *Multicast.* Multicast is a special service in which a switch with N duplex ports accepts a signal from one ingress and replicates that signal to a specified subset of the other $N - 1$ nodes (or the subset may possibly include the node of origin). Each such pattern of a single ingress to multiple egresses is called a *multicast tree.* Each ingress may have multiple multicast trees in service at any time (at least for TDM and for cell and packet switches), so any overall switch may have many multicast connections in progress at any time. Multicast can occur at the physical signal, the TDM, or the cell and packet levels. Although multicast typically uses less bandwidth than broadcast, it is considerably more difficult to implement in cell and packet switches, for two reasons: (a) more sophisticated control is required to tailor the egress signals to the desired multicast "footprint," and (b) when many signals are being multicast at the same time, complex patterns of loading at egress ports can occur, complicating QoS assurances. Multicast is less challenging in physical circuit and TDM switches, but issues remain regarding the discovery of open paths in these switches.

3. *Bicast.* Bicast is a restricted form of multicast in which a signal is emitted twice. Bicast is an important special case because it is used in many traffic protection schemes. Each stream of customer traffic is replicated and sent on two independent transmission paths. The receiving node receives both streams. One, referred to as the *working stream*, is normally taken as the correct version of the customer's traffic. The other is referred to as the *protect* (or *protection*) *stream*. When the working stream is damaged (by bit errors) and the receiver detects this damage by internal consistency checks on the stream, the receiver switches to accept the protection stream instead, assuming that it is better than the working stream and better than some minimum quality level. As we shall see, a requirement to support bicasting has significant implications for the design of physical circuit switches.

9.6 SWITCHING IN ONE OR MORE STAGES

For reasons that will emerge as we progress through these switching chapters, the ideal switch has a single-stage of switching. For present purposes we represent a single-stage switch as a single box, as in Figure 9.4. Notice in the figure that the duplex ports are separated into two parts: ingress and egress. This diagram style will be useful as we proceed. But keep in mind that the ingress and egress portions of a port are almost always built together as a single unit. Perhaps the most fundamental requirement of any general switch is that any ingress can reach any egress. Assuming that the box labeled "single-stage switch" in Figure 9.4 can make any connection, it is clear that the ingress nodes can connect in some fashion to any egress node.

Single-stage switching architectures have tremendous advantages on many issues that are discussed in subsequent chapters. However, available technology limits our ability to build arbitrarily large single-stage switches, and we are sometimes forced to make use of various multistage switching architectures. Figure 9.5 illustrates a representative multistage switch architecture. Multistage switches such as the three-stage switch illustrated in the figure also

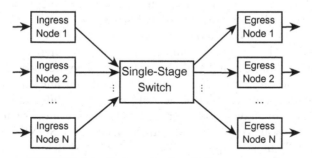

FIGURE 9.4   Representative single-stage switch.

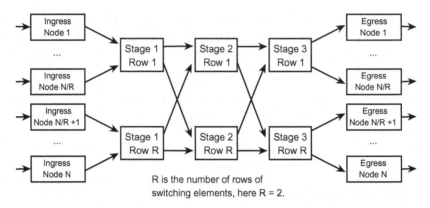

FIGURE 9.5 Representative multistage switch (a three-stage switch with two rows).

should permit any input to be connected to any output. Assuming that each of the six (2 × 3) switch elements can connect any input to any output, it is clear that any (single) input can be connected to any set of outputs. Although this is a good start, we shall discover blocking issues within such switch fabrics as we study them in detail.

9.7   COST MODEL FOR SWITCH IMPLEMENTATIONS

As we progress further into the design of switches, we become more and more interested in the achievement of switching capabilities at low cost. To have concrete means for the comparison of alternative switch architectures, we develop an approximate cost model for each switch architecture. We make the assumption that all switches are to be implemented in CMOS VLSI. This assumption is fair because (1) CMOS VLSI is the dominant implementation technology for most digital products at this time, and (2) most modern switches are implemented in CMOS VLSI.

In any switch implementation, there are some unavoidable costs that do not vary with the choice of switch architecture. An example is link and protocol termination in the ports of the switch. For instance, a SONET/SDH switch must terminate the SONET/SDH signal as it arrives at the switch and relaunch a valid SONET/SDH signal leaving the switch. This involves a variety of requirements, including framing, error checking, and pointer processing. Implementation of these requirements requires a significant amount of VLSI real estate. However, from the point of view of switching architecture, such blocks or megacells are an unavoidable overhead and not part of the switching architecture. Another way to say this is that the costs of these SONET/SDH termination megacells must be paid regardless of what switching architecture is used. We drop all such protocol termination costs in our costing model, because

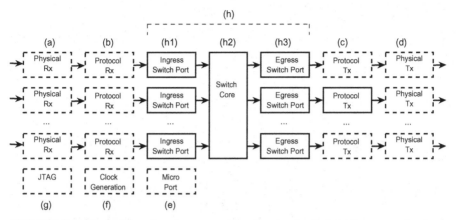

FIGURE 9.6 Typical functional components of a simple digital telecommunications switch.

the first role of this costing model is to compare the effectiveness of a variety of switch architectures. Similarly, any VLSI chip has various overheads, such as a microprocessor port for control, and test features for manufacturability. We ignore all such components that do not vary with the switch architecture itself or which are too deeply ingrained in the switching logic to separate (e.g., is the circuits used to generate clocks that control a large switching structure). Of course, to arrive at realistic VLSI cost estimates from these switching architecture models, it is necessary to add back in all these fixed costs that we have proposed be ignored. Nevertheless, this focused model will help us better understand the pure switching issues.

Figure 9.6 presents a very high level view of a typical switching VLSI product. The components represented include (a) receiver physical layer ports, (b) receiver protocol blocks, (c) transmitter protocol blocks, (d) transmitter physical layer ports, (e) microprocessor access and control blocks, (f) clock generation blocks, (g) test access blocks (e.g., JTAG), and (h) internal switching mechanisms, which are our primary focus in these chapters. These mechanisms of interest typically divide into (h1) ingress switch port blocks, (h2) core switching blocks, and (h3) egress switch port blocks. The overall traffic flow in Figure 9.6 is from left to right.

9.8 CROSSBAR SWITCH CONCEPT

Recall that physical circuit switches allow each egress port to select and emit the physical layer signal arriving at any of the ingress ports. Figure 9.7 illustrates the standard abstract view of such switches. In the figure, N ingress ports on the left of the diagram accept incoming physical layer signals and propagate these signals on the horizontal channels toward the right of the figure. Egress

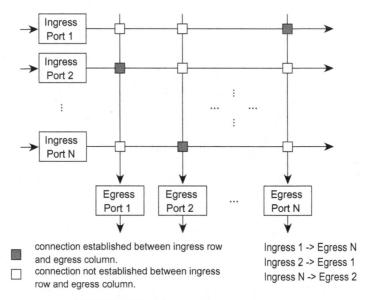

FIGURE 9.7   Abstract view of switching.

ports are arranged in a row at the bottom of the figure. Each egress port has an input channel rising above it. The horizontal ingress channels and the vertical egress channels are conceptual "bars" that "cross" in a regular Cartesian pattern. Each intersection of ingress and egress channels represents an opportunity to make a connection between the ingress channel and the egress channel. Each egress channel must be connected to one (or zero) ingress channels. Thus, each egress channel and its connected egress port carry or copy the signal (or no signal) from one selected ingress port.

The first implementations of this class of switch were called *crossbar switches*. This name has stuck as the general term for this class of switches, and we use it here. The crossbar concept is quite simple, but we shall see it grow in flexibility and features as we progress through the chapters on switching. A crossbar switch allows each egress port to select any ingress as its source. Clearly, this permits any permutation of inputs to outputs. Recall that a permutation is a one-to-one mapping in which the resulting vector contains all the elements of the argument vector, but in some arbitrary order. For $N = 4$, the ingress (argument) vector is $[1,2,3,4]$; the egress (result) vector is any rearrangement of the ingress vector.

An example egress vector is $[3,1,2,4]$; it implies that egress 1 (the first position in the vector) is connected in ingress 3 (the value in the first position in the vector), that egress 2 is connected to ingress 1, and so on. Other permutations selected are $[4,2,1,3]$ and $[1,2,3,4]$. Any egress port may not select an ingress to connect to, or copy. We represent this formally by extending the set of ingress port symbols to include ϕ, which represents the condition of no input being selected: $\{1,2,\ldots,N,\phi\}$. Thus, egress selection vectors include such instances as $[3,\phi,1,2]$ and $[\phi,\phi,\phi,\phi]$. Often in practice, when an egress selects ϕ, the protocol and the switch have some null signal value to emit, as opposed to leaving that output floating.

An immediate consequence of the crossbar definition (each egress selecting its own ingress) is that the crossbar is capable of multicast and broadcast. *Broadcast* means that every egress emits a copy of the identical signal. An $N = 4$ crossbar in broadcast mode, with ingress 1 as the source to be replicated on each output, has an output vector of $[1,1,1,1]$. *Multicast* means that an ingress is copied to some subset of the egress ports. An $N = 8$ crossbar with two multicasts in progress (one from ingress 3, the second from ingress 5) and with several unicast connections in progress is described by the egress vector $[3,2,5,\phi,5,1,3]$. Figure 9.8 illustrates an $N = 4$ crossbar with only unicast traffic ($[2,4,3,1]$) and the same crossbar with one multicast and two unicasts ($[2,4,2,1]$).

The crossbars we have examined so far are symmetric in the sense that they have equal numbers of ingress and egress ports. That symmetry can be broken by separating N into N_i and N_e for the number of ingress and the number of egress ports, respectively. This gives us the ability to increase the number of paths that a set of signals can take by having $N_e > N_i$ [Figure 9.9(a)] or to reduce the number of paths a set of signals can take by having $N_i > N_e$ [Figure 9.9(b)]. The $N_e > N_i$ case is described as a crossbar with *speedup*; the $N_i > N_e$ case is described as a crossbar with *slowdown*. The degree of speedup or slowdown is the ratio N_e/N_i. Symmetric crossbars have a speedup of 1. Crossbars with speedup and slowdown will be useful in several ways in the sections and chapters to come.

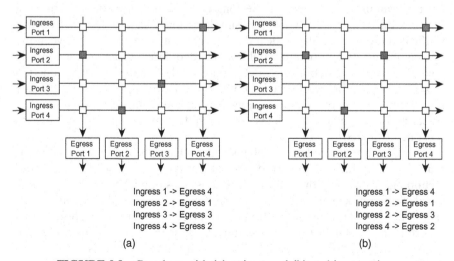

Ingress 1 -> Egress 4
Ingress 2 -> Egress 1
Ingress 3 -> Egress 3
Ingress 4 -> Egress 2

(a)

Ingress 1 -> Egress 4
Ingress 2 -> Egress 1
Ingress 2 -> Egress 3
Ingress 4 -> Egress 2

(b)

FIGURE 9.8 Crossbars with (a) unicast and (b) multicast settings.

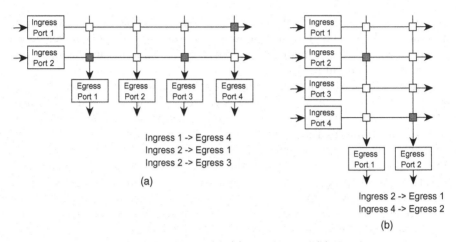

FIGURE 9.9 Crossbars with (a) speedup and (b) slowdown.

Thus far in our model of crossbars, we assume that the ingress and egress "channels" carry the complete signal passing from an ingress port to an egress port, but we have not discussed how this signal is carried. For the moment we take the naive view that these channels carry an exact analog copy of the signal that arrives at the ingress port and that exactly that same signal departs on the egress port. Of course, many optical, electrical, and practical issues will cause us to abandon this assumption in many applications as we proceed.

9.9 OPTICAL CROSSBAR SWITCHES

Photonic methods are used for the long-distance transmission of almost all network traffic. It would be a lovely simplification if all switching of network traffic could remain in the optical domain, as many conversions between the optical and electrical domains could be avoided. Much research has recently gone into the design of optical switches that could fulfill the switching function in such all-optical networks. Some initial optical switch products are already in service in commercial systems. However, the area remains predominantly a research area, as many problems remain and many alternative approaches must be evaluated.

There are a few standard problems with which optical switching solutions must deal and which are in part responsible for the limited use of these optical technologies. Probably the most important issue is optical power control. In a standard network, with optical transmission and electrical switching, each signal received at a switching station is converted to a digital electronic system before being switched. This process fully restores the signal in the sense that a firm decision between 0 and 1 is made, any degeneration is removed by forcing the signal to a pure 0 or 1, and the output power level of the signal is restored to optimal levels when the digital signal is reemitted as an optical signal. Thus, signals are cleaned up fully at each switch-to-switch hop.

There is no convenient (albeit expensive) opportunity of O-E-O conversion to restore signals in an all-optical network. Thus, signal quality degrades with distance. In addition, the power levels of incoming optical signals may vary from channel to channel, depending on the distance since these signals were last restored (by optical amplification or by a full O-E-O). Thus, signals emitted also vary in their power levels. In a general network, as optical signals are routed over arbitrary paths, it becomes impractical to effectively manage the power levels of all the optical signals. Forced restoration at each switching node avoids these problems.

There are a number of interesting proposals for all-optical switches. We present briefly only the simplest and most straightforward solution, leaving other proposals for your further study. MEMS (microelectrical mechanical system) devices provide the basis for the most straightforward optical switches. In these devices, small mirrors lie on the surface of a silicon die or wafer. Each mirror can be pushed into a vertical position by the MEMS mechanisms. The simplest MEMS optical crossbar switch is shown in Figure 9.10. Optical signals are constrained to cross the array of mirrors from left to right, across the rows. In each row, the leftmost mirror that is standing (as opposed to lying flat) will reflect the optical signal through a right-angle turn to be emitted in that column. Thus, with one mirror standing per row and one per column, the device acts as a permutation (unicast) switch, with each ingress signal row being sent to a distinct egress signal column. Switch configuration changes are made by raising and lowering mirrors.

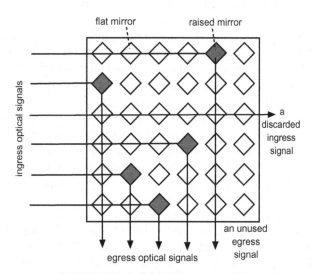

FIGURE 9.10 MEMS optical switch.

A significant limitation of the MEMS optical switch architecture shown in Figure 9.10 is that it is not capable of any form of multicast or broadcast; the mirrors support only unicast operation. The most immediate consequence of this is that entirely different mechanisms would have to be devised for network protection, as the basic bicast opertion is not available. The bicast problem is quite difficult, as bicasting would divide the optical power in two halves. Thus, some form of amplification is clearly required. If an optical fiber contains multiple signals (colors), and if it is desired to switch these components separately, it is necessary to pass the signal through a prism to separate the colors before entering the optical MEMS switch and to recombine the colors as desired after passing through the optical switch. Optionally, it is possible to switch all colors as a unit in one row to one column.

At least conceptually, MEMS and other optical techniques produce attractive solutions. Some physical implementation aspects of optical switching devices were already discussed in Chapter 3. Much work remains to determine how widespread they may become in practical networks. However, as we examine TDM and cell and packet switches, we will discover that their requirements place them further and further from solution by all-optical methods.

9.10   DIGITAL ELECTRONIC CROSSBAR SWITCHES

Although MEMS switches are highly attractive for low-layer optical networks and will probably come to play an important role in the future, there is no doubt that crossbar switches implemented in some form of digital electronics now dominate the crossbar product space. There are a number of reasons for this dominance of digital electronic implementations:

1. The appropriate technologies are mature and economical. CMOS is the most common implementation technology for most digital crossbars, due to its widespread economical and technological dominance of the industry. There are applications for which the greater speed of more exotic technologies such as GaAs are appropriate, but they are the exception.

2. There is often a need to improve the quality of a signal before retransmission on the egress side of a switch. Required improvements can include removing jitter from a received clock, sharpening bit transitions, correcting errors, adding control information, and boosting power levels. All of these improvements are practical today only through digital implementation.

When a digital electronic implementation is used, there is commonly an issue regarding the relative rates of the signals passing through the crossbar and the maximum rate at which signals can reasonably be transmitted across the silicon dies implementing the switch. With common protocols and signaling rates such as SONET's OC-48 (in which bits are emitted and received at 2.48832 Gb/s) and common implementation techniques such as CMOS, it is not possible or sensible to transmit the full signal rate in a single CMOS signal. Instead, it is necessary to delay the incoming signal and capture some number of bits to be sent in parallel between the ingress and egress ports. This approach permits reasonable CMOS clock rates. For example, an OC-48 signal might be sent across a switch's CMOS die with $b = 8$ bits in parallel, with a clock rate of 2.48832 Gb/s ÷ 8 = 311.04 Mb/s, thereby allowing the use of a 311.04-MHz clock in CMOS. The alternatives are to work harder in the physical design of the CMOS part to achieve full OC-48 line speed with $b = 1$, to reduce the rate of the signal being switched (e.g., a 100-Mb/s Ethernet signal) or to use a more aggressive implementation technology (e.g., GaAs).

Figure 9.11 illustrates our first view of the implementation of a digital crossbar. The arrangement of ingress and egress ports remains as before. The ingress channels and the descending egress channels are taken to be metal wires. Thus, each ingress broadcasts its signal to all egresses. Each egress has a selection multiplexer, which allows it to accept whichever of the ingress signals is required. This is the simplest instance of a general pattern of "broadcast by ingresses; selection by egresses."

Figure 9.11 does not include the necessary control components which deliver addresses of required ingress ports to each egress multiplexer. This logic is straightforward and is derived immediately from the egress selection vectors discussed in an earlier section. Figure 9.12 adds this control feature. The lower left portion of the figure represents a vector of registers, each of at least $\log_2 N_i$ bits. Each register drives the control inputs of one of the N egress selection multiplexers. Clearly, this is a direct implementation of the selection vectors discussed earlier in the chapter. Where it is important that a null input (ϕ) be selectable at each egress, $\log_2(N_i + 1)$-bit registers are required in the vector, and the multiplexers must accept an $(N + 1)$th input of ϕ.

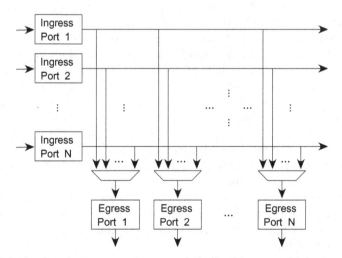

FIGURE 9.11 Crossbars are most commonly built with one multiplexer per egress.

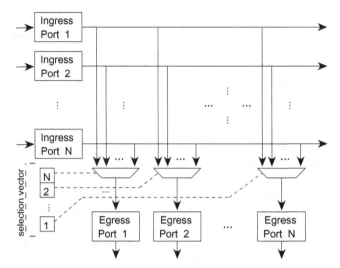

FIGURE 9.12   Digital crossbar with an egress selection vector.

The abstract digital crossbars shown in Figures 9.11 and 9.12 assume that each ingress port drives a single channel that is broadcast to all N egresses. In general, this is not practical. The capacitive load seen by a single ingress port (by the final gate or buffer driving a metal trace that fans out to all N egresses) is generally far too high to allow the crossbar to operate at an acceptable speed. The electronic speed of this broadcast function can be improved dramatically by the use of fanout trees of gates or buffers, which spread and repeat the ingress signal to all N egresses. Figure 9.13 presents an abstract representation of this need. In the figure, fanout trees are represented by triangular symbols. They are driven by each ingress signal, and they emit N copies of these signals, one to each egress. As we shall see in later designs, the cost of fanout trees can be significant and cannot be ignored.

Figure 9.13 remains an abstract representation of the switching logic of digital crossbar. An objective of this book is to develop a useful model of the implementation costs of each type of switching solution studied. To permit the development of this cost model, and to enhance our understanding of switch implementations, we need to delve into how digital crossbars are implemented at the logic gate level. Figure 9.14 illustrates this next level of detail. This figure retains the ingress fanout tree and the egress multiplexer, but expands both into networks of standard gates. We examine both of these structures in turn. The ingress fanout trees provide a distributed amplifier to allow one ingress signal to drive N egress ports. This is a standard, elementary problem in digital circuit design. The electrical size (i.e., the resistance of their output stages in CMOS) of the amplifying buffer cells and the number of outputs (downstream buffers or egress port multiplexers) is determined by a simple analysis in digital design courses.

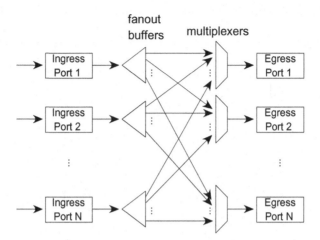

FIGURE 9.13 Crossbars require fanout buffers.

For the purposes of modeling costs of switch architectures, we require a good approximation but do not wish to be highly dependent on particular technologies. The size of each fanout buffer and the number of nodes driven by each fanout buffer is a level of detail best avoided in our model. Consequently, we assume that one "small" buffer is used to drive two loads, and that the fanout trees required by our digital crossbars are simple binary expansion trees (e.g., the root of the tree is one node, and each node of the tree drives two loads, such that the numbers of nodes per layer of the tree are in the sequence $1,2,4,8,\ldots,N$). When N is some power of 2 ($N = 2^D$ for some integer D), the tree is balanced (i.e., all root to terminal leaf paths are of equal length). Such a tree has $\sum_{i=1}^{D} 2^D = N - 1$ nodes, which we approximate as N. In Figure 9.14 each ingress port has a tree of depth $D = 2$, with $N = \sum_{i=1}^{D} 2^D - 1 = 3$ nodes. The N outputs at the base of the fanout tree drive the N egress ports of the crossbar.

At the egress side of these digital crossbar switches we find an $N{:}1$ multiplexer. Much as digital designers prefer to avoid packing a $1{:}N$ fanout in one large buffer, it is standard practice to implement $N{:}1$ multiplexers as fanin trees of smaller multiplexers. In fact, most digital design is done based on standard gate libraries, and it is rare for such libraries to include a multiplexer greater than 8:1. The optimal fanin ratio in multiplexer trees is dependent on the underlying technology. Again, we prefer to avoid such detailed technology dependencies in our efforts to model switch fabrics. Consequently, we assume only a 2:1 multiplexer, and build larger multiplexer fanin trees from these small elements. Although this design is typically not optimal with respect to time, it is not far off the optimal implementation size in a VLSI circuit area. It is a good approximation for our purposes. We assume that the 2:1 multiplexer is composed of two AND-2 gates and one OR-2 gate (or DeMorgan

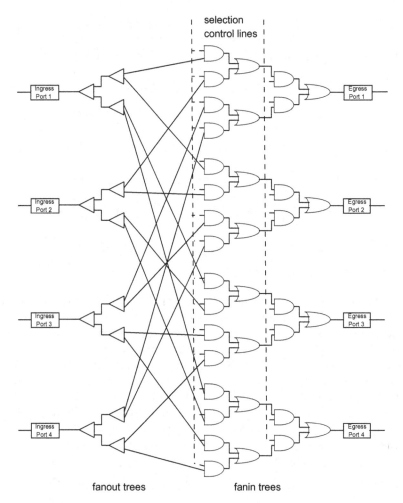

FIGURE 9.14 Crossbars as interlaced fanout and fanin trees.

equivalents), as shown in Figure 9.14. Each fanin tree is of depth D and contains $N = \sum_{i=1}^{D} 2^D - 1$ two-input multiplexers. In an asymmetric crossbar, with $N_i \neq N_e$, it is also necessary to break the symmetry between the D values in the equations for the two trees. In the ingress broadcast fanout tree, D must become $D_i = \log_2 N_e$; in the egress multiplexer fanin tree, D must become $D_e = \log_2 N_i$.

9.10.1 Control of Digital Crossbar Switches

As we have discussed, a crossbar switch must have sufficient control to select which of the ingress ports it is to copy at any time. This leads to the selection vector shown in Figure 9.12. It is natural to use binary-encodings of these choices, because these control values immediately drive multiplexers, which naturally implement decoding of binary-encoded values. Looking at any egress fanin tree, it is clear that each bank of two-input multiplexers requires one bit of control, and that a total of $\log_2 N_i$ such bits are required. It is conventional to assign the highest-value bit to the rightmost (closest to the egress root of the fanin tree) and to proceed assigning successively lower-valued bits to successively leftward banks of multiplexers.

If these control bits were semistatic, changing only infrequently and tolerating relatively long periods to complete a change, it would be appropriate simply to wire each control bit to the multiplexer(s) it controls. However, it is commonly necessary for these fanin trees to be switched within the time required to transmit a single bit on the fanned-in data path. This requirement places the propagation of control information from the registers containing the values to the trees of multiplexers directly in the critical path of the device. Consequently, one generally cannot ignore the need to provide buffering fanout trees to drive the multiple multiplexers of each tree. For the same motivations, we use a fanout tree equivalent to the ingress data path fanout trees described above (binary fanout trees).

It is slightly more complicated to account for the costs of these control fanout trees, because each bit requires a differing degree of fanout. Since each two-input multiplexer requires control inputs (control and control bar), we assume the use of a final buffer that both carries out the fanout amplification and produces both forms of the control value. Thus, even the highest-order bit requires one fanout cell. From this starting point, we can conclude that there must be $D = \log_2 N_i$ control bits per egress, that these control bits (in sequence from high order to low order) require fanout trees of depth $[1, 2, 4, \ldots D]$, that these trees have buffer counts of $[2^1 - 1, 2^2 - 1, 2^3 - 1, \ldots, 2^D - 1]$, and that the total buffer count of each tree is $2^{D+1} - D - 2$, which can be approximated conservatively by 2^{D+1}.

Figure 9.15 shows a crossbar with dual-output selection control pages. Each output has two registers which can drive the output selection multiplexer. A single fabric-wide bit determines which of these two selection registers is active for each page. At any time one set of registers is "active" and makes an effective selection of the outputs. The other set of "standby" registers can be programmed with some new set of connections. When the fabric-wide control page bit is flipped, the active and standby pages switch their roles. In synchronous crossbars it is necessary that the switchover occur during a single bit time, such that the inputs change from one bit to the next. This is commonly the critical timing path in a VLSI implementation of a crossbar. Other crossbars, and other applications, do not impose this strict requirement (either the signals are asynchronous or some control protocol allows any bits scrambled during a page switch to be ignored).

9.10.2 Cost Model for Digital Electronic Crossbar Switches

A major theme of the switching logic chapters of this book is to develop simple models for the cost of the various types of switch discussed. We begin this

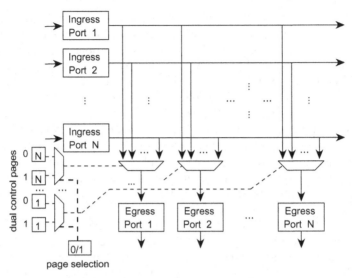

FIGURE 9.15 Dual-output control pages.

theme by providing equations for the costs of physical signal crossbar switches. Keep in mind that these equations are useful only as rough guidelines; any implementation will have many constraints and/or features that cause the implementation costs to vary significantly from this simple model. In particular, the model equations assume that the fanin and fanout trees are binary. As this is unlikely to be the case in a practical switch, these key costs will vary somewhat from this model. Also, and critically important for these simple physical circuit crossbar switches, the cost of the ingress and egress ports will dominate the cost of the crosspoint matrix. Yet this model considers only the crosspoint matrix and ignores the port costs. Thus, before using this model in practice, it is necessary to add implementation costs for the ports, the microprocessor interface, any built-in self-tests, and clock generation and distribution. Thus, the value of this model for physical circuit crossbars is clearly very limited, yet it is the beginning of a development that will become quite important and relevant when we study TDM switches in Chapter 10 and cell and packet switches in Chapter 11.

The overall model produces a result in terms of the number of primitive logic gates required for the switch fabric. A *primitive logic gate* is roughly equivalent to a 2-AND or 2-OR gate. The datapaths of the crossbar are given in the following equation, which assumes that the crosspoint is built of N binary fanout trees coupled with N binary multiplexer fanin trees. N_i is the number of ingress ports, N_o is the number of egress ports N_o/N_i is the speedup of the device, although commonly $N_i = N_o$), and b is the number of bits carried in parallel through the crossbar (e.g., $b = 8$).

$$xbar(N_i, N_o, b) = N_o b(N_i - 1) \times mux(2) + N_i b(N_o - 1) \times buffer(2)$$

The costs of the assumed primitive components (a multiplexer built of two 2-ANDs, one 2-OR, and a simple amplifying buffer) are given by the following two equations:

$$mux(2) = 2 \times and(2) + or(2)$$
$$and(2) = or(2) = buffer(2) = 1$$

The control registers that drive the output multiplexers and the fanout trees required to drive all the multiplexer control inputs are given by the equation

$$xbarCntl(N_i, N_o, b) = 2N_o \times \log_2 N_i \times registerBit + 2N_o 2^{\log_2(bN_i)+1} \times buffer(\text{?}$$

The cost of a *registerBit* varies greatly with CMOS libraries, but *registerBit* = 16 is a good estimate. The data path and control cost for a crossbar are thus

$$ClosCost(N_i, N_o, b) = xbar(N_i, N_o, b) + xbarCntl(N_i, N_o, b)$$

9.10.3 Growth Limits of Digital Electronic Crossbar Switches

In Section 9.10.2 we developed an abstract cost accounting for digital crossbar switches. One might expect the costs reflected by that model to limit the size of practical digital crossbar switches as N grows. This is generally not the case, however, and certainly is not the case where high-speed differential signaling physical layer protocols such as CML or VLDS are used. Recall that complete switches require implementations of ingress and egress ports. Depending on the type of design (whether it is reclocking and/or retiming, and how complex the protocol may be), the port implementations vary in size. But with current technology it is certain that the port implements will far exceed the crossbar switch in area on a CMOS die. In other words, we run out of space to implement the ingress and egress ports before the cost of the switching core becomes too high. This relaxed cost of switching will change dramatically when we examine other types of switches in later chapters.

9.10.4 Commercial Examples of Electronic Crossbar Switches

Vitesse Semiconductor Corporation offers a family of CMOS crosspoint (crossbar) switch devices. Overall, this family of devices represents the state of the art in this type of switch. In this section we introduce a few members of this product line to give the reader some understanding of the state of the art in crossbar switches. Vitesse's devices range from the 4×4 (four inputs and four outputs) VSC3104 device to the 144×144 VSC3140 and VSC3144 devices. The first observation to be made is the width of the range of this family: In terms of input and outputs, the family covers a range of 144/4 = 36; in terms of the fundamental gate cost of the necessary crossbars on these dies, the range is $144^2/4^2 = 1296$. Nevertheless, we can assume that both dies are dominated by the areas required for their input and output ports, as even the crossbar for the 144 × 144 devices comprises a quite modest number of CMOS gates.

All of these devices are strictly nonblocking. When used as single-stage switches, this nonblocking property is assumed by the overall switch; when used in Clos-like networks, the blocking status of the overall fabric depends on the design of the Clos and the traffic model (unicast, bicast, multicast). Switch settings are determined by programming a pair of registers associated with each output. At any time, one set of these registers controls the switch and determines which input signal each output will replicate. The second set of registers is used to program the next set of inputs per output; all outputs can be switched from one register set to the other register set at once, allowing nearly instantaneous change from one connection pattern to another. These (and other) switch settings are controllable by a flexible microprocessor interface.

Most of these switches are asynchronous, which means that each ingress/egress stream can have an independent bit rate and phase. Some devices include third-generation input signal equalization (ISE) Eye Opener technology. The purpose of this circuit is to improve the quality of the signal received by "opening the eye", as discussed elsewhere in the book. Additionally, the output level of each transmitter can be programmed independently.

Other features include a built-in self-test capability and control software. A representative power consumption is 16W by the VSC3144 (144 × 144 asynchronous crosspoint switch). Notice that this is 111 mW per channel.

Any of the devices in Vitesse's crosspoint switch family can be used to build the single- or multistage fabrics described in this chapter. The Vitesse crosspoint switches have no knowledge of the protocol being carried. This frees them to be used with a great variety of protocols (e.g., from SONET, to storage network protocols, to Ethernet) but requires that any protocol checking or processing be offloaded onto other devices in any system.

9.11 MULTISTAGE CROSSBAR-BASED SWITCHES

When the needs of an application for physical circuit switching exceed the capacity of the largest feasible single switch element (e.g., larger than the 144 × 144 devices from Vitesse), it becomes necessary to consider the use of multistage networks. There are many forms of multistage networks, but in the domain of physical circuit switching, the networks defined by Clos in 1952 remain the standard. In this section we begin our examination of Clos's multistage networks.

Figure 9.16 illustrates Clos's multistage network definition. There are three stages: stage 1 is known as the *ingress stage*, stage 2 as the *middle stage*, and

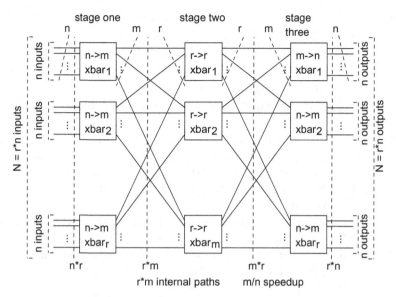

FIGURE 9.16 Clos network structure.

stage 3 as the *egress stage*. Each stage consists of some number of crossbar elements. In a nutshell, each stage 1 crossbar spreads a portion of the overall ingress load over all the elements of stage 2, each stage 3 crossbar collects a portion of the overall traffic from the stage 2 devices, and the stage 2 devices make the required connection from the stage 1 devices to the stage 3 devices. Now let's expand on these roles.

1. Each stage 1 device accepts the traffic from n sources and redistributes this traffic to m egress links, which are connected to stage 2 devices. These internal connections are implemented by an internal crossbar switch.

2. Each stage 3 device accepts the traffic from m stage 2 devices, rearranges it as required using an internal crossbar switch, and delivers the rearranged traffic to the n connected egress ports.

3. Each stage 2 device has a link from each stage 1 device and to each stage 3 device. Thus, each stage 2 device can establish connections from stage 1 to stage 3 thus completing the connection patterns required.

The integers n, m, and r determine the overall size of the Clos network.

4. There are r stage 1 devices. As each accepts n links, stage 1 as a whole accepts $N = rn$ ingress links.

5. There are r stage 3 devices. As each emits n links, stage 3 as a whole emits $N = rn$ egress links.
6. Thus, the overall fabric accepts and emits $N = rn$ links.
7. Each of the r stage 1 devices emits m links, so stage 1 as a whole emits rm egress links to stage 2.
8. Each of the r stage 3 devices accepts m links, so rm egress links connect stage 2 to stage 3.
9. Since each stage 1 device has m egress links, and there are exactly m stage 2 devices, each stage 1 device is connected to each stage 2 device with one link; since each stage 2 device has r ingress links and there are exactly r stage 1 devices, each stage 2 device is connected to each stage 1 device with one link.
10. Since each stage 3 device has m ingress links, and there are exactly m stage 2 devices, each stage 3 device is connected to each stage 2 device with one link; since each stage 2 device has r ingress links and there are exactly r stage 3 devices, each stage 2 device is connected to each stage 3 device with one link.

The overall pattern allows the signal from any ingress link to be routed via its stage 1 crossbar to some stage 2 crossbar, for forwarding by the stage 2 crossbar to the required stage 3 crossbar, for connection to the required egress link.

11. Each ingress signal must pass through a specific stage 1 crossbar.
12. Each egress signal must pass through a specific stage 3 crossbar.
13. The Clos structure allows any stage 1 to connect to any stage 3 device though any of the stage 2 devices, assuming no conflicting traffic.
14. Establishing connections in a Clos network is a matter of finding a stage 2 device that has a free path (link) both from the required stage 1 device and to the required stage 3 device.

Any Clos network is described by three numbers n, m, and r. The form $C(n, m, r)$ is used to represent any Clos network. Figure 9.16 represents the general concept of $C(n, m, r)$. Figure 9.17 represents a particular Clos network, $C(2, 3, 2)$, with $n = 2$, $m = 3$, and $r = 2$. Something interesting is happening in Figure 9.17: $m > n$. When $m > n$, the Clos network exhibits speedup. This means that there are more paths between stage 1 and stage 2 and between stage 2 and stage 3 (rm) than there are into stage 1 or out of stage 3 (rn), and $rm > rn$. Speedup is a slightly strange term to use in this situation, as the "speed" of any switched physical signal is certainly not changed. Instead, speedup in a Clos means that there are more than the minimal number of paths connecting stages 1 and 3. The minimal number of paths is certainly $N = rn = rm$, but with speedup we break this symmetry and have $(N = rn) < rm$.

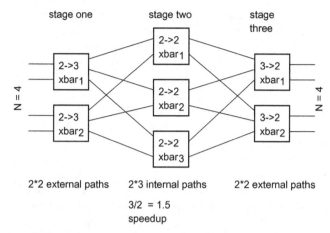

FIGURE 9.17   Clos network C(2, 3, 2) with speedup = 1.5.

The practical importance of speedup is that it both makes the search for an open path through stage 2 more likely to succeed because there is a larger pool of possible open paths, and it allows a Clos fabric to survive a failure in a link or even a stage 2 device, as there are more likely to be alternative paths.

Speedup is given the numerical value m/n. Although it is possible for a Clos network to have $n > m$, this represents an architecture with slowdown that cannot possibly route all $N = rn$ ingress or egress links. The Clos network shown in Figure 9.17 has a speedup of 1.5.

9.11.1   Routing and Blocking in Clos Networks

As we just discussed, routing a new connection through a Clos network is a matter of finding a stage 2 (or center-stage) device that has a free path from the required stage 1 device and to the required stage 3 device. Any such open path found can be used to route the new connection. In Figure 9.18(a) we see a Clos network with three established connections: A → B, though center-stage element <2,3>, C → D, through center-stage element <2,1>, and D → A, through center-stage element <2,2>. Unused links are drawn as lightweight dashed lines.

Suppose that a request for a new connection from B to C arrives. A search must be undertaken for an open path from B's stage 1 element <1,1> to C's stage 3 element <3,2>. Element <1,1> has two available links to the center stage (to elements <2,1> and <2,2>), so our open path search algorithm defines a variable that is assigned the list of these (partially) open middle-stage devices: *IngressOpens* = {<2,1>, <2,2>}. Element <3,2> has two available links from the center stage (from elements <2,2> and <2,3>), so the open path algorithm defines a variable for these (partially) open middle-stage devices: *EgressOpens*

= {<2,2>, <2,3>}. If we compare *IngressOpens* and *EgressOpens* for overlap (checking for identical elements), we produce *OpenCenters* = {<2,2>}, which is the list of all middle-stage devices that are open with respect to both the ingress and egress paths. There being only one item in *OpenCenters*, there is only one open path, which is $B \rightarrow <1,1> \rightarrow <2,2> \rightarrow <3,2> \rightarrow C$, or the path from the request ingress port B through the fixed ingress element <1,1>, through the only open center-stage element <2,2>, to the fixed egress element <3,2>, and hence to the egress port C requested. Figure 9.18(b) illustrates the successful addition of B $\rightarrow$ C to the connection configuration of Figure 9.18(a).

Unfortunately, the universe is not always such a compliant place. As shown in Figure 9.19, sometimes new connections between an otherwise idle ingress port and an otherwise idle egress port cannot be completed. In the figure we have Clos C(2, 2, 2). This Clos has the initial connections A $\rightarrow$ B and D $\rightarrow$ C. Consider what happens when we attempt to connect B $\rightarrow$ D (B's ingress and D's egress being "otherwise idle"). We find that *IngressOpens* = {<2,2>} and *EgressOpens* = {<2,1>}. Finding the intersection of these two lists, we find

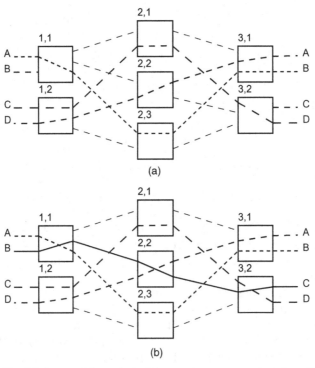

FIGURE 9.18 (a) Successful routing of four unicast connections in a Clos network with 1.5 speedup; (b) connection B $\rightarrow$ C added.

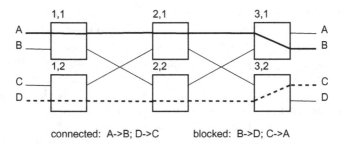

connected: A->B; D->C   blocked: B->D; C->A

FIGURE 9.19   Unicast blocking in a Clos(2, 2, 2) network.

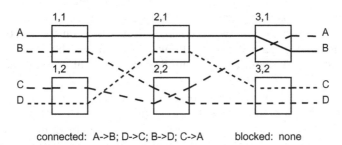

connected: A->B; D->C; B->D; C->A   blocked: none

FIGURE 9.20   Rearrangement allows blocking to be avoided.

OpenCenters = {}. This informs us that there is no complete open path from B's ingress element <1,1> to D's egress element <3,2>. The requested connection is blocked.

Notice that this Clos is also blocked to C → A. Again, we arrive at *Open-Centers* = {}. This is an unfortunate situation. Our Clos could be connecting hugely expensive optical links from B and C and to D and A. Customers of this switch know very well that their links are idle and that the links they want to connect to are idle. They suspect our Clos-based switch of fraud or incompetence. How can their expensive optical links be useless to them due to some failure of a switch to perform?

This problem must be addressed. We shall quickly find a solution to this particular problem, but the general issue of blocking within switches will be with us for the remainder of the book.

Now let's address the blocking issue we found in trying to add the connection B → D to the Clos in Figure 9.19. Look at Figure 9.20, which contains both the established connections from Figure 9.19 (A → B and D → C) and the two apparently blocked connections from Figure 9.19 (B → D and C → A). How did we get this blocked set of connections to fit? The solution was to rearrange existing traffic to allow new traffic to fit. Notice in Figure 9.20 that the D → C connection has been moved from the path *D* → <1,2> → <2,2> →

<3,2> → C to the path D → <1,2> → <2,1> → <3,2> → C. This rearrangement then allows both B → D and C → A to be added to the Clos.

When we move an existing connection from one middle-stage path to another to make room for a new connection, it is important that we not disrupt the connection being moved. Otherwise, each time a connection was moved to allow a new connection, the existing service would suffer one or more bit errors, and that customer's QoS would be compromised by the addition of the new service. Thus, each crossbar must be capable of changing its settings between bits without corrupting either the last bit from the old connection or the first bit from the new connection. In addition, all inputs to the crossbar must be aligned temporally (such that all can be switched hitlessly between bits at the same instant). Another requirement is that the three columns of the Clos must change their settings in a carefully controlled and aligned sequence such that switch changes occur between the same two bits in each customer stream in each column of the fabric. This is a demanding requirement which significantly complicates the implementation of Clos networks.

Notice that we saw no initial blocking in Figure 9.18 in a Clos with a speedup of 1.5, then saw initial blocking in Figure 9.19 in a Clos with no speedup (because $n = m$). Is it the case that we incur blocking only in Clos networks without speedup? Figure 9.21 will dispel this hope. The Clos in Figure 9.19 has a speedup of 1.25, yet is blocking (without rearrangement) for several candidate connections. Yet once again we find that by rearranging the existing

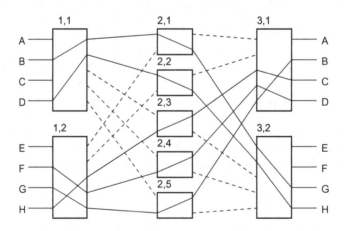

connected: B->G; D->H; F->D; G->B; H->C

blocked: C->A

also blocked: A->A; E->E; E->F

available: E->A; F->A; A->E; A->F; C->E; C->F

FIGURE 9.21   Unicast blocking in a Clos(4, 5, 2) network with 1.25 speedup.

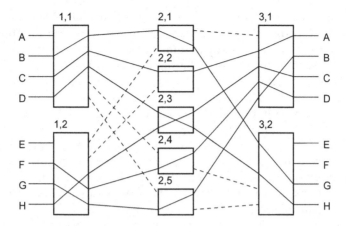

connected: B->G; D->H; F->D; G->B; H->C
newly connected: C->A

FIGURE 9.22 Rearrangement allows a new connection in a Clos(4, 5, 2) network.

connections shown in Figure 9.21, we can add the desired new connection shown in Figure 9.22.

We have seen that partially loaded Clos networks can appear to be blocked to new connection requests, but that in some examples at least, rearrangement of some of the existing connections can make space for the new connection. This process seems somehow related to the possible need to repack a suitcase to accept some new object.

Important issues for switch and network designers arise out of this process of rearranging existing connections to accept new connections. At least the following issues must be addressed:

- Is it always possible to arrange existing connections to permit the acceptance of an arbitrary new connection, up to the involvement of all ingress and egress ports? Or, is it possible that some combinations of traffic are inherently difficult and refuse to permit any rearrangement that allows some particularly troublesome new connection to be added? If it is possible for a switch to block the use of network links, the entire network must be aware of the potential problem.

- Whether or not it is possible to block a Clos, are there constructive algorithms to route N connections or as many connections as possible? How complex are such algorithms?

- What is the worst-case time to add a new connection?

- If it is necessary to rearrange existing connections to make a new connection, can this be done "hitlessly" (i.e., without disruption of the existing connections)?

Without satisfactory answers to these questions, it would not be sensible to proceed with Clos-based designs. To begin to answer these questions, it is necessary to introduce a new way to think about the connections that exist on a Clos. This discussion begins with the simplest nontrivial Clos (Clos(2, 2, 2)), but the ideas generalize to all Clos networks. Figure 9.23 illustrates a fully loaded Clos(2, 2, 2). Notice that all connections can be divided into two groups: those that follow a path through the upper center stage (<2,1>), and those that follow a path through the lower center stage (<2,2>).

Figure 9.24 takes the idea of the two types of center-stage paths (upper and lower) and formulates an abstraction of connections between stage 1 and stage 3 devices based on which center-stage layer they transit. The layer of center-stage connections used is indicated by the type of connecting lines used in Figure 9.25 (and subsequent figures). In Figure 9.25, lines with short dashes represent connections through the upper center-stage device (<2,1>), and lines with long dashes represent connections through the lower center-stage device (<2,2>). Thus, in Figure 9.24 we find that connection A → D goes from <1,1> to <3,2>, by transiting the lower center-stage device <2,2>.

Each of the stage 1 and 3 devices in the network in Figure 9.23 has one connection to each center-stage device. In Figure 9.24, this turns into the rule that each first or third stage device can have only one line of each center-stage type. Thus, if node <1,1> had two short-dashed lines, that would indicate that

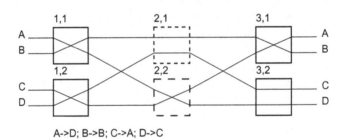

A->D; B->B; C->A; D->C

FIGURE 9.23   Full set of connections on a Clos(2, 2, 2) network.

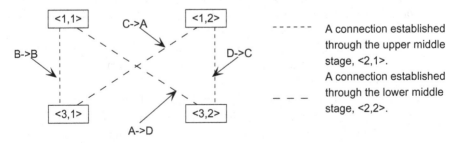

FIGURE 9.24

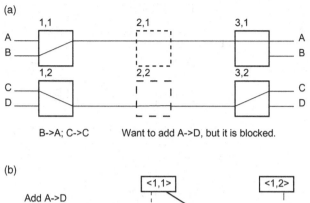

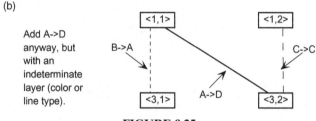

FIGURE 9.25

it was trying to use the upper center stage twice, and that would be an illegal (impossible) connection. Another (obvious) constraint on these connectivity diagrams is that stage 1 (3) devices cannot have connections to other stage 1 (3) devices. Happily, Figure 9.24 is legal.

Consider Figure 9.25. Part (a) shows a Clos with two connections. It is noted that there is a wish to add connection A → D but that that connection is blocked. (One could get from A to the lower center-stage device, but not from there to D; one could get from the upper center-stage device to D, but it is not possible to get from A to the upper center-stage device.) The situation in Figure 9.25(a) is represented in Figure 9.25(b), where we see B → A from <1,1> to <3,1> on the upper center stage (line type) and C → C from <1,2> to <3,2> on the lower center stage. Figure 9.25(b) also shows the new connection desired from A to D as a heavy solid line. Let us assume that this connection has not yet been assigned a center-stage layer; thus far, it is only being considered.

If A → D is to be accepted, it must be assigned to one center-stage layer or the other. So we bite the bullet and make an arbitrary assignment (to the lower layer). This assignment immediately creates a conflict at crossbar <3,2>; crossbar <3,2> is now doubly connected to the lower center-stage crossbar, which we know to be illegal (Figure 9.26).

Figure 9.27 illustrates the next step in our developing algorithm. The conflict at <3,2> is resolved by flipping the previously existing connection involved in the conflict (C → C) and moving it to a new layer, in this case to the upper layer. In general, we flip the previous connection to any otherwise unused layer. In this case there is only one, but in general we can have *m* center-stage

Pick an arbitrarion layer (colour or line type) for A->D.

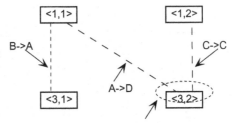

There is an illegal assignment: <3,2> has two connections on the same layer (longer dash line type, or the lower center-plane layer).

FIGURE 9.26

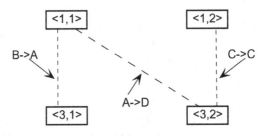

At the node with the newly generated conflict in layer assignments, flip some other edge to another layer (colour) to resolve the conflict. In this case, flip C->C to the upper layer.

FIGURE 9.27

layers and hence a choice from as many as $m - 1$ layers. The only important thing in this choice is that we chose some layer that is not otherwise engaged at the node that had the conflict.

Having made this single flip of C $\rightarrow$ C, the new state of our algorithmic diagram (graph) contains no center-stage layer assigment conflicts. This situation indicates that the new connection has been accepted without blocking, after rearrangement. The flipping of C $\rightarrow$ C was the rearrangement that allowed A $\rightarrow$ D to be added. Figure 9.28 represents the actual connections through the Clos(2, 2, 2), as determined by the algorithmic steps taken in the discussion above.

The example just presented is limited in that the Clos is quite small and only one rearrangement was required. It could be that the flip of C $\rightarrow$ C caused a conflict at <3,1>. If it had, this conflict would have been resolved in the same manner but by flipping the other edge involved in the conflict. If one randomly

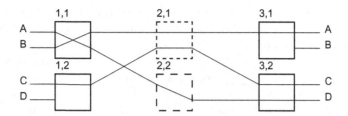

B->A; C->C are maintained, with C->C on a new
center-stage layer. A->D is successfully added.

FIGURE 9.28

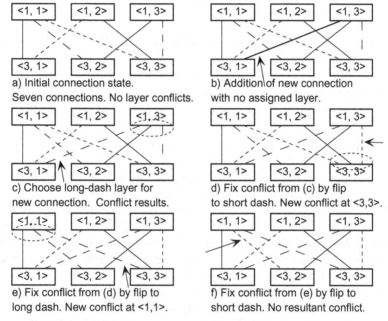

a) Initial connection state.
Seven connections. No layer conflicts.

b) Addition of new connection
with no assigned layer.

c) Choose long-dash layer for
new connection. Conflict results.

d) Fix conflict from (c) by flip
to short dash. New conflict at <3,3>.

e) Fix conflict from (d) by flip to
long dash. New conflict at <1,1>.

f) Fix conflict from (e) by flip to
short dash. No resultant conflict.

FIGURE 9.29 More extended example of conflict resolution in a Clos(3, 3, 3)
network.

flipped either one of the two conflicting edges, the algorithm could repeatedly
flip the same edge (or varying sequences of edges), and never terminate. Thus,
the algorithm continues tracing out a chain of conflicts until no more conflicts
exist.

Figure 9.29 presents a more complex example of traffic rearrangement in a
Clos to allow the admission of a new connection. A sequence of four connection changes is required before a conflict-free configuration is reached. The

possibility of longer chains of connection flipping re-raises the questions that introduced this discussion: Does the algorithm always succeed by finding a way to make any number of (unicast) connections in bounded time?

Proof of the foregoing points is presumed to be beyond the appropriate scope for many students of this book. However, students with a strong background in discrete mathematics and algorithms should be able to solve this problem. Some hints are provided: Consider that the edges in the abstract representation may describe a loop (or multiple loops); some edges may not be involved in loops but may be isolated segments or acyclic chains of segments; which cases are potentially difficult, and how can you show that the more difficult cases are always resolvable in finite time by this algorithm?

For those without the background to solve this problem, we cheat and give the answers:

1. Any possible set of unicast connections can always be packed into a Clos (with speedup ≥ 1.0).

2. Algorithms based on the ideas presented above always find a solution.

3. The worst-case performance for such algorithms is $O(N)$; the time cost of finding a solution grows as fast as the number of ports in the Clos.

There are practical consequences of the use of this sort of "open-path algorithm" which are discussed later in the chapter.

[The abstract representation of the connections through the Clos is derived directly from the *graphs* of graph theory. Each Clos device is represented by a *node* of graph theory; each connection is represented by an *edge* of graph theory. In this case, there are two *colors* of edges (one representing the upper center-stage layer, the other representing the lower center-stage layer). To help keep the price of this book down, different dashed-line types have been used to represent the normal colors of graph theory.]

There are important applications that require Clos-like networks to achieve a sufficiently large switch capability but which cannot accept the delays associated with an open-path algorithm such as the one we have been discussing. Such applications require Clos fabrics that cannot block and which do not require rearrangements to accept a new connection. Such a switching fabric is called a *strictly nonblocking fabric*, whereas Clos's with less speedup are called *rearrangingly nonblocking fabrics*.

Happily, Clos's original work provides a strictly nonblocking solution for unicast traffic. Strictly nonblocking for unicast traffic means that a new connection between an otherwise idle input port on stage 1 can always be connected to an otherwise idle output port on stage 3 without disruption of existing unicast connections. An approximate, simple statement of Clos's result is that if a Clos network has a speedup of 2.0, it will be strictly nonblocking for unicast traffic. In more detail, Clos's statement is that a fabric requires

$2n - 1$ center-stage paths per stage 1/3 device to be strictly nonblocking for unicast, where n is the number of inputs to stage 1 and the number of outputs from stage 3. Clos's proof that such fabrics are nonblocking is simple. Refer to Figure 9.16 as you read this proof, as $N, n, m,$ and r are defined in the figure. Each stage 1/3 crossbar has n sources/sinks of signals (physical ports). Consider any stage 1 crossbar and any stage 3 crossbar to which a new connection is to be added. They each have no more than $n - 1$ connections immediately before the new one is added. We assume that there are $n - 1$ preexisting connections, because filling the last possible connection is the most difficult case. Each of stage 1's $n - 1$ connections will consume one middle-stage path; each of the stage 3's $n - 1$ connections will consume one middle-stage path. In the worst case, there is no overlap between the $n - 1$ ingress connections and the $n - 1$ egress connections, so a total of $2n - 2$ middle-stage paths may be in use. The new connection requires that one more middle-stage path be available if it is to be completed; thus, we require $2n - 1$ middle-stage paths between these two particular stage 1 and stage 3 devices to allow this new connection to be made. As all middle-stage devices are equivalent, any free middle-stage device is appropriate for our new connection. Of course, the same holds for any choice of stage 1 and stage 3 devices. Thus, we must have $m \geq 2n - 1$, which dictates that we have at least $2n - 1$ middle-stage crossbars. Each middle-stage crossbar connects to $r = N/n$ stage 1 and 3 devices, so we have a total of $(2n - 1)r = (2n - 1)(N/n) = 2N - N/n$ links to and from the middle stage. This results in an overall speedup of $2 - 1/n$.

In practice, strictly nonblocking Closes are usually built with $m = 2n$ and a speedup of 2.0. Figure 9.30 illustrates a Clos with a speedup of $m/n = 2.0$, which meets Clos's criterion and provides strictly nonblocking unicast connection making.

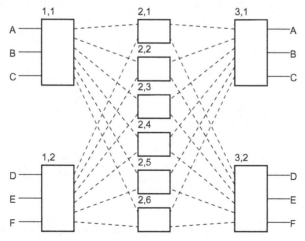

FIGURE 9.30 Clos(3, 6, 2) network with 2.0 speedup.

9.11.2 Multicast in Clos Networks

Thus far, we have examined only unicast routing in Clos networks. Many of the more important practical applications of physical signal switching require more than unicast. The most common motivation for requiring more than unicast is protection. In Chapter 13 we discuss protection requirements and mechanisms in some detail. For the moment we notice only that protection mechanisms generally require that two copies exist of every signal, in order that downstream systems or components can switch from a corrupted copy of the signal to a valid copy. Creating two copies of the signal requires a multicast capability at numerous points in a network. Multicasting exactly two copies is referred to as *bicast*. While two copies is the simplest form of multicasting that we can face, we shall discover that it creates severe difficulties for Clos networks. On the other hand, as we saw earlier, any degree of multicast (including bicast) in a single crossbar switch is straightforwardly easy to provide. Let us proceed to examine the difficulties raised by multicast and a variety of solutions to these problems.

Consider the problem depicted in Figure 9.31. We require a bicast connection from A to B and C (denoted as A → B, C). We assume that a crossbar switch within a Clos network is capable of internal multicast and see how this capability can be harnessed to provide multicast across the entire Clos. Figure 9.31 shows two ways of implementing A → B, C:

- *Stage 1 multicast.* Any signal to be multicast arrives in a particular stage 1 device. In Figure 9.31 the signal from A arrives in element <1,1>. That crossbar can carry out the multicast, thereby emitting two copies of the signal, independently, toward two stage 2 devices. This is illustrated by the solid line connections.
- *Stage 2 multicast.* It is possible not to perform any multicast in stage 1 and wait until one copy of the signal arrives in stage 2. Here, the multicast can

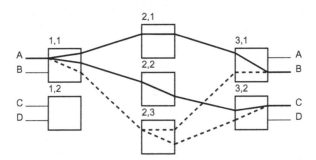

FIGURE 9.31 Multicast opportunities in multistage switch fabrics.

result in independent signals that are routed independently to the required stage 3 devices. This is illustrated by the dashed line connections.

There are two other standard strategies for implementing multicast in Clos networks:

- *Stage 3 multicast.* In the special case when all of the multicast egress ports required are attached to the same stage 3 device, that device can carry out the entire multicast, with the signal being carried only once through stages 1 and 2. (This is not illustrated in Figure 9.31 because the two multicast targets in that example are not attached to the same stage 3 device.)
- *Multiple-branching multicast.* It is possible to combine the foregoing three techniques arbitrarily to build a branching tree of multicasts to eventually cover all of the multicast targets required. Figure 9.31 illustrates an example of this approach.

In Figure 9.32 we see the multicast A → A,B,C,F implemented by one bicast in each of the three stages. Obviously, there are a number of alternative branching trees that can cover the five nodes required.

Given this generality of methods of implementing multicasts, why should it be hard to accomplish, and how can one chose between methods? As we begin to answer these questions, we must examine the limitation of the various techniques.

1. Stage 1 multicast can strand input ports or at least reduce the speedup seen from a stage 1 device. Suppose that a Clos has $n = m$ (the fabric has no speedup). Then if one signal consumes two or more outputs from the device, at least one other ingress signal cannot be emitted. Where $n < m$ (the fabric has speedup), each stage 1 replication of a signal reduces the available speedup.

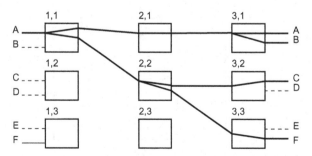

FIGURE 9.32 Use of multiple levels of branching to accomplish a multicast connection.

Eventually, added replications will consume all speedup and leave the case in which other ingress signals are blocking in stage 1.

2. Stage 2 multicast has a comparable drawback. Each replication within these r^2 devices limits our ability to use the ingess ports. This drawback is less severe, as other connections may be routed to other stage 2 devices.

3. Stage 3 multicast is painless, but is available only randomly. We would expect only $1/r$ of all bicast to be satisfied in stage 3.

4. Multibranching multicasts appear to have some ability to "steer around trouble." This is true, but the degree of complexity required in an automated routing algorithm goes up dramatically when multibranching trees are employed. We shall see that the complexity of routing algorithms is a significant negative to many practical applications of these switches, so we must avoid indulging in overly complex open-path algorithms.

So we have concluded that multicast routing decisions can be both painful (isolated capacity) and difficult (managing multiple levels of branching). It turns out that much worse problems lurk beneath this seeming innocent request to share information. Examine Figure 9.33, which shows three simultaneous bicasts: $B \rightarrow A,G; E \rightarrow B,E; I \rightarrow D,I$. All three bicasts are accomplished in stage 2. Let us assume that this system prefers stage 2 bicasts. We find one bicast per stage 2 element. Now try to add one more bicast: for instance, $F \rightarrow C,H$. $F \rightarrow C,H$ cannot be added in device <2,1> because that device does not have a free path to <3,1>, which is required to reach C. $F \rightarrow C,H$ cannot be added in device <2,2> because that device does not have a free path to either <3,1>, which is required to reach C, or to <3,3>, which is required to reach H. $F \rightarrow C,H$ cannot be added in device <2,3> because that device does not have

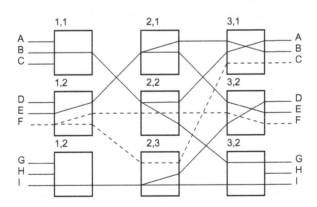

connected: B->A,G; E->B,E; I->D,I.
F->C,H has been added in stage 1.

FIGURE 9.33 All further stage 2 multicasts are blocked in a Clos(3, 3, 3) network.

a free path to <3,3>, which is required to reach H. Thus, F → C,H, as a new stage 2 multicast, is blocked.

Figure 9.34 presents a connection graph for the situation in Figure 9.33. The notation for connections has been enhanced to permit the representation of center-stage multicasts. Recall that center-stage multicasts must occur on the same center-stage crossbar device. Thus, each leg of each of the three bicasts must be on the same layer (or line type, in our connection graph representations). This added constraint is indicated by horizontal bars connecting the two legs of each bicast (or x legs of each x-cast).

Inspection of Figure 9.34 quickly shows that no new (stage 2) multicasts can be added to this Clos(3, 3, 3). There is no single line type (layer) that is free at any pair of stage 3 devices. Thus, further center-stage multicast is not possible. [It is interesting and instructive to attempt to add another bicast to this situation, but we leave this as an exercise. *Hint*: Add the new bicast; pick a line type for both legs of the bicast; look for a resulting conflict at both destinations of the bicast; resolve any conflict(s); continue. In this continuation, can you show that you have rereached the same situation and hence are in an endless cycle? Can you show that the cycle is unavoidable?] F → C,H cannot be added as a stage 2 bicast, but Figure 9.35 shows that it can be added with a stage 1 multicast.

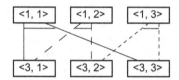

FIGURE 9.34   Connection graph of Figure 9.33.

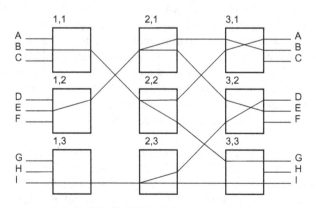

connected: B->A,G; E->B,E; I->D,I.
all other possible stage 2 multicasts are blocked.

FIGURE 9.35   Adding a stage 1 multicast.

It is possible to build strictly nonblocking Clos fabrics for bicast traffic. As was pointed out earlier, the standard techniques used for protection of customer traffic require that two copies of a customer's traffic be forwarded on separate links (these are called working and protection copies). Obviously, bicast is required to generate these two copies, to support this approach to protection. Thus, it is of commercial importance that nonblocking bicast Clos fabrics exist.

In such a bicast protection scheme, each Clos fabric receives both the working and the protection copies of each customer signal. The ingress stages of the switch pick the best copy of the customer's signal (normally, the working copy, but it is possible to switch quickly to the protection copy if the working copy becomes corrupted and the protection copy still has adequate quality). Thus, in a sense, the switch immediately throws away half of its input! But its next operation is to make two copies of (to bicast) the better of the two received sources of each customer's signal and to direct them to the appropriate output channels. Thus, the switch carries the full received bandwidth across its central stages, but this traffic consists of only half of the received traffic, with all of the retained half doubled by their bicast connections.

Let us take the time to study the design of appropriate bicast Clos fabrics, because a seemingly subtle point arises that forces us to design two quite different switches, depending on the choice made on this seeming small point. To my way of thinking, this is a beautiful example of complexity arising in a system but being apparent only when the entire problem is understood. It would be easy for an engineer who specialized in Clos fabrics to miss the subtle point if he or she did not consider the larger picture. So we will develop this bicast fabric as if we didn't know the subtle problem, then expose the problem, and then redevelop the fabric to meet the additional challenge.

Our first step is that we must add a functional unit before the stage 1 crossbars which evaluates both the working and protect signals for each customer traffic stream and picks the better of the two sources for forwarding to the Clos switch. This is really just a multiplexer, with sophisticated signal quality measurements and control logic to determine when to switch sources. We shall not concern ourselves with the details now, but just notice that either the working or protection source can be selected as desired. These multiplexers are shown in the leftmost column of Figure 9.36. Each of four customer services (A to D) has a working and a protection source; a decision is made between these alternative sources, and the winner is passed to stage 1 of the Clos.

Each stage 1 element of the Clos thus receives only $n/2$ traffic streams, but each of these traffic streams received is bicast, resulting in n traffic streams emerging from stage 1. At this point we ignore the fact that the $n/2$ pairs of streams are related and route each of the n streams independently.

We now consider how many middle-stage links we need for each first-stage crossbar in order to remain strictly nonblocking for this traffic pattern. Just

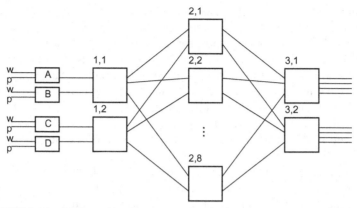

FIGURE 9.36 2.0 Speedup Clos network suitable for traffic protection bicast.

before the last possible customer stream is added to a stage 1 device, that device is receiving $(n/2) - 1$ traffic streams and emitting $2[(n/2) - 1] = n - 2$ streams. The added customer service will increase the number of traffic streams received to $n/2$ and two more streams will be emitted, bringing the number up to n. Consequently, we must consider the case in which there are $n - 2$ stage 1 to stage 2 streams, and two more such steams are about to be added. Thus, the ingress load is $(n - 2) + 2$. There are two cases for the middle stage to egress stage paths: the two new streams go to separate stage 3 devices, or the two new streams go to the same stage 3 device. These cases reduce to the same single case once the first stream is added, so we consider only the addition of the second of the new bicast streams. The worst case for the stage 3 device(s) to which the last branch of this new bicast is directed is when it already has $n - 1$ traffic streams. If we are adding one new stream to each of two stage 3 devices, each can have $n - 1$ existing connections and each needs one free connection. If we are adding two new connections to the same stage 3 device, it can have $n - 2$ previous connections and needs two free connections. Combining these two requirements and focusing on the addition of the last connection, we find that there must be at least $(n - 1) + (n - 1) + 1 = 2n - 1$ center-stage paths. In Figure 9.36, with $n = 4$, we use $2n = 8$ center stages.

In summary, we have halved and then doubled our bandwidth in (or just before) the first stage. The resulting bandwidth coming out of the first stage is n. Unsurprisingly, we require $2n$ center stages. This seems to be very good news: We seem to have accomplished full bicast for customer service protection without enlarging our Clos network. While the total traffic through the middle stages of the Clos remains unchanged (N streams), we have definitely added an important new service: traffic protection. But we have missed an important systems requirement. The companies that use such switches in their telecommunications networks (i.e., the "telcos") view the N inputs to their (Clos-implemented) switch as independent and equal. When a new customer service is added with working and protection paths, a telco does not want to be limited in the choice of which of the N ports can be used. More important, they do not want to add the working and protection pairs for a given customer service to the same place on the switch. Instead, they need to maintain maximal independence between working and protection to avoid single failures which kill both sides of the service. It is generally an absolute requirement that working and protection pairs be separated onto different customer line cards or switch shelves.

Now look again at Figure 9.36. We have assumed that the working and protection sides of each customer's service arrive at the same selection multiplexer just before each stage 1 crossbar. This is convenient for switch designers, but it implies a coupling or colocation of working and protection streams, which is unacceptable to our customers, who insist on allowing the working and protection sides to arrive at different and arbitrary stage 1 devices. This implies that selection between working and protection must be a little more subtle, in that it operates across two stage 1 devices because we can no longer assume that the working and protection streams arrive at a common device. This matter is not critical, and we shall ignore it here. It is still possible to measure the quality of both streams, to select the better of the two and bicast it, and to kill the poorer of the two. But this new customer-driven freedom means that we can no longer assume that the n streams arriving at each stage 1 device will be halved before being doubled. It may happen that one (unlucky) stage 1 device carries nothing but working traffic (or protection traffic that is being used due to failures in its working half)! All of that working traffic must be doubled by bicast operations. Thus, any stage 1 device can consume as many as $2n$ middle-stage links. (The total traffic across the middle stages remains N, but instead of being equally divided over the r stage 1 devices, the traffic may be lumpily allocated across stage 1 devices, leading to some devices being fully loaded and some being empty. Nevertheless, we must provide separately for the worst cast of maximal traffic occurring at any stage 1 devices.) The stage 3 devices still require only n middle-stage devices, but the overall requirement for middle-stage links has gone up to $3n - 1$ [$2n - 1$ for stage 3, $n - 1$ for stage 1, and 1 for the last stream added (of the two)]. The Clos illustrated in Figure 9.36 does not meet this requirement. It is necessary to increase the number of middle stages to 12 for that example, to provide for independent placement of bicast protection streams.

9.11.3 Implementation Costs of Clos Networks

In Section 9.10.2 we developed a cost model for crossbar switch fabrics. As Clos networks are simple replications of crossbars, it is easy to extend that model to give fabric costs for Clos networks:

$$ClosCost(n, m, r)$$
$$= r \times crossBarSwitch(n, m, b) + m \times crossBarSwitch(r, r, b)$$
$$+ r \times crossBarSwitch(m, n, b)$$

To add in the input/output costs, one simply adds in the receivers in each stage (N, $(m/n)N$, and $(m/n)N$) and the number of transmitters in each stage [$(m/n)N$, $(m/n)N$, and N]. The total number of chips is $r + m + r$. The total number of port-through-fabric-to-port links is $N + (m/n)N + (m/n)N + N = 2(1 + m/n)N$.

9.12 DESIRABILITY OF SINGLE-STAGE FABRICS AND LIMITS TO MULTISTAGE FABRICS

Recall that we opened the issue of multistage networks (Clos networks) when we realized that there was an implementation associated with single-stage switches. Nevertheless, we must keep in mind that single-stage non-blocking switches are, for several reasons, greatly superior whenever they can be built:

1. Single-stage switches are strictly nonblocking for any traffic pattern (when designed appropriately). On the other hand, three-stage Clos networks raise the issue of blocking. In general, blocking can be avoided by building more speedup, but this clearly raises implementation costs.

2. Even when built to nonblocking standards, Clos networks require an algorithm to find open paths. Nonblocking single-stage switches require no such algorithm.

3. When a Clos that permits initial blocking is used, it is necessary to have a more complex open-path algorithm that can rearrange traffic when necessary. Nonblocking single-stage switches never impose this requirement.

4. When a Clos that permits absolute blocking (unfixable by rearrangement) is used, network-wide open-path searches may fail. This greatly raises the complexity of finding open paths across entire networks. Again, nonblocking single-stage switches never raise this complexity.

5. Clos networks are more expensive. In terms of input/output of customer streams, Clos networks require twice as many hops to get from the receiving line card, to stage 1, to stage 2, to stage 3, and back to the transmitting line card as do single-stage networks, which require only two hops, one from the receiving line card to the switch and one from the switch to the transmitting line card.

Thus, we would certainly use nonblocking single-stage physical signal switches whenever feasible, but we must recognize that regardless of the implementation technology (CMOS or MEMS), there are hard physical limits.

How large can Clos networks be built? In simple terms, there are three limits: the size of the largest individual crossbar that can be implemented in CMOS, the aggregate size of multiple chips that can be packaged on printed circuit cards and placed on shelves, and the number of input/output streams that can be routed on the printed circuit cards and the backplanes between cards.

It is possible to go beyond Clos networks to build systems even larger than three-stage Clos fabrics can support. The expansion pattern is accomplished by adding new pairs of stages, one more ingress stage before the middle stage and one more egress stage after the middle stage. Thus, we can go beyond three-stage Clos fabrics with five-, seven-, . . . stage switches. With today's technology and today's switch size requirements, such monsters are generally not required. It is relatively easy to build sufficiently large switches with single-stage techniques, or failing that, with three-stage Clos designs. In addition, the wide range of multistage switch fabrics presented in the literature can be used in multistage switches.

KEY POINTS

- Switches can be divided into three classes: physical circuit switches, logical circuit (or time-division-multiplexing) switches, and cell or packet switches.
- A variety of quality of service measures are used to characterize the service experienced by a network customer.
- Although the basic switching operation is unicast (one-to-one), important special services include broadcast, multicast, and bicast.
- Crossbars capture the most straightforward means of implementing switching services. There are straightforward and practical ways to implement crossbars in digital logic.
- A cost model for crossbar switches will indicate that there are practical limits to the maximal size of these switches.
- When crossbars become too expensive, multistage switches are used. The Clos switch fabric is the design most commonly used in very large commercial switches.
- Clos fabrics introduce the possibility of blocking and the need for rearrangement. An open-path algorithm is available for conflict resolution.
- Blocking properties of Clos switching under multicast and bicast traffic must be considered, including the use of bicast in protection schemes.
- The implementation costs of Clos fabrics need to be reviewed.

REFERENCES

Benes, V., *Mathematical Theory of Connecting Networks and Telephone Traffic*, Academic Press, New York, 1965.

Clos, C., A study of non-blocking switching networks, *Bell System Technical Journal*, vol. 32, no. 5, pp. 406–424, 1953.

Hui, J. Y., *Switching and Traffic Theory for Integrated Broadband Networks*, Kluwer Academic Publishers, Norwell, MA, 1990.

Hwang, F., *The Mathematical Theory of Nonblocking Switching Networks*, World Scientific, Singapore, 1998.

Tung, M. F., An introduction to MEMS optical switches, http://132.236.67.210/engrc350/ingenuity/Tung_MF_issue_1.pdf, 2001.

10

TIME-DIVISION-MULTIPLEXED SWITCHING

10.1   INTRODUCTION

This chapter is the second in a sequence of three chapters on the switching technologies used in various layers of transport networks. Chapter 9 provided an overview of switching issues and considered the switching of complete

Network Infrastructure and Architecture: Designing High-Availability Networks,
By Krzysztof Iniewski, Carl McCrosky, and Daniel Minoli
Copyright © 2008 John Wiley & Sons, Inc.

physical signals. This chapter covers the switching of logical circuits or time-division-multiplexed (TDM) signals. Chapter 11 covers the switching of packets and cells.

Although the Internet seems to be moving toward packetized traffic of the TCP/IP protocol suite, it would be a mistake to assume that only packet (or cell) switching techniques are required and that TDM is a sunset technology that could slowly be retired as it is replaced. Instead, there are compelling reasons to continue to use and to understand TDM protocols and systems. The irreplaceable advantage of TDM is that large physical pipes can be subdivided to serve multiple purposes, with the multiple logical subchannels of TDM serving as parts of different routes or as different services and priorities. TDM thus removes the need to switch an entire stream at a packet or cell level when a stream of packets or cells is merely being transported. Thus, we believe that TDM techniques will continue to have a robust and interesting future as the Internet evolves.

We begin this chapter with a review of TDM structures. Many of the details of SONET/SDH, the most common TDM structure, were presented in Chapter 4. In the present chapter we deliberately avoid most of the details of SONET/SDH, concentrating instead on the switching of arbitrary TDM signals.

An abstract definition of TDM switching is presented. This simple definition leads to the straightforward idea of a central memory architecture for TDM switching. These central memory switches are strictly nonblocking (SNB) for all traffic patterns, whether or not multicast is involved. Naturally, there are no open-path algorithm (OPA) issues to discuss, because paths do not need to be found, as they are obvious. At this point, TDM switching looks like a breeze. But an analysis of the limitations of central memory switching soon raises dark clouds of technological pessimism. We come to understand that central memory switching is severely limited and limiting, and alternatives must be found for larger commercial switches.

Alternatives to central memory switching fall into two groups. The first group of alternatives maintains the SNB property and remain OPA-less. These switches are the ingress-buffered switch, the egress self-select switch, and bit-sliced versions of either of these switches. We discuss their operation, hardware structures, and costs. The second set of switches we examine in response to the limits of central memory switches are multistage switch fabrics. Three are examined in varying levels of detail: the time–space–time switch, the space–time–space switch, and the multistage memory switch. For each of these types of switch fabric, we discuss their operation, present their hardware architecture, discuss their blocking and rearranging properties, and study appropriate OPAs for use with them.

10.2 TDM REVIEW

Recall that time-division multiplexing divides a physical signal into some number G of subchannels. Each subchannel is composed of a repeating

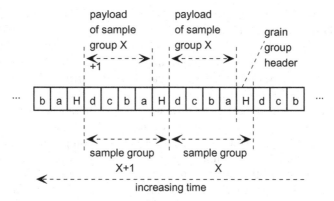

FIGURE 10.1 Time-division multiplexing of a single physical channel.

pattern of samples from a common source. For instance, subchannel A might consist of the sample sequence $\dots, a_x, a_{x-1}, \dots, a_2, a_1, a_0$, where a_0 is the first sample in time sequence, and increasing subscript values indicate increasing time in the subchannel. Each sample a_i is of some fixed number of bits B (commonly, $B = 8$). A group of samples, one from each subchannel, forms a sample group. For instance, where we have a TDM signal with four subchannels called A, B, C, and D, a sample group might consist of the temporally ordered group $[d_i, c_i, b_i, a_i]$. Sample groups may have a header, which may support alignment, error detection and correction, and OA&M (operations, administration, and maintenance) features. Within each sample group, the optional header and each subchannel sample has a fixed location. The use of these fixed locations within sample groups makes it trivial to find the various components once the sample group boundary has been identified. Figure 10.1 illustrates the structure of a TDM signal with a header and four subchannels.

As the term *time-division multiplexing* implies, the time required to transmit a sample group is an important characteristic of a TDM signal, as this time is what is subdivided among the various channels. The number of bits in a sample group and the transmission rate in bits per second determine the sample group transmission time and its corresponding sample group frequency. TDM structures can occur on several levels. In SONET/SDH, the primary physical signal, STS-x, is divided into x subchannels, each consisting of an STS-1 of 8000 frames/s $\times$ 9 rows $\times$ 9 columns = 6,480,000 samples/s, where each sample is $B = 8$ bits, the aggregate subchannel bandwidth is 51.84 Mb/s, and the aggregate channel bandwidth is $x \times 51.84$ Mb/s. SONET/SDH requires no sample group header, as all alignment, error detection, and OAM functions are embedded in the subchannels. An STS-3 signal consists of 3-byte interleaved subchannels; for instance: $\dots, c_3, b_3, a_3, c_2, b_2, a_2, c_1, b_1, a_1$. Both the overall physical signal and each component STS-1 are generally referred to as *broadband signals*, due to their relatively high bit rates.

In SONET/SDH it is possible to subdivide each STS-1 into a secondary layer of TDM structure. The general idea is that each SONET/SDH frame of nine rows and 90 columns is divided into a number of subchannels, each of which consists of some fixed set of columns. Thus, each TDM-defined STS-1 can contain its own set of subchannels, called *virtual tributaries* and typically consisting of several million bits per second. This level of TDM transmission is referred to as *wideband*. Further, each virtual tributary can be subdivided into multiple *narrowband* subchannels, each consisting of 64 kb/s and used to carry one voice channel. Thus, we see a three-level decomposition of the physical channel into smaller and smaller portions of bandwidth.

For the purposes of this chapter, the importance of nested TDM structures is that the number of subchannels in SONET/SDH applications can vary tremendously, depending on the level of TDM structure that is being considered: A single OC-48 signal can be viewed as one physical layer channel, 48 broadband channels, $28 \times 48 = 1344$ wideband channels, or $24 \times 28 \times 48 = 32{,}256$ narrowband channels. As we shall see in this chapter, this varying granularity of TDM signals (the number G of defined subchannels) has a huge impact on the difficulty of constructing feasible and economical switching solutions. For the remainder of this chapter, the details of the TDM structure will not be important. What is most important is the total number of subchannels that must be supported. We shall refer to the number of subchannels, G, as the *granularity* of the problem, and we shall refer to each subchannel (whether it be a broadband STS-1 or a narrowband DS0) as a *grain*.

10.3   TDM SWITCHING PROBLEM

By ignoring the details of the TDM structure, it is easy to give a general statement of the TDM switching problem. Some number N of physical signals arrive at a switching site. Each physical signal is divided by TDM techniques into a common number of grains, G (we restrict ourselves to a common number of grains for simplicity, but generalization to varying G values for each port are straightforward). The total number of ingress grains is NG. There are also N egress physical channels, each divided into G grains, yielding NG egress grains. The TDM switching problem is that each of the egress grains must be fillable from any of the ingress grains. *Fillable* means that the incoming sequence of grains on the selected ingress subchannel will be emitted on the egress subchannel. A TDM switch may add only a limited number of grain group latency periods to the signals as they are received, switched, and transmitted.

Figure 10.2 illustrates this situation for $N = 2$ and $G = 4$. The ingress grains on port 1 are called a_2, b_2, c_2, and d_2; the ingress grains on port 2 are called e_2, f_2, g_2, and h_2. The same grain names, one grain group earlier (i.e., a_1 instead of a_2), are shown in their new position in the egress grain groups on the right-hand side of the switch. In the figure, ingress subchannel 'a' in the first tem-

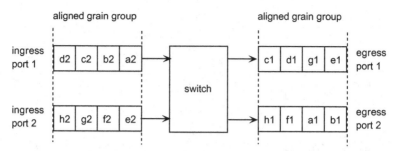

FIGURE 10.2 TDM switching problem.

poral position on port 1 is being switched to egress port 2 in the second temporal slot. (The second grain of the 'a' connection, 'a2', appears in the first position of ingress port 1. The first grain, 'a1', has been switched to a new position on another port.)

Notice that the grains a_* move backward in time in terms of their egress time position in their egress sample group (when they are in the second temporal slot) relative to their ingress time position on ingress port 1 (where they occupy the first temporal slot). Now consider ingress grains b_*, which are switched to egress port 2 in the first temporal position. These samples move forward in time in terms of their positions in their ingress and egress sample groups. It is clear that moving a sample forward in time (i.e., to an earlier time) is difficult to accomplish with current technologies!

Since any ingress gain must be switchable to any egress gain, and an ingress grain must arrive before it can be emitted by an egress port, it is clear that all ingress grains of a particular grain group must arrive before any of the egress grains can be emitted. A second key observation is that the ingress ports must accumulate the next grain group as it arrives while the egress ports are selecting their egress grains from the last ingress grain group. These observations lead to a standard approach to TDM switching:

1. All ingress ports align their incoming TDM signals to a common grain group boundary.
2. The ingress ports save all incoming grains in grain group memory 1.
3. At each common grain group boundary, the entire contents of grain group memory 1 is copied to grain group memory 2. Thus, a complete set of NG ingress grains is available in memory 2.
4. On each common grain group boundary, all the egress ports select their desired grains from among the NG grains stored in memory 2.

This approach allows each egress port to select arbitrary grains from the last ingress grain group without creating problems regarding the temporal movement of grains within grain groups. An alternative to this approach is to

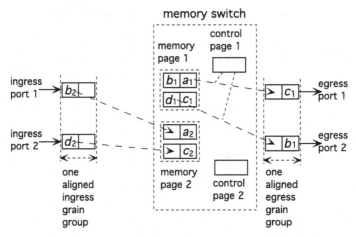

FIGURE 10.3   Conceptual memory TDM switch.

swap two memories between the two roles: filling by ingress ports, and selection by egress ports. In this style, the two memories exchange roles on each common grain group boundary. The TDM switch illustrated in Figure 10.3 uses this approach. At the time of the drawing, the ingress ports are just moving their second grains (b_2 and d_2) into the second temporal columns of their rows of memory page 2. At the same time, the egress ports are selecting grains to fill their second grain slots (egress port 1 selects grain a_1, egress port 2 selects grain d_1). Egress selection is controlled (or programmed) by the memory control page 1 which gives the addresses in memory page 1 that are to be read to fill each egress slot.

After these two ingress grains are moved to memory page 2 and the two egress grain slots are filled from memory page 1, we have arrived at the end of the switch's common grain group boundary (because $G = 2$ in this example). On the temporal boundary (i.e., logical clock edge) between this common grain group and the next common grain group, the role of the two memories will be swapped, so memory page 1 will be filling and memory page 2 will be used to fill the egress ports' time slots. Several clarifications regarding the operation of this switch are required. These issues are addressed in the following two subsections.

10.3.1 Temporal Alignment

The incoming signals may arrive with any temporal alignment of their grain groups, as they have come from arbitrary locations over arbitrary lengths of fiber (or copper). The switch in question picks an arbitrary grain group boundary, then each ingress port uses a delay FIFO to align its incoming signal's grain group boundary to this common boundary. The switch then extracts exactly aligned grain groups from these FIFOs, according to the switch's local clock. If all the remote transmitters that provide these ingress signals to the switch in question share exactly the same clock, and transmit exactly the same bit rate, this simple FIFO alignment technique is satisfactory. However, it is never the case in practice that all the remote transmitters share exactly the same clock. For many practical reasons, these remote transmitters will have clocks with very small errors in their rates. If such clocking errors were to go uncorrected, eventually any finite alignment memory in the switch in question would be exceeded (FIFO overflow or underflow), and an alignment error would be forced [some port(s) must either drop an ingress frame or must reuse an ingress frame in order to keep aligned with the local common clock]. SONET/SDH addresses this problem with the ability to make positive and negative slips. These slips allow compensation for clocks that are either too fast or too slow. Refer to Chapter 4 for more detail regarding this temporal structure and mechanism.

10.3.2 Dual Control Pages

Figure 10.3 shows two control pages, with page 1 actively selecting grains for the egress ports from memory page 1. When the roles of the memory pages are reversed, control page 1 will remain in control—it will then be used to control the selection of grains from memory page 2. Control page 2 will remain in the background until users of the switch want to make some connection change (change the pattern of selection of grains by the egress ports). When a switching change is required, a control system (e.g., a microprocessor bundled with the switching hardware) will program control page 2 with the desired settings, then will cause the switching gear to start using control page 2 instead of control page 1. This change of control pages occurs on the boundary between two of the common grain group times in the switch. Consequently, the switch's output will change from one connection pattern to the next exactly on some grain group boundary (instead of potentially corrupting a grain group by using one control page for part of the grain group and another control page for the rest of the grain group). This simple feature is essential for reliable switching of TDM networks.

10.3.3 Strictly Nonblocking Design

The TDM switching architecture we have examined in preceding sections cannot block. There is no set of connection settings that cannot be achieved. This property arises trivially from the ability of the egress ports to read any of the ingress grains. This property holds for unicast (each ingress grain is used once, to complete a one-to-one connection) as well as for arbitrary multicast (any ingress grain can be read repeatedly to fill any number of egress grain slots). Of course, any multicast requires that some ingress gains not be used (switched to egress grain positions), assuming that the number of ingress grains is the same as the number of egress grains. As a corollary to this SNB property, TDM switches based on this general design do not require any sort of open-path algorithm to establish new connections.

10.3.4 Varying Port Configurations

In our discussion above, we have assumed that the switch has equal numbers of ingress and egress ports ($N_{in} = N_{out}$) and that all ports have the same grain count, G. Neither of these assumptions are required. It is possible, although perhaps not useful, to have $N_{in} \neq N_{out}$. It is very common to have G vary by port. Consider the case of a SONET/SDH switch with several major WAN connections at OC-192 and a collection of smaller MAN connections of OC-48, OC-12, and OC-3. These ports have granularities of $G = 192$, $G = 48$, $G = 12$, and $G = 3$. This could create problems for TDM switching were it not for the temporal and bandwidth structure of SONET/SDH. For all SONET/SDH OC-X signals, X copies of a 9×90 byte frame arrive each 125 µs, and consequently one grain of each of the X STS-1s arrives every 125 µs/($9 \times 90 \approx$ 0.15432 µs. If $X = 3$, the grains arrive in about 51.44 ns each; if $X = 48$, the grains arrive in about 3.215 ns each; but in both cases, a complete grain group for each port arrives in about 154.32 ns. Thus, a SONET/SDH switch can be set to have a common grain group clock running at 6.48 MHz (the inverse of this common grain group time). Each OC-X port will require X grain positions in each of the two shared memories.

10.4 CENTRAL MEMORY TDM SWITCHES

Figures 10.4 to 10.6 illustrate three common variants of simple memory-based TDM switches. The switch in Figure 10.4 has two physically separate memories,

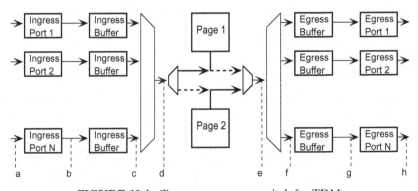

FIGURE 10.4 Two-page memory switch for TDM.

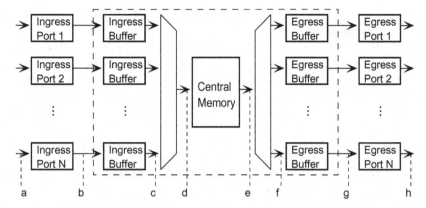

FIGURE 10.5 Two-ported single memory switch for TDM.

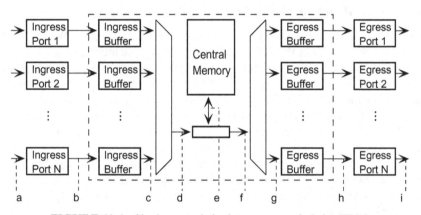

FIGURE 10.6 Single-ported single memory switch for TDM.

called page 1 and page 2. At any time, one memory is connected to the ingress ports and the other is connected to the egress ports. All the ingress ports save their most recent sample in their ingress buffer. During each common grain group period, the mux accepting inputs from the N ingress ports moves from one ingress buffer to the next, accepting grains and transferring them to the small mux facing the dual pages of memory. This memory interface mux remains set for each grain group, steering each complete ingress grain group into one of the two pages of memory. Thus, the page of memory selected is time divided among the N ingress ports and must run at N times the rate of each port. At the same time, the egress ports are fed from the other page through the other memory mux (under the control of a connection-setting memory that is not shown in the figure). Like the input side, the large egress mux is used to read one sample at a time and fill the egress sample buffers within each egress sample time (and the memory serving the egress ports must also run at N times the rate of each port). Each memory must have NGB bits to store a complete grain group.

Figure 10.5 illustrates the use of a dual-ported memory instead of two physically separate memories. This memory has a write port dedicated to the ingress side and a read port dedicated to the egress side. The memory must contain two full pages of grain groups, the same amount of memory as was required by the switch in Figure 10.4. Each port of this memory must run at N times the rate of any single port (assuming that all ports operate at the same rate). The central memory must support bits to store two complete grain groups. Figure 10.6 illustrates a TDM switch design with one single-ported central memory. This single port alternates between writes serving the ingress ports and reads serving the egress ports. It, too, must contain two logical pages of memory ($2NGB$ bits).

10.4.1 Cost Model for Central Memory TDM Switches

What are the implementation costs of this style of TDM switch? We can identify three separate costs, those for the data paths, sample memories, and control memories. To allow more concrete results from this analysis, we consider the two-memory design shown in Figure 10.4. We consider N ports, of G grains each, with B bits per grain. The common grain clock runs at C hertz. The data paths consist of the ingress port mux from N ports to the ingress memory mux and the dual structure on the egress side. The costs of these paths are quite modest for practical switches: $2 \times N{:}1$ mux/demux at B bits and $2 \times 2{:}1$ mux/demux at B bits. Each sample memory must be capable of storing NGB bits. Each control memory (assuming dual control pages) must be capable of storing one address in the range 0 to $NG - 1$ for each of NG egress samples, a total of $NG \times \log_2 NG$ bits.

Equations are given below to summarize these key costs. First we define *fanout* and *fanin*, which give the costs for our binary amplifier fanout trees and our binary multiplexer fanin trees. Recall that a binary tree to/from N nodes requires $N - 1$ fanout or fanin elements:

$$fanout(N) = (N-1) \times buffer(2)$$

$$fanin(N) = (N-1) \times mux(2)$$

The following three functions define the costs of this central memory architecture. Notice that data path and control memory return cost in gates and data memory returns costs in SRAM bit cells.

$$datapath(N, G, B) = B[\,fanin(N) + fanout(2) + fanin(2) + fanout(N)]$$

$$datamemory(N, G, B) = 2NGB$$

$$controlmemory(N, G) = 2NG\log_2 NG$$

In practice, both memories require extra bits per word for error checking and/or correction.

Given the simplicity of the data paths and the high density available in modern integrated memories, the hardware costs of this switch architecture are never an important issue. For example, one could build a 1024-ported OC-192 switch core (with 10 Tb/s of duplex bandwidth) with only

$$datapath(1024, 192, 8) = 49{,}152 \text{ gates}$$

$$datamemory(1024, 192, 8) = 3{,}245{,}728 \text{ bits}$$

$$controlmemory(1024, 192) = 7{,}077{,}888 \text{ bits}$$

10.4.2 Limits of Central Memory Designs

The switch mentioned briefly at the end of the preceding section is enormous—it carries 1024 10-Gb/s signals, for 10 Tb/s! This is far beyond what can be built with today's technology on a single VLSI substrate. If the design is broken up into multiple VLSI chips, the number of interchip communications paths is excessive. Clearly, there are practical limits to the sizes of such switches which are unrelated to the quite optimistic costs considered in the preceding section. We now turn our attention to two key limitations to the size of feasible central memory TDM switches.

Our first concern is the amount of input/output that can pass through a single chip. As we discussed in an earlier chapter, high-bandwidth switches can only be built using high-speed differential serial links (today, although some day we hope that direct optical links will become feasible). High-speed differential serial ports range in bandwidth from 1 to 10 Gb/s at this time, although a speed of 2.48832 Gb/s exactly supports SONET/SDH's STS-48 bit rate and is a highly practical choice of link speed. The transmitters and receivers for such serial links require considerable VLSI real estate, but these costs are highly proprietary and cannot be discussed in a publicly available text such as this. However, it is widely known that such links have a relatively high power requirement of about 150 mW per 2.5-Gb/s duplex serial link.[†] This cost is adequate to place reasonable bounds on the size of the largest single-chip TDM switches, as summarized in the equation

$$serialPower(N) = \frac{N}{2.5 \times 10^9} \times 150 \text{ mW}$$

Thus, the example chip discussed at the end of Section 10.4.1 would have consumed 614,400 mW or 614.6 W for its serial input/output links, which—

[†]150 mW per 2.5-Gb/s link is a typical number for VLSI implementation of chips available in the marketplace today. With CMOS scaling and future design innovations, that number will drop to 100 mW, perhaps down to 50 mW.

alone—is an absurd power budget. Of course, this power estimate ignores the power costs of switching, which can be expected to be of the same order of magnitude. Suppose that we take the currently reasonable assumption that serial link power cannot exceed 20 W per VLSI die. This translates to a limit of 133 2.5-Gb/s links, or 332.5 Gb/s of total duplex bandwidth. These serial link power costs apply to all switches regardless of whether they switch physical channels, TDM, or packets/cells. If we require one-third of the power budget for the switching logic (an optimistic estimate), the maximum-size single-chip switch should be about 96 2.5 Gb/s ports, or 240 Gb/s.

Returning our focus to central memory TDM switches, we find that while the cost of the appropriate hardware is not generally an issue, the speed requirements of these designs is a critical issue for all but the very smallest designs. Consider a SONET/SDH VT1.5 (the VT1.5 is a commonly used virtual tributary in SONET) wideband switch with N OC-48 ports. Each STS-1 carries 28 VT1.5 subchannels, so the switch has $G = N \times 48 \times 28$ grains. This switch can be built as a "SONET column switch." Virtual tributaries, such as the common VT1.5, occupy a fixed set of columns in their hosting STS-1. Each VT1.5 occupies three such columns. A SONET column switch treats each STS-1 column as a grain. Thus, a column switch can be programmed to switch a complete VT1.5 by moving the three columns which make up that VT1.5. SONET column switches are convenient for wideband switching because all of the virtual tributary formats and positions can be accommo- . dated by selecting the right combination of STS-1 columns. In a column switch, during each complete SONET row arrival, taking $125 \mu s/9 = 13.88 \mu s$, the full set of $G = N \times 48 \times 90$ eight-bit grains must be switched. This gives us a grain group clock rate of $C = 1/13.888 ms = 72 kHz$. Assume the dual sample memory design illustrated in Figure 10.5. Thus, each memory must perform $(N \times 48 \times 90)C$ eight-bit reads or writes per second. For a modest $N = 16$, the memories must carry out 9.953×10^9 operations per second. With current technologies, rates of no more than one-twentieth of this require-ment are possible. Memory speeds are not increasing quickly, yet much bigger switches are required. As a result, the simple central memory-based TDM switch designs we have examined thus far are practical only for small products of NGC, such that $NGC <$ maximum-memory rate. If we take the maximum memory rate as 500 MHz, we could build a broadband (STS-1) central memory switch (with $C = 8 kHz \times 9 \times 90 = 6.48 MHz$) with a total of $NG \leq$ 500 MHz/6.48 MHz = 77 STS-1's, perhaps configured as four STS-12's and nine STS-3's. Such a switch is much too small to be considered for transport applications.

So we are forced to depart from a clean, simple central memory model to obtain the switching performance that is required by today's networks. Note the relationship between this unfortunate situation in TDM switching and the need to move to multistage switching for physical circuits in Chapter 9. We shall look in two directions: cleverer internal switch architectures, which avoid the performance pitfall of central memories, and multistage networks, which allow bigger switches to be composed of smaller switching elements.

10.5   INGRESS-BUFFERED TDM SWITCHES

In our first attempt to find alternatives to the simple central memory TDM switch, we examine ingress-buffered TDM switches. Here, we attack the infeasibility of the central memory by dividing that memory into N components, one associated with each ingress port. Our hope is that each memory only has to run at $(1/N)$th the speed of the single central memory, but the reality will be a bit more complex.

As illustrated in Figure 10.7, each ingress port has $2G$ grain buffers (of B bits each). Each ingress port supplies grains to a mux, which directs each arriving grain to the left-hand buffer of one double grain buffer. The mux rotates through this set of G grain buffers once per grain group time, such that each grain occupies the same one of the G grain buffers in each grain group time period. On grain group boundaries, the complete set of G recently supplied grains is copied to the right-hand buffers of each of the G double buffers. This right-hand set of buffers is stable for the next grain group period and is available for the egress ports to read.

Each right-hand buffer broadcasts its current grain to all N egress ports. Each of N ingress ports thus emits GB bits to each of N egress ports. Each egress port thus receives NG grains of B bits each. During each egress grain period, each egress port uses G B-bit N:1 muxes to select all the G grains for the desired port, and then one B-bit G:1 mux to select from the port selected the grain that it should emit during that time period. (In fact, this combination of fanout trees and selection muxes is one way to implement an N-ported read memory.)

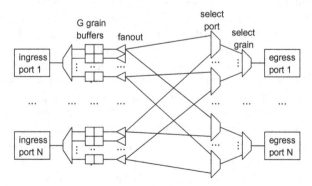

FIGURE 10.7   Ingress-buffered memory switch for TDM.

This switch architecture is clearly SNB, as all grains are presented to all egress ports for independent selection. Arbitrary multicast is possible; no OPA is required. The gate costs of this design are straightforward to evaluate. There are two pages of grain buffer flops; data paths consisting of ingress demuxes, fanout trees, and $NG:1$ muxes:

$$grainBuffers(N, G, B) = 2NGB$$

$$datapath(N, G, B) = B[N \times fanout(G) + NG \times fanout(N) + N \times fanin(NG)]$$

$$controlMemory(N, G) = 2NG \times \log_2 NG$$

Let us consider the speed demands of this design. Writing to the ingress buffers must occur at the rate at which samples arrive on a single port. For OC-48, samples arrive at 311.04 MHz, which can be implemented in present-day technology. For OC-192 ports, the required rate of 1244.16 MHz exceeds practical CMOS limits (for large switches, which consume large VLSI areas), but the STS-192 can be decomposed into four STS-48's by a thin layer of very fast logic. Then the switch accepts the four STS-48 inputs in place of the single STS-192.

Consider what happens at the common grain group boundary within this switch. All left-hand ingress grains replace all right-hand ingress grains, and then all of the fanout networks broadcast these new grains across the switch to all the egress ports. While on average only one-half of the bits actually change, we must consider the worst case, in which all NGB bits change from 0 to 1, or from 1 to 0. Since each bit is fanned out to N ports, we find that N^2GB destination gate capacitances, plus all the intervening wire and fanout buffers (which are usually a larger capacitive load), must be switched. Working out the details of the power consumed by this event is more appropriate for an advanced VLSI course than for this course, but the power consumption is a serious problem. All power consumption and consequent heat generation is bad, as heat dissipation is generally a critical issue in large-bandwidth switches. But worse, this high level of power consumption is a radical departure from the normal power consumption for this switch architecture. Between these ingress broadcast events, the switch is merely accepting new grains into ingress buffers and selecting available grains at each egress port. These local events required very little power, comparatively speaking. The result is that once per grain group the switch experiences a sharp spike in power consumption. These spikes tend to disrupt the chip's power supply (pulling V_{dd} down and/or V_{cc} up). These disruptions can cause bit errors in the switch, which are unacceptable. An alternative is to build the chip's power supply rails and bond-outs to supply the currents required by the grain group spike. This is a difficult path as well, as the resources required for the power distribution are quite large and expensive, and there remain severe dynamic current changes. As an overall result, this switch architecture is also limited in aggregate NGC; the limit is much larger than the limits of the central memory switch (and are dependent on complex details of VLSI technology); but we are still left requiring much larger switches.

10.6   EGRESS-BUFFERED SELF-SELECT TDM SWITCHES

Our goal with this architecture is to overcome the power spike problem introduced by the ingress buffered TDM switch of Section 10.5. Our initial stab at this solution will be to move all buffering from the ingress side to the egress side, and have the ingress ports continually broadcast (fanout) their stream of grains to all egress ports. The egress ports are left to capture the grains. If each egress port captures all ingress grains, we find that the write side of each egress memory is a duplicate of the central memory design that got us into severe trouble. Rather than reexperience that limitation, we change the egress memory design such that it captures in memory only those grains that it requires given the current connection requirements (Figure 10.8). Clearly, each egress port can require only G grains per grain group, saving us a factor of N in memory costs. Although each egress requires only G grains, it must still be able to see all NG arriving grains in order to make an SNB selection. The solution is to provide G independent grain-capture devices at each egress. Each such device

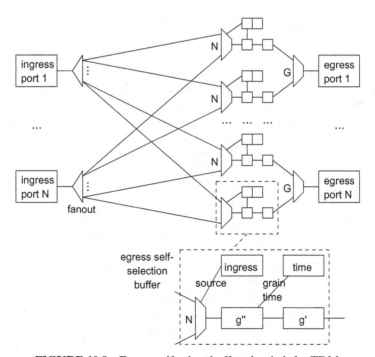

FIGURE 10.8   Egress self-select buffered switch for TDM.

receives the current grain from all N ingress ports; it is programmed to know which port it wants to take its grain from and what time (in the grain group cycle) that grain will arrive. Then, when the right time comes, it grabs its required grain using its $N:1$ mux. After a complete grain group, all G self-select buffers at each egress port have made their selection. On the next grain group boundary, they copy their G recently selected grains into a duplicate right-hand grain buffer, where these selected grains are fed to the egress port while the next set of grains are selected by the mechanism described above.

This architecture is SNB, as all egress grain slots (each with its own self-select buffer) sees all NG grains by looking at all N ingress ports over all G time slots. Only having to make one selection, it cannot fail by blocking in any way. Consequently, this architecture does not require an OPA. The important hardware costs of this egress self-selection architecture consist of datapath gates, flops to hold the grains, and the switching control information which specifies when in the grain group and from which ingress port an egress should capture its grain.

$$grainBuffers(N, G, B) = 2NGB$$

$$datapath(N, G, B) = B[N \times fanout(NG) + NG \times fanin(N) + N \times fanin(G)]$$

$$controlMemory(N, G) = 2NG(\log_2 G + \log_2 NG)$$

10.7   SLICED SINGLE-STAGE SNB TDM FABRICS

Even with the best internal architecture for an SNB TDM fabric, the needs of transport networks continue to exceed the limits imposed by VLSI technology (the critical limits are VLSI die area due to gates, flops, and RAMs; and power consumption). Simply put, the biggest switching problems exceed the capacities of the most advanced VLSI technologies. As we shall see in the following sections, there are multistage architectures for the TDM switching problem, but as we saw with physical circuit switches in Chapter 9, these switches introduce the possibility of blocking and the need for open-path algorithms. In this section we apply a new technique to extend the range of single-stage SNB fabrics for the TDM switching problem.

This new technique can best be introduced by examining how large electronic memories are commonly built for computers. Suppose that we require a large memory of 256-bit words for a computer. The large word size is chosen because memory contents are transferred to a processor cache, and system performance will be improved by these larger transfers to the cache. However, memory chips (e.g., SDRAMs) are not manufactured with 256-bit words. Such chips are not manufactured because they would make poorer use of VLSI real estate and packaging options, due to the large number of inputs and outputs, and would cost more per bit of storage. Instead, suppose that we limit ourselves to a more economically feasible memory chip with a 32-bit word. With such a memory chip, we must either (a) read eight successive 32-bit words to accumulate our 256-bit cache word (a solution that negates our attempt to improve system performance by use of a wide cache fetch word size), (b) use eight such 32-bit chips in parallel, with each chip contributing 32 bits to the overall 256-bit word expected by the cache system (this solution allows all 256 bits to be read at once, but forces us to use eight memory chips), or (c) use some combination of (a) and (b): for example, four 32-bit chips providing 256-bit cache lines in two read/write operations. Solution (b) is commonly referred to as a *bit-sliced memory* (here the slicing is by 32-bit segments, but the underlying idea is the same). The problem of producing a large system (here, a 256-bit memory subsystem) is solved by the use of parallel components (here, 32-bit memory chips) to build up the solution required. In general, bit slicing allows us to preserve performance, albeit at a cost in replicated hardware.

This general approach is applied to our problem of building large and larger SNB TDM switches. In our prototypical SONET/SDH TDM switching problem, each switch port provides or consumes data in 8-bit chunks. In all earlier switches in this chapter, all 8 bits of these SONET/SDH samples were presented to one port of a switch chip in sequence over a serial link. This corresponds to an unsliced memory solution in which the entire word is provided by a single chip. Using the bit-sliced memory concept, we slice our TDM switch fabric into multiple chips and have each port send only a subset of the 8-bit samples to (or receive from) each slice of the fabric. Suppose that we slice the fabric into two chips and send/receive 4 bits to/from each slice of the fabric. The hardware required for this solution is illustrated in Figure 10.9, for an arbitrary number of slices, Z.

Figure 10.10 illustrates the operation of this concept. In Figure 10.10(a) we see that two ports each carry two 8-bit grains. Port 1 carries the 8 bits of grains A and B; port 2 carries the 8 bits of grains C and D. In both cases, these 16 bits are shown arriving in serial format, as from a SONET/SDH link. We

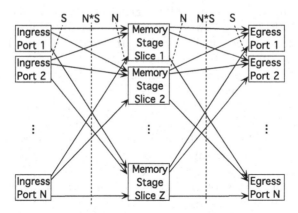

FIGURE 10.9 Stack of bit-sliced memory switches.

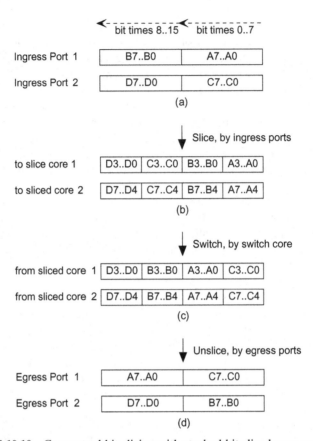

FIGURE 10.10 Conceptual bit-slicing with stacked bit-sliced memory switches.

required that both ports be implemented on a shared VLSI chip in order that all of the arriving/departing bits of these grains can be manipulated arbitrarily. In Figure 10.10(b) we see how these four grains (A to D) are rearranged by our assumed dual-ported chip before being transmitted to our sliced switching core. The first 4 bits of each byte grain are assembled in one sequence, to be transmitted to the upper core switching chip; the second 4 bits of each byte grain are assembled in another sequence, to be transmitted to the lower core switching chip. Each $4 \times 4 = 16$-bit sequence is ordered such that the bits of grain A are transmitted first, followed by the bits of grain B, . . . , followed by the bits of grain D. Thus, considering the two port-to-fabric links as a unit, all the bits of grain A are transmitted in the first 4-bit times to the sliced core fabric, followed by the other grains in sequence.

From the point of view of the sliced switching core, each grain arrives as a unit, sliced over the dual-element central fabric. These two fabric elements are conventional single-stage TDM switches, except that they are designed to operate with 4-bit samples instead of SONET/SDH's 8-bit samples. If these two fabrics are given identical switch settings, then the two four-bit groups (nibbles) of each grain will be switched separately, but identically, to the same egress port. These egress grains will be sent over two serial links to a common egress port, where they will arrive simultaneously over the two links in four bit-times, as shown in Figure 10.10(c). The egress paths of the dual port chip accepts grains in this sliced format and rearranges them into the standard SONET/SDH format of eight bits before transmitting them on the egress port data paths, as shown in Figure 10.10(d).

Since each sliced switching chip is assumed to be SNB, the resulting sliced system is also SNB, as we simply replicate the switch settings from one switching plane to another. What has been gained by this solution? Put simply, we have managed to use two VLSI chips to attack our switching problem instead of one chip, without giving up our SNB property. If we can show that this arrangement allows a switching system with twice the throughput of a single chip to be built, we shall have an ideal solution. Before we jump to this conclusion, we must examine in more detail the costs of these bit-sliced SNB TDM switches.

From the point of view of the switching core, two things have changed: The size of the grains to be switched has been divided in two, from 8 bits to 4 bits, and the number of grains to be switched has doubled (suppose that we were dealing with STS-48 links, each carrying forty-eight 8-bit grains; our slicing solution changes this to ninety-six 4-bit grains). We characterize this change by a slicing factor, Z, where here, $Z = 2$. Our new grain size, B', is B/Z; our new grain count, G', is GZ; our required number of core switching chips is Z. Clearly, we can generalize this solution by allowing Z to be any integer that evenly divides B. For $B = 8$ we can support a Z value of 1, 2, 4, or 8.

The choice of Z is nontrivial. As Z grows, the effective number of grains per chip, G', grows, so the costs of implementing each individual switch increase, even though B' is decreasing. This effect is due to the importance of G in the equations in Section 10.6. Detailed cost modeling shows a diminishing return in supported bandwidth as Z increases to B. For instance, with $B = 8$, $B' = 2$ and $B' = 4$ are practical, but $B = 8$ is much less desirable and would be used only if the marginal gain in overall bandwidth was worth yet another doubling of the number of switching chips.

Interestingly, this is not the end of this slicing game. While the $B = 8$ limit of SONET/SDH seems to limit the possibilities of slicing, in fact it does not. It is possible to have the port devices save up two temporally adjacent grains from each subchannel, making a total of 16 bits, which can then be sliced with (reasonable) Z values of up to 8. Or more than two temporally adjacent grains can be accumulated to make a yet bigger bit pool to slice over large values of Z.

10.8 TIME–SPACE MULTISTAGE TDM FABRICS

Requirements of real networks commonly exceed the largest single-stage switches that can be built in available technology, even with the cleverest architectures and implementations available, especially for highly granular switches. This is not to say that single-stage switches have no role. In fact, they dominate in the sense that most TDM switches are built with single-stage, SNB architectures. But there are always core network applications that exceed the limits of single-stage architectures. For such applications, multistage networks must be considered.

In this section we introduce time–space switching, which is the natural multistage switching concept for TDM signals. Time–space switching has a clear relationship with the multistage Clos networks we examined in Chapter 9, where Clos networks were used to switch physical circuits. Time–space switching fabrics add the concept of temporal switching and arrive at a new switching paradigm.

We begin our examination of time–space switching at the conceptual level. There are multiple ways to define and build time–space switching fabrics, but the most common and perhaps the simplest such fabric is the time–space–time fabric. Time–space–time fabrics consist of three switching stages, of which the first and last are time switching stages and the second is a space switching stage. *Time switching* applies to a single TDM signal, consisting of multiple subchannels or temporal grains. A time switch has the ability to reorder temporally the grains within each grain group. For instance, a group of grains $(a \ldots d)$ might enter a time switch in alphabetical order (d, c, b, a) and be switched to a new temporal order, (b, c, a, d). In time–space–time switching, each time stage repeats the same reordering operation to each grain group, at least until some new end-to-end connection is required by the network.

Space switching applies to multiple TDM signals (say, N signals). Each such signal enters a space stage with a common temporal alignment of grain groups and contained grains (i.e., all signals present their ith grains at time gxi, where $g = [0, \ldots, G - 1]$ is the number of the grain and x is the temporal duration of each grain). Each set of N grains can be switched from any ingress port to any egress port, as a permutation. Each of the G grains in each ingress grain group can be switched in this port-to-port manner, one at a time. Space switching is accomplished conceptually by a crossbar with varying per-grain-time control.

Thus, a time–space–time switching fabric gives us the ability to (1) reorder each ingress TDM signal temporally in a set of N first-stage time switches, then (2) switch spatially each set of temporally aligned grains emitting from the first-stage time switches in a set of G N-to-N permutations, then (3) reorder temporally each of the N TDM signals emitted from the central space stage in a final stage of N time switches.

Notice that a time–space–time switch with N ports and G grains per grain group has NG grains. If we view time in the horizontal dimension (with time

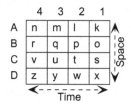

FIGURE 10.11 Time–space matrix.

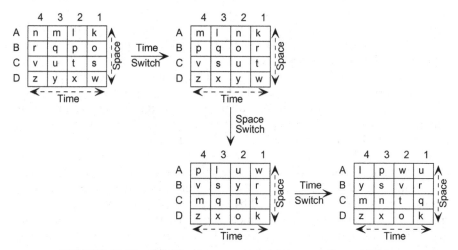

FIGURE 10.12 Combined conceptual time–space–time switching.

increasing from right to left), and space in the vertical dimension (with port number increasing from top to bottom), then $N \times G$ matrices can be used to represent the temporal and spatial position of all of the NG grains of a time–space–time switch. Figure 10.11 shows the temporal–spatial position of $NG = 4 \times 4 = 16$ grains, named $[k \ldots z]$. The temporally first grain on port A is k. The third grain on port D is y. This representation of time and space is used to explain time–space switching.

A complete time–space–time switch has N first-stage time switches (one per ingress signal), one space switch (with G different temporal settings), and N third-stage time switches. Each of these switching stages has the ability to reorder our temporal–spatial matrices on one dimension. N time stages allows each (temporal) row of the matrix to be rewritten (reordered). One space switch, operating in G time periods, allows each column of the matrix to be rewritten (reordered). The combined effect of all three reorderings is shown in Figure 10.12. The upper-left matrix represents the time–space position of the 16 grains as they enter the overall time–space switch (after temporal alignment to a common clock in the ingress line cards). After passing through the first temporal switching stage of N time switches, we get to the upper-center matrix, which may have arbitrary temporal rearrangements of each temporal row of the upper-left matrix. Then, after G time steps through a single space switch, we get to the lower-center matrix, which may have arbitrary spatial rearrangements of each spatial column of the upper-center matrix. Finally, after passing through N final space stages, we get to the lower-right matrix, which may have arbitrary temporal rearrangements of each temporal row of the lower-center matrix.

Consider the trajectory through time and space of some representative grains in Figure 10.12. Grain u appears in time 3 on port C in the upper-left matrix; u is moved to time 2, still on port C by that port's ingress time switch; u is then moved to port A, still in time 2, by the central space switch; finally, y is moved to time 1 (still on port A) by that port's egress time stage. Thus, grain u has moved from <C, 3> to <A, 1> in three separate switching steps. Grain z appears in position <D, 4> and remains in that position through all three stages of switching, to emerge from the overall time–space–time fabric without having changed its position. Grain t appears in position <C, 2>, is moved to <C, 1>, remains in <C, 1>, then is moved back to <C, 2>, to remain in the same position on egress from the switch as it was in on ingress. Thus, z and t experience no net movement in space or time, but follow different internal switching patterns. The traffic carried in Figure 10.12 has several distinguishing properties: Every ingress time–space position is filled by some named grain; every egress time–space position is filled by some named grain; and there is a one-to-one relationship between the ingress matrix and the egress matrix. Such switching loads are referred to as *saturated unicast* or *permutation loads*.

In earlier chapters we discussed the need for multicast switching. Time–space–time switches have the ability to perform multicast; in fact, they have three opportunities for multicast, one in each switching stage. Figure 10.13 illustrates these three multicast opportunities: We see grain p being multicast from <B, 3> to <B, 2> and <B, 4> in the first time stage for port B. Although no other traffic is shown, notice that port B has three free time slots in the upper-left (ingress) matrix but only two free time slots in the upper-center (after ingress time switching) matrix. Clearly, if port B had a full ingress load, one of its ingress grains would have to be dropped to allow grain p to be multicast (bicast). Grain s is also being bicast. The central space switch is also performing multicast: Grain s at <D, 3> is being bicast to <A, 3> and <B, 3>; grain q at <C, 1> is being bicast to <A, 1> and <B, 1>. The final time stage is also carrying out some multicasts: We see s being bicast on port A and separately on port D, and we see r being multicast on port C.

This pattern of multicast causes the one ingress copy of p to occupy two egress positions, having been bicast in the first time stage. Grain q occupies two egress positions, having been multicast in the space stage. Grain s occupies six egress positions, having gone from one ingress grain, to two copies after the first time stage (via one temporal bicast on port D), to three copies after

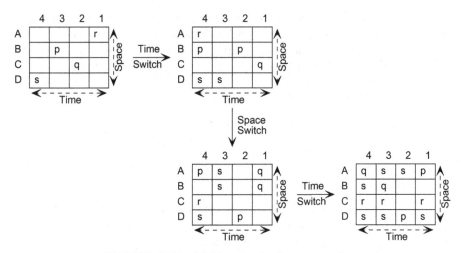

FIGURE 10.13 Multicasting in time–space–time.

the space stage (after one spatial bicast in time 3), and finally, to six copies after the final time stage (via a bicast on port A and *a* 3-cast on port D). The more complex trajectory of *s* shows that any grain can be replicated in a three-stage fanout tree, with any copy being replicated up to $G - 1$ times in any time stage and up to $N - 1$ times in the space stage.

10.8.1 Architecture and Costs of Time–Space–Time Switch Fabrics

Figure 10.14 illustrates the high-level architecture of a standard time–space–time switch fabric. Each ingress and egress port is linked to a central space switch by a time-switching stage. Time–space–time switches are composed of $2N$ time stages for G grains of B bits each, and one space stage of N ports of

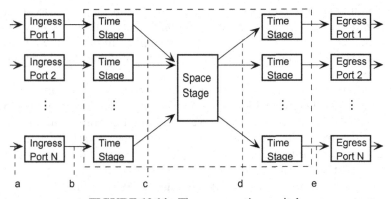

FIGURE 10.14 Time–space–time switch.

B bits each, connected as shown in the figure. Switching control features are not shown, but are straightforward: Each time stage requires $2G$ control words of $\log_2 G$ bits each; the space stage requires $2N$ control words of $\log_2 N$ bits each. Thus, the total important gate and flop costs of this architecture are

$$datapathFlops(N, G, B) = 2 \times 2NGB$$

$$datapathGates(N, G, B) = xbar(N, G, B) = B[fanout(N) + fanin(N)]$$

$$controlFlops(N, G) = 2 \times 2NG \log_2 G + 2NG(N-1)mux(2)$$

10.8.2 Blocking and OPA

The example of time–space–time switching we examined in Figure 10.12 shows that it is possible to switch a complete ingress load of grains to new egress positions. However, our experience in Chapter 9 with routing issues causes us immediately to ask two questions: (1) Is it possible to route arbitrary patterns of traffic through such a switch? and (2) Is there an algorithm to find such routings? Happily, when the traffic includes no multicast, the answer to both questions is "yes," but let's examine these issues in some detail. In the pages that follow we look at time–space–time routing from two points of view. Useful algorithms can be constructed from either point of view, but the first is probably more useful in explaining the routing issues, and the second is probably more useful in producing a useful algorithm.

The first point of view focuses on the space stage as the critical resource of the overall switch and the set of input-to-output mappings that must be threaded through the critical central space stage. We observe that the space stage gives us G opportunities to carry out N-to-N permutations. Clearly, any (unicast) traffic pattern on this switching fabric must fall within the limitations imposed by this set of G permutations. So this first view of the problem focuses explicitly on making effective use of the space stage. The goal is to avoid wasting switching opportunities in the space stage (i.e., using some time step to switch fewer than N grains through the space stage). Let us represent the overall switching problem as a set of input-to-output mappings, as shown in the stylized Venn diagram in Figure 10.15, which represents a complete unicast load in that each of NG ingress time slots is mapped to one of the NG egress time slots. The items in the Venn diagram are the individual port and time slot mappings.

From a mathematical point of view, we can consider the mapping of the Venn diagram in Figure 10.15 as a single permutation from NG input opportunities to NG output opportunities. We could number both input and output opportunities from 1 to NG and consider permutations on this sequence of numbers. The individual mappings in Figure 10.16 are the equivalent problem. We know that we must be able to decompose the set of mappings into G individual unicast mappings from the N ingress ports to the N egress ports. Let us begin by finding and extracting such an N-to-N unicast mapping from

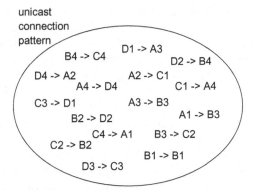

FIGURE 10.15   Representative set of TDM connections.

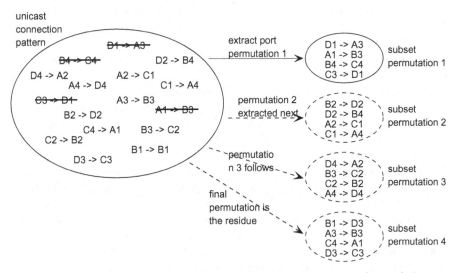

FIGURE 10.16   Extracting subpermutations for a time–space–time switch.

our complete Venn diagram. Any such subset must consume each ingress port once and must consume each egress port once. Figure 10.16 illustrates this process. These subsets are found by search. We keep a list of unused ingress and egress ports. At each step we select the next unused ingress port. Then we look through the remaining mappings in the Venn diagram until we find one that uses the ingress port selected. If that mapping requires an unused egress port, we remove that mapping from the Venn diagram, add it to the current subset mapping, and mark the ingress and egress ports used. If that mapping requires a previously used egress port, we continue looking for a mapping from the ingress selected to an unused egress.

In our example the first permutation for this space switch includes the input-to-output mappings D1 → A3, A1 → B3, B4 → C4, and C3 → D1. The set of required ingress ports is {A, B, C, D}; the set of required egress ports is also {A, B, C, D}. Figure 10.16 goes on to show the selection of the subsequent $G - 1$ permutations. Notice that after the selection of the third permutation, the remaining mappings are B1 → D3, A3 → B3, C4 → A1, and D3 → C3. This residual set of mappings is itself a permutation set, so the last complete permutation exhausts the original set of mappings which were to be routed through our time–space–time fabric.

Notice that we have paid no attention to the ingress and egress times of the various mappings as we assigned them to times to cross the central space stage. For instance, the first permutation we selected has mappings from time slots {1, 3, 4} to time slots {1, 3, 4}, yet presumably this permutation will be assigned to the first time slot through the space stage. It is necessary to align the inputs of these four mappings to the time they all pass through the space stage. Thus, we must begin in the first time stage with the mappings D1 → D1, A1 → A1, B4 → B1, and C3 → C1. This aligns all four connections to move through the space stage together in time slot 1. The space stage then must move each ingress port's grain to the correct egress port, by continuing the overall mappings begun in the first time stage: D1 → D1 → A1, A1 → A1 → B1, B4 → B1 → C1, and C3 → C1 → D1. Now all the grains are on the desired egress ports but may not be in the correct time slot for emission. The final time stages adjust each grain to the proper time slot by adding one more mapping to the grain-path mappings we have accumulated in this paragraph: D1 → D1 → A1 → A3, A1 → A1 → B1 → B3, B4 → B1 → C1 → C4, and C3 → C1 → D1 → D3. In this manner, the ingress time stages are used to move grains to the space-stage permutation time to which they have been assigned, and the egress time stages are used to move grains from these space-stage permutation times to their assigned output time slots. Once the space-stage permutation assignments are made, the required settings for the time stages are obvious and always nonblocking. The determination of each time adjustment requires only a very small fixed amount of computation time.

Any such set of NG input-to-output permutations mappings can always be decomposed into G N-to-N permutations. The proof of this property is beyond the level of sophistication in discrete mathematics assumed by this book, but as some classes using this book can assume the necessary background, a proof of this property will be left as an exercise. For the moment, however, we merely observe that the set of G permutations, each permutation consumes a weight of 1 for each ingress and egress port, and that the sum of such weights over all G space-stage permutations adds up to the weight of G time slots per port assumed by the TDM model. This algorithm requires that we find G subset permutations. Each such mapping requires that we find N input-to-output mappings. Finding each such input-to-output mapping requires, as a worst case, that all the remaining input-to-output mappings be checked. On average, there are $(NG)/2$ remaining input-to-output mappings to check. As a consequence, the cost of this algorithm grows as $NG(NG/2)$ or, $N^2G^2/2$. A serious limitation of this algorithm is that it assumes that the switching setting effort will be redone each time any connection is changed; that is, the algorithm is not suitable for rapid incremental switching changes. This limitation of our first algorithm is added motivation to look for a second approach to this problem.

Our second approach to an open-path algorithm for a time–space–time fabric is strongly related to the algorithm established in Chapter 9 for finding open paths in Clos networks. Recall that we developed a graph algorithm that represented first and third Clos stages as nodes and represented each center stage as a unique line type. The fact that each first stage could communicate only once to each center stage switch was represented by the rule that each first-stage device could use each line type only once; similarly, the fact that each last stage could communicate only once from each center stage switch was represented by the rule that each last-stage device could use each line type only once. Each line type represented a complete path through one of the central switch elements.

This Clos OPA problem is adapted to time–space–time OPA needs. The classic time–space–time architecture has only one central stage—the space switch—so does not seem to fit the Clos OPA model of allocating paths to and from multiple center stages. But the TDM division of the links to the single central space stage in time–space–time fabrics provides an opportunity to state a strong correspondence between the two OPA problems. With time–space–time switches, each time slot opportunity through the central space stage is considered to be an alternative path and an alternative line type in the graph model of the algorithm. During each such time slot, the central space stage is reconfigured by its control memory; each such configuration allows a permutation of grains to pass through the space switch; each such permutation is considered to be a separate routing path controlled by the line types of the basic OPA algorithm. Thus, we were left with a graph edge type of problem: of finding an assignment of connections to line types such that neither the input nor the output line type restrictions were violated.

The first- and third-stage time switches are the nodes of the graph. The times through the center switch stage become the line types of our graph problem. Clearly, each time stage can communicate with the center space stage only once per time. Hence, the time stages can terminate each edge type only once, and there must be G line types.

Figure 10.17 is a representative example of this algorithmic view of time–space–time routing. There are $N = 3$ ports and $G = 3$ time slots. We see five ingress-to-egress mappings assigned to the space stage [e.g., in time step 1, ingress time stage 1 sends a grain to egress time stage 3 (the solid line)]. As only five of nine possible connections are made through this space stage, we expect to see idle paths—in this case, time slot 1 has only one assigned mapping, and each of the other time slots have two of three opportunities in use.

This algorithm is intended to be incremental; that is, new connections can be added without starting over as was necessary with our last algorithm. Let

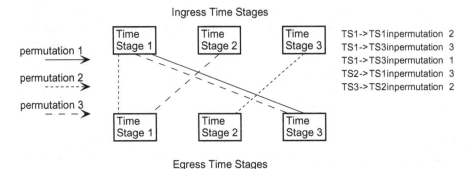

FIGURE 10.17 Abstract graph model for time–space–time open-path algorithms.

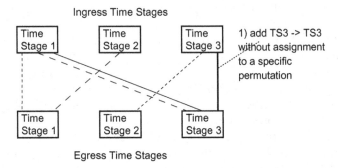

FIGURE 10.18 Adding the new connection.

us examine the addition of a new ingress-to-egress mapping to the connection status represented by Figure 10.18. The new connection goes from ingress time stage 3 to egress time stage 3. This new connection is added in the figure as a tentative new line type (heavy solid line). With this new connection/line type added, our immediate goal is to find one of the nontentative line types in this problem which is unused at ingress 3 and at egress 3. If such a line type exists, we immediately assign this type to our new, tentative connection, and are done. However, in this case, each of the three legal line types is used at either ingress 3 (short dash) or egress 3 (long dash and solid). Figures 10.18 to 10.25 walk through the algorithm in detail. In the first step, the new connection from time stage 3 (TS3) to time stage 3 is added without being committed to any permutation (line type). In these figures a heavy solid line is used to indicate such uncommitted connections.

With the new, uncommitted connection in place, examine the two time stages at the ends of the new connection for any line type that is not used by either time stage (Figure 10.19). Should such a line type be available, it and the space permutation that it represents are selected for the new connection,

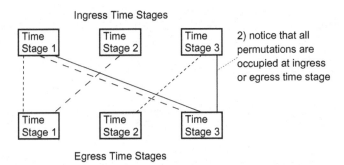

FIGURE 10.19 Checking for a free line type (space permutation).

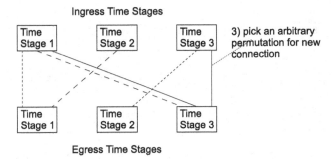

FIGURE 10.20 Picking an arbitrary line type when no free line type exists.

and the algorithm is done. In our running example, however, we find that all available line types are used either at the ingress or egress time stage.

When the two end stages of the new connection have consumed all line types between them, as shown in Figure 10.20, an arbitrary choice is made for the line type of the new connection (in this case, the light solid line type is selected). The reuse of a line type forces a conflict over line type assignment at one of the two endpoints of the new connection (Figure 10.21). The algorithm determines which end has the conflict. In this case, egress time stage 3 has the conflict, brought about by reuse of the light solid line type at the node.

The line type conflict at egress time stage 3 is resolved by arbitrarily switching the connection with the previously existing use of the conflicted line type (from time stage 1 to time stage 3) to some new line type. This new choice of line type is selected from the set of line types not otherwise used by egress time stage 3, in this case the short dashed line type. It is important that this changed line type move away from the newly added connection instead of again changing the line type of the new connection (Figure 10.22). If the algorithm did not continue away from the point of the initial conflict in this

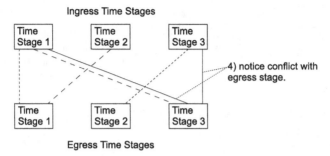

FIGURE 10.21 Discovering where the conflict occurs.

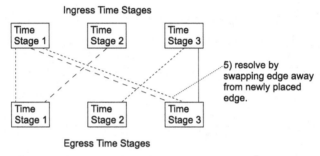

FIGURE 10.22 Resolving the conflict by swapping a line type away from the problem.

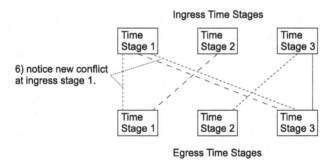

FIGURE 10.23 Checking for and finding a new conflict.

manner, it would be subject to nontermination, due to continual reflipping of the same edges or edges.

The line type change at egress time stage 3 to the short dashed line type causes a conflict at ingress time stage 1 (Figure 10.23). This new conflict at ingress time stage 1 is resolved in the same manner as the earlier conflict at

Ingress Time Stages

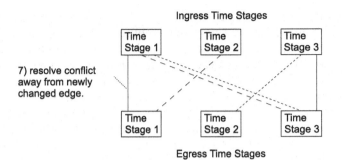

7) resolve conflict
away from newly
changed edge.

Egress Time Stages

FIGURE 10.24   Resolving the new conflict away from the original problem.

Ingress Time Stages

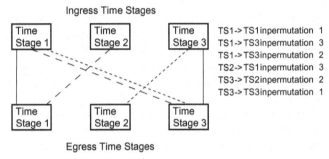

Egress Time Stages

FIGURE 10.25   No remaining conflict—OPA success.

egress time stage 3: A new, otherwise unused line type at ingress time stage 1 is used for the conflicting edge farther away from the original conflict (Figure 10.24). This new change leads to no further conflicts at either end of the newly changed edge from ingress time stage 1 to egress time stage 1. This means that the algorithm is complete, and the set of connections has no conflict in line type or in use of the permutations scheduled through the central space stage (Figure 10.25).

This algorithm follows a chain of connected lines, alternating between ingress and egress time stages. There are at most NG such edges, so the costs of this algorithm are bounded by $O(NG)$. In practice, this algorithm terminates much faster than $O(NG)$, as it is unusual for a conflict chain to exist throughout the entire NG length.

We have observed that a time–space–time fabric has one multicast opportunity per stage, allowing multicast trees of up to depth 3. However, the addition of multicast severely complicates the clean picture of an efficient, rearrangeably nonblocking switch fabric that we have seen thus far. The complications due to multicast are quite similar to the complications we found in physical circuit multicast. From the point of view of the open-path algorithm, a multicast through the space stage represents two or more edges that must be switched jointly from one line type to another. This means that the frontier of swapping line types can grow wider, and in fact can wrap back on itself to form loops in the edges which cause the algorithm to loop unendingly.

It is true that multicast can be carried out in the first stage, via time. However, this option is available only if there are otherwise unused time slots from the time stage in question to the space stage. Additionally, if all endpoints of a multicast are different time slots on the same egress port, the multicast can certainly be accomplished in the egress time stage (there must be unused incoming time slots, as we are limited in the number of output time slots in the egress link, and the multicast will consume some of these). Overall, the use of multicast in these simple time–space–time fabrics is a very challenging issue. Any degree of multicast can cause blocking, but unless the extent of multicast is high, the probabilities of blocking are low.

We will now examine two methods of improving the OPA routing performance of time–space–time networks in the face of some degree of blocking. The first technique is to add speedup to the center of the fabric. The ingress time stages accept G grains. Normally, they also emit G grains, but it is possible to add extra bandwidth at this point. Where S is a speedup factor ($S > 1$), we can give each stage one time-stage SG egress time slot by increasing the transfer rate on these links by S. Thus, the ingress time stages receive G grains but emit SG grain slots, allowing added flexibility for paths through the space stage. The space stage must run at a clock rate S greater, to honor SG time slots. The egress time stages must accept all SG ingress grains and emit G grains.

Double speedup could be provided by giving each ingress time stage two egress links, each capable of carrying G grains. This requires twice as many ports on the space stage and two links into each egress time stage. Small amounts of speedup greatly reduce the probability of blocking. This same speedup also decreases the OPA algorithm time to completion by increasing the probability of finding spare time slots within the fabric. Nevertheless, speedup does not absolutely preclude blocking until the amount of speedup is $S = N$ for $1:N$-casts.

10.8.3   Space–Time–Space Switching

To this point in the chapter we have concentrated on time–space–time fabrics. The time–space switching paradigm is much more general than is apparent from having reviewed this one form of switch. There are two principal ideas that generate alternative network forms: Time and space stages can be systematically exchanged to form new fabrics, and deeper networks of five or perhaps more stages can be used. We examine briefly the dual fabrics with exchanged time and space stages. Although deeper networks have been used in commercial products in recent decades, there appears to be little interest in these forms

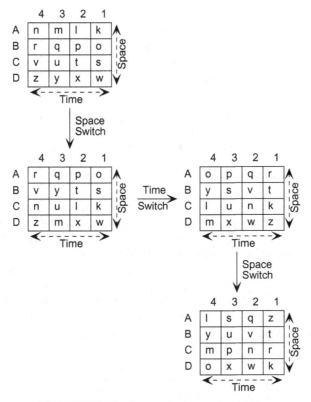

FIGURE 10.26 Space–time–space switching.

today, due to their costs and the ability of current VLSI technologies to construct huge three-stage networks.

In a space–time–space fabric, as the name indicates, there are space stages in the first and last positions, with a set of time stages sandwitched between them. Switching is accomplished by G space permutations, followed by N time permutations, followed by G final space permutations. Figure 10.26 illustrates this conceptual switching pattern. The OPA algorithm for space–time–space switches is a simple dual of the algorithm for time–space–time switches. Recall that the time–space–time algorithm focused on the allocation of paths through the central space stage, where there are G N-to-N space permutations. The time stages are used to line up permutation groups for the center stage, and finally, to move grains to their correct output time slots. The space–time–space focuses on the allocation of paths through the central time stages, where there are N G-to-G time permutations. The ingress space stages are used to line up temporal permutation groups for the center stages, and finally, to move grains to their correct output ports. All of the ideas developed above regarding

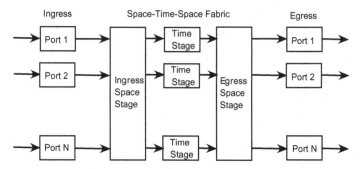

FIGURE 10.27   Space–time–space switch fabric.

time–space–time switching apply to space–time–space switching with N and G, and space and time interchanged. The physical hardware required for a space–time–space switch is illustrated in Figure 10.27.

The cost of a space–time–space fabric is easily derived from the analysis of the costs of a time–space–time switch. Space–time–space fabrics have two NN crossbars and $N\,GG$ time stages (whereas the time–space–time fabrics had $2N$ GG time stages and one NN crossbar). Depending on the relative costs of time stages (principally, memory) and crossbars (principally, VLSI gates and wires), one or the other approach may appear to be more economical in a given technology.

However, time–space–time fabrics have an important advantage that is difficult for space–time–space to overcome. It is natural to place both ingress and egress time stages on the port cards in most designs that have multiple line cards and one switch fabric card. This has two important advantages: the overall size of the switch core (now just the single space stage) shrinks substantially, allowing the central switch core to reside on one circuit card; and the fabric now requires only one duplex link between the line cards and the switch card. Compare this to a space–time–space fabric (which is too large to fit on a single printed circuit card): In this configuration, the initial and final space stages cannot be placed on the line cards, as their inputs/outputs cross all line cards. Similarly, although the central time stages could be placed on the line cards, there is little motivation to do so, and generally, they would have to exist on core switch cards with the space stages. The number of intercard links then must be one from the line card to the first space stage, one from there to the time stage, one from there to the second space stage, and one from there back to the line card. Thus, four intercard links are required, compared to only two such links for time–space–time switches. Thus, time–space–time fabrics have generally dominated.

Nevertheless, there is a significant difficulty in having time stages on the line cards: Changing switch settings to implement new connections requires that the control system make nearly simultaneous changes to settings on the ingress line card, the central space stage, and the egress line card. This introduces control and protection complications. In some time–space–time switching products, these difficulties force the time stages into the central fabric.

In time–space–time or space–time–space switching systems that do not push the achievable density of circuit boards, it may be possible to place the entire fabric on a single core switch card, in which case the relative advantage of time–space–time disappears, and the raw hardware comparison of these two approaches should dominate. For many technology cost combinations, space–time–space solutions may be less expensive when they fit on a single printed circuit card.

10.9 MULTISTAGE MEMORY SWITCHES

A second major approach to building networks for switching TDM signals is three stages of memory switches, as shown in Figure 10.28. Each memory stage has multiple ingress and egress ports. Their numbers may be balanced, biased toward speedup (fewer ingress than egress ports), or biased for slowdown (fewer egress than ingress ports). In all cases, these switching elements are SNB switches. They may be built using any of the SNB architectures explored earlier in the chapter: memory, ingress buffered, egress self-select, or others.

If should be pointed out that a memory switch can be viewed as a switch that can switch in both time and space at once (by changing both port and time by reading from an arbitrary location in a conceptually two-dimensional memory, where the two dimensions represent time and space in our TDM switching problem). Thus, we can consider working with time, space, or memory switching elements. Our earlier pessimistic conclusions regarding the upper

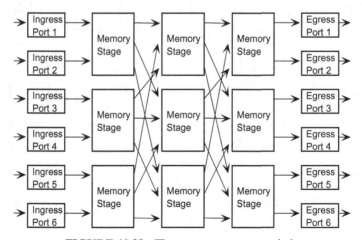

FIGURE 10.28 Three-stage memory switch.

practical limit of memory switch sizes still apply, but here we consider the use of many small (and practical) memory switches to compose a much larger three-stage switch). Switch fabrics are designed much like a Clos network: some degree of speedup, S, from first stage to second stage, and then the same degree of slowdown, $1/S$.

Recall from our consideration of time–space–time blocking and OPA issues that a grain had to pass through the central space stage at a particular time and that our difficulty in finding an open path centered on finding a time that the grain could get from an ingress to an egress time stage. With the use of memory switch elements in the center stage, this restriction to a particular time is removed. If a grain is routed through a particular center stage, it does not matter when in the grain group period it arrives or when it departs. The memory switch can move the grain in both time and space, so it can compensate for whatever choices are made. Hence, the routing problem becomes just an issue of finding a center stage with any available grain time from the ingress memory switch and with any available grain time to the egress memory switch. This is a much easier problem and results in an OPA that runs faster for unicast and multicast and blocks significantly less often for multicast.

A practical OPA for a three-stage memory switch can focus on the number of available grains on the various links from stage 1 devices to stage 2 devices, and from stage 2 to stage 3. Figure 10.29 represents the state of such a switch. The numbers on each link are the number of free grains on those links. The goal in Figure 10.29 is to add a new connection from A to B, from ingress port 1 to egress port 6. Clearly, it is necessary to find an open path from ingress 1's first-stage memory switch to egress 6's third-stage memory device. Such a middle-stage memory switch has a nonzero count from the first-stage switch and to the third-stage switch. From the first-stage switch, we can get grain A

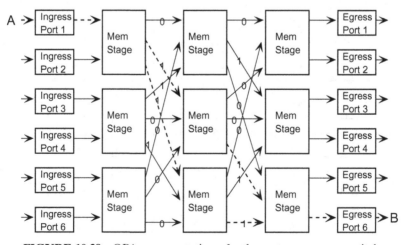

FIGURE 10.29 OPA representation of a three-stage memory switch.

to the second or third middle-stage switches; we can get to the third-stage memory switch from the second or third middle-stage switches. The intersection of these two sets is the second middle-stage switch, which is assigned to this new connection. This is the essence of the OPA for three-stage memory switches. It is a slight generalization of the OPA for Clos networks: With Clos networks, a link was available or not; with three-stage memory switch fabrics, a link has from 0 to G free paths. But in the final analysis, a link is or is not available to a new grain.

There are a variety of options in designing an OPA for a three-stage memory fabric: When there are multiple middle-stage paths, which should be taken? One approach is to try to balance the use of all links, so one adds the new call to the least used pair of middle-stage links. Another approach is to try to balance the use of middle stages, so one adds the new call to the least-used middle stage. Alternatively, there is benefit in adding new connections on maximally used resources in order to maximize the density of use of some resources, leaving other resources maximally free. Studies have shown that all of these alternatives are useful. What is less useful is random assignment. However, the differences are subtle, as open-path identification in three-stage memory switches generally works very well, especially in fabrics with some degree of speedup. Speedup can be obtained by extra stage $1 \rightarrow 2$ and stage $2 \rightarrow 3$ links, or by temporal speedup on these links.

Alternative Multistage Switching Fabrics The literature of switch and network research has presented many alternative networks, principally since Clos's work in the 1950s. Although many of these switching networks contain interesting ideas, and a few have been used in practice, the use of switching technologies in transport networks has been dominated by various SNB memory switches, time–space–time switches, and three-stage memory switches. Without intending any criticism of these other approaches, we leave the study of alternatives to another book.

10.10 SUMMARY

TDM (logical circuit) switching is conceptually very simple. The initial obvious architecture and hardware solutions for TDM switching are also very straightforward. Happily, these central memory switches are strictly nonblocking and require no OPA. But these simple models do not scale well, as larger TDM switches are required. Technological limits to the speed of memories; the number of gates, flops, and RAM bits that can be used economically on a single VLSI substrate; and the amount of power that can be dissipated on a single VSLI substrate combine to limit the utility of the central memory TDM switch.

A variety of responses to these limitations are examined in the chapter. The first group attempts to retain the SNB and OPA-less properties of the ideal switches as long as possible. These switches include the ingress buffered switch, which gets around central memory switch problems but raises serious power system issues; the egress self-select switch, which retains the scaling advantages of the ingress buffered switch but also resolves that switch's power system issues; and the bit-sliced architectures, which extend the range of either the ingress buffered or egress self-select architectures.

Admitting that the SNB solutions above will fail to provide sufficiently large switches for current and expected Internet needs, the chapter presented and analyzed several multistage TDM switching solutions. Time–space–time, space–time–space, and multistage memory switches were studied. For each, the conceptual model and architecture were presented, the costs were examined, and the blocking issues and suitable OPAs were examined.

KEY POINTS

- There are a number of SNB and OPA-less architectures for TDM switching.
- These SNB or single-stage architectures are all based on the simple central memory model; the more feasible solutions differ in architectural detail but retain the obvious SNB properties.
- The simple central memory design is severely limited in the bandwidth it can absorb but is the idea switch for very small problems (problems too small to be of interest to transport networks).
- The ingress-buffered switch scales well and retains good block/OPA properties but suffers from the critical flaw that it requires a severe spike in power consumption once per grain group. This power spike is very difficult or expensive to design around with current VLSI technologies.
- The egress self-select switch retains the highly favorable scaling and blocking/OPA properties of the other single-stage networks. In addition, it solves the power spike problem of the ingress-buffered TDM switch. Consequently, this architecture is highly attractive for a wide variety of applications.
- Either the ingress-buffered or the egress self-select architecture can be generalized and increased in scalability using the bit-slicing technique. This technique slices the TDM grains and transmits the multiple slices from each grain simultaneously to a stack of bit-sliced SNB switches, which themselves switch the multiple portions of the grain in parallel, to a common egress port where they are reassembled into the original unsliced grain for emission on the proper outgoing link. This technique allows a substantial increase in the range of achievable SNB switches in any technology.
- Time–space–time switches represent another way to expand the scale of feasible switches. These fabrics have three switching stages: time, preced-

ing space, preceding a second time stage. These fabrics are rearrangably nonblocking for unicast traffic but blocking for multicast traffic. Happily, OPAs exist that rearrange unicast calls efficiently. These algorithms have been adapted for use with limited degrees of multicast (although details were not studied in detail in the chapter).

- Space–time–space switches are a dual of time–space–time switches, have the same essential blocking issues, and have similar OPAs. Which of these two models is superior depends on technologies and the size of the overall switching solution. But in general, time–space–time is more useful because it can be built with one-half of the blackplane input/output that a space–time–space switch may require.

- Three-stage memory switches are similar to time–space–time and space–time–space switches, but support much simpler and hence more reliable and rapid discovery of open paths.

REFERENCES

Clos, C., A study of non-blocking switching networks, *Bell System Technical Journal*, vol. 32, no. 5, pp. 406–424, 1953.

Hui, J. Y., *Switching and Traffic Theory for Integrated Broadband Networks*, Kluwer Academic Publishers, Norwell, MA, 1990.

The 5ESS switching system, *AT&T Technical Journal*, vol. 64, no. 6, pt. 2, 1985.

11

PACKET AND CELL SWITCHING AND QUEUING

Network Infrastructure and Architecture: Designing High-Availability Networks,
By Krzysztof Iniewski, Carl McCrosky, and Daniel Minoli
Copyright © 2008 John Wiley & Sons, Inc.

11.1   INTRODUCTION

This chapter is the third in a sequence of three chapters on switching technologies. In Chapter 9 are provided an overview of switching issues and considered the switching of complete physical signals. In Chapter 10 we examined the switching of logical channel [or time-division-multiplexed (TDM)] signals such as SONET/SDH. This chapter covers the switching of packets and cells. After briefly defining the packet–cell switching problem, we present a series of queuing models and policies for switches. With the ideas of queues in hand, the chapter will turn to study a series of progressively more sophisticated architectures for packet and cell switching. With each solution, we study blocking properties, multicast, scalability, arbitration methods, and cost.

11.2   PACKET–CELL SWITCHING PROBLEM

In earlier chapters of the book we examined existing packet and cell formats and protocols in some detail; in this chapter we ignore most of the details, and present a more abstract view of switching of packets and cells. Our reason for this approach is that the essential ideas of this type of switching easily apply across a wide range of protocols. The details of packet and cell formats are just details. They can be worked out with each new protocol encountered, as long as the general view is well understood. A second reason to examine this material from an abstract point of view is that there are many packet–cell protocols and there is no sensible single choice to study.

Transport packets are varying-length sequences of bytes. Any packet protocol has some minimum length of packet; we refer to a packet of this length as a *minigram*. Similarly, there is some maximum length; we refer to a packet of this length as a *maxigram*. For most protocols, any length between these two is legal; for some protocols, only multiples of some number of bytes are legal lengths (e.g., multiples of 4 bytes). Packets may describe their own lengths, or their length may be determined by the mechanisms of some physical layer coding or carrying protocol [e.g., Ethernet packets embedded in their native physical layer or in the generic framing procedure (GFP)]. Cells are fixed-length sequences of bytes. Cells do not carry any indication of their own length, as that length is implicit [e.g., asynchronons transfer mode (ATM)].

The term *frame* is generally used as a synonym for packet, as in Ethernet and in frame relay. We adopt a nonstandard use for frame in this book: We use it to refer to either a packet or a cell when the differences between the two do not matter. As we shall see in later sections of this chapter, the most important practical differences between packets and cells have to do with the design of the data paths that handle them.

From an abstract point of view, a switch or router must derive at least the following five pieces of information from each incoming frame and from tables stored within the switch or router: the length of the incoming frame; whether the incoming frame has been corrupted; which egress port the frame should be directed to; what new label, if any, should be put on the frame when it exits the egress port; and with what priority the frame should compete with other frames for the required egress port.

Frame Length Frame length can be determined in several ways. The most common are by an explicit-length field in the frame header, by the embedding of the frame in a lower-layer protocol that controls and specifies frame starting positions (e.g., packet ower SONET), or by an assumed length (e.g., the 53-byte cell length of ATM). In some protocols or combinations of protocols, there are two independent sources of frame-length information. When this is the case, they are both used as a self-checking source of length information.

Self-Checking Most frame protocols carry redundancy fields that allow the correctness of the received frame to be probabilistically verified. In some protocols, only the header information is covered because it is assumed that the payload is protected by some end-to-end, higher layer of protocol (e.g., ATM); in other protocols, the entire frame is self-checked (e.g., TCP). Also it is common that the frame header carry information that can be checked for internal consistency: for instance, only a subset of the bit combinations of certain command or type fields may be used, in which case a check can be made for illegal codes. Failure of these self-checking mechanisms as the frame is received causes the frame to be dropped, counted, and reported as a statistic to the OA&M system, and possibly negatively acknowledged.

Next-Hop Determination In any N-ported frame switch, each arriving frame must specify in some manner at which of the N egress ports it should be directed. There are a large number of ways in which this is done:

1. Ethernet switches use a 48-bit unique Ethernet address as an argument to a table-lookup function that returns the desired egress port. The number of possible addresses is huge ($2^{48} \approx 2.8 \times 10^{14}$), so Ethernet switches use sparce table-lookup techniques such as content addressable memory (CAM), hash tables, or binary trees.

2. IPv4 routers use a 32-bit address as their table-lookup argument. Commonly, routers will have the final routes for a relatively small number of addresses, and default routes for the rest of the huge address space. Again, sparce table-lookup techniques are useful.

3. ATM cells carry much smaller routing fields. Depending on where in the ATM network the cells are found, their effective routing fields can be 12

bits (VPIs in the NNI definition) or 24 bits (VPI and VCI in the UNI definition). Direct memory-lookup techniques are appropriate for ATM.

4. When MPLS frames are created by label edge routers (LERs), arbitrary aspects of the incoming frame are inspected to determine a forwarding equivalence class and the initial MPLS label to be appended in the new MPLS header. This process can be complex, so it is restricted to occurring only once per packet lifetime in an MPLS network.

5. Once MPLS frames have been classified and tagged by an LER, they carry a 20-bit label which is used to determine the next hop (and the next label) in each label swap router. These fields are intended to be small enough to allow direct memory lookup to determine the egress port.

Next Label   Some protocols use only end-point labels. Such labels are held constant during the lifetime of the frame as it is forwarded from switch to switch or from router to router. IP, with its destination address label, is the most important example of this sort of protocol. With such protocols, the switch does not find a new label for each frame emitted. However, some protocols determine a new label for each switch-to-switch hop that a frame takes. MPLS is an example of this approach. New labels are found by table lookup on the old, incoming label. The reason to swap labels is to avoid having to reserve labels all the way across a potentially large network. Since labels are unique only to individual switch pairs, each pair of connected switches can make their local agreements without concern for use of the label elsewhere in the network. This allows label reuse and the considerably smaller labels required by MPLS.

Class of Service   Many of the protocols that are of interest in the transport networks have some mechanism of specifying or determining a class of service for each frame. The details of these mechanisms vary widely and are sometimes quite complex. The key objective is to distinguish classes of traffic that require different levels of services. These different levels of service translate, through the queuing and arbitration mechanisms that will be studied in later parts of this chapter, into different priorities in being routed through the frame switch.

11.3 TRAFFIC PATTERNS

Frames bring two radical changes to the traffic models and blocking issues we saw with physical and logical circuit switches.

1. Traffic is now broken into relatively small segments, each of which is self-addressing. Thus, any frame switch must respond continually to incoming connection requests (i.e., frame destination addresses) in the data flow. With physical and logical circuits, there is no comparable segmentation of the data stream, and no addresses are recognized in the data stream.

2. In general, there is no overall discipline to control the flow of frame traffic. In physical and logical networks and switches there is an implied call acceptance control (CAC) mechanism that does not allow a new call (connection) to use a physical or logical network link unless that network resource is unused. If a network resource is in use, the attempted new call is blocked. If the required network link is available, the attempted new call is granted. In most packet protocols, and certainly in IP, there is no notion of call acceptance control. Each host is free to send an IP datagram to any IP address at any time. A consequence of this freedom is that loads throughout a network are unpredictable. It is always possible for the aggregate ingress flow to a switch or router to exceed the egress capacity of some link(s). Protocols, switches, and routers must be designed to cope with such traffic overloads. There are subsets of protocols that are protected from overload conditions throughout a network by CAC mechanisms for selected classes (e.g., in ATM). These mechanisms can be effective, but the overall goal of today's transport networks of carrying IP means that ultimately, the network is faced with traffic with little or no effective flow control (or at least that IP is not an effective source of the required control, which must then be either abandoned or sought elsewhere).

In practice, switches can be exposed to a nearly infinite variety of traffic load conditions. It is useful, however, to examine queuing systems when presented with two types of loads: benign loads and hotspot loads.

- *Benign loads.* A benign (or ideal) load consists of permutations of frames such that each ingress and each egress is kept equally busy. This is the simplest conceptual load model that can be used to analyze a switch or queuing system.
- *Hotspot loads.* A hotspot (nonbenign) load has some concentration of bandwidth at some subset of egress ports, such that the sum of bandwidth directed at these outputs exceeds their capacity for some period of time. The degree of concentration of traffic at hotspot egresses and the duration of that concentration determine the degree of difficulty presented to the queuing system or switch. In general, we allow hotspots to be arbitrarily severe when examining their effects on queuing systems.

11.3.1 Realistic Loads

Neither of the above load models is intended as a realistic representation of actual network traffic. Actual behavior is much more complex, with dynamically changing requirements and traffic loads coming in periodic bursts. These simpler benign and hotspot load concepts are used only to study the potential behaviors of switches.

Responses to realistic loads are important. These responses are studied in academic and industrial research. However, from the switch architecture point of view, there is a case to be made that these realistic loads are of only secondary importance. Transport quality switches must not allow congestion at one egress port to affect other flows, with the single exception of lower-priority flows to the congested egress port. To ensure this level of quality, it is necessary to consider arbitrarily severe hotspot loads. If transport-quality switches are made to be completely (or very nearly) nonblocking under severe hotspot loads, they will behave as well as possible under realistic loads. Traffic may be delayed due to congestion at its egress port, but never due to artificial limitations in the switch and never due to congestion on crossing traffic. When this is the case, all delays due to congested egress ports simply build the depth of queues which feed that port. At some point, the queues will fill and frame loses will be forced. So the behavior of a (nearly) ideal switch is limited primarily by the depth of its queues and, of course, the relative bandwidths of its various ports. Realistic traffic patterns can and should be used to tailor the depths of queues and to size telecommunication links, but not to determine the nature of the switching architecture, which in the world of transport networks should be essentially ideal. Realistic traffic patterns are also important to the design of policies regarding the sharing of buffers between otherwise independent queues.

As hinted in the preceding paragraph, it is possible to adapt to realistic loads by tuning the depth of queues in switches or by tuning the bandwidth of various network links. One must be careful, however, in excessive increases to queue depths:

1. Increasing depth adds cost, of course.
2. Increasing depth also adds to the maximal latency traffic can experience. Excessively late delivery is a compromise of overall quality of service, and is particularly painful when it is accomplished by paying substantially more for queue memories throughout a network.
3. Finally, research in recent decades on the bursty nature of network traffic indicates that the gains to be had from increasing queue depths are small, as the highly bursty nature of data traffic tends periodically to overwhelm even deep queues.

11.3.2   Responses to Congestion

Any switch subject to more ingress traffic for a particular egress with an arbitrary frame protocol has only a few options: (1) send back pressure signals to all sources of a traffic overload, to slow or stop the incoming flow; (2) hold excess traffic in a queue until the temporary excess of traffic passes and the excess traffic can safely be emitted; (3) misdirect some of the traffic by sending it out some egress link other than the ideal egress link; or (4) drop the excess traffic.

Serial RIO (sRIO) is an example of a protocol that employs a form of back pressure. sRIO allows frames to take forward hops only when it is known that there is a free buffer for that frame and another buffer for each higher-priority frame that may come along. Although this is useful in local domains, it is not a good network solution, as it is prone to developing congestion trees (which are discussed later in this book). In addition there are multipath protocols that allow packets to take any one of multiple possible output links. There may be a preferred forward link, but if it is unavailable, or its egress queue is overly long, a packet can be forwarded on some other link. This solution allows packets to flow around congestion but leads to misordering of packets when they reach their destinations. Use of IP prohibits both the back-pressure solution and the misdirection solution, leaving us with only the queue and drop options. (Recall that IP has essentially no discipline. In TCP/IP, it is TCP that provides effective control with its receiver windowing mechanism, but this applies only end to end and gives no protection to hop-by-hop internal network behavior.)

11.4 LOGICAL QUEUE STRUCTURES AND THEIR BEHAVIOR

As we have seen, switches and routers hold temporarily unforwardable frames in queues, to await emission on the required egress link. These queues, their locations, their policies with respect to use by ingress, selection by egress, and sharing are deeply important to the design of switches and routers.

11.4.1 Queues

Each queue is a first-in first-out (FIFO) data structure . All frames contained are sorted from the youngest frame at the tail to the oldest frame at the head (Figure 11.1). New youngest frames are added to the tail of a queue by an enqueue event. The oldest remaining frame is removed from the head of a queue, by a dequeue event. There are many ways to implement queues,

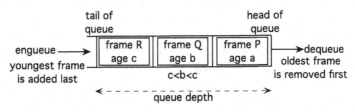

FIGURE 11.1 Packet–cell queue.

including linked lists and pointers into circular memory buffers. These details are not particularly important to our examination of switch architectures. If one proceeded from the concepts covered in this book to designing a more detailed switch microarchitecture, these issues would become vital. But for now we simply assume the use of any reliable queue implementation. It should be pointed out that we restrict all queues to finite size. This is an obvious restriction for anyone concerned principally with constructing hardware, but it is also an important basis of some of the performance statements that will be made of the various queuing systems. The principal point is that any queue will eventually overflow if subject consistently to more enqueues (insertions) than dequeues (removals).

11.4.2   Flows and Logical Queues

A *flow* is an ordered stream of frames from some ingress, I, at some priority class, C, to some egress, E. We use the notation $F_{I,C,E}$ to denote a particular flow (e.g., $F_{1,2,3}$ represents the flow from ingress 1, at priority 2, to egress port 3). Commonly, different flows occupy the same physical path in a switch or a network. We denote this with a simple extension to our flow notation—when one of the subscripts of a flow is replaced by an asterisk, we refer to the combination of all flows with any value for that subscript. For example, $F_{1,*,3}$ denotes all the flows (at any class) from ingress 1 to egress 3, and $F_{*,*,3}$ denotes all the flows (from any ingress at any class) to egress 3. Each queue is used to hold a particular flow or set of flows. We use similar notation to represent queues: $Q_{I,C,E}$ denotes a queue that holds the flow from ingress I, at priority C, to egress E; $Q_{*,2,7}$ denotes a queue that holds the flow from all ingresses, at priority 2, to egress 7; and $Q_{*,*,*}$ denotes a single queue that contains all flows through a switch.

11.4.3   Queuing Systems

Queuing systems, as found inside switches, are combinations of ingress and egress ports, multiplexers (muxex), and demultiplexers (demuxes), queues, communications paths, and control signals. Each of these components has important properties which affect the overall queuing system.

1. Ingress ports inject a frame into the queuing system whenever a frame arrives at the port. There is little or no buffering in the ingress port, so the frame must be passed on to the rest of the queuing system nearly immediately. If the rest of the queuing system cannot accept a frame from an ingress port, the frame will be lost.
2. Communications paths have no memory. They serve only to connect the other components of the queuing system. They move frames from one location to another as a sequence of bytes or other small chunks of frames.

3. Queues are the only components that offer storage of frames. A frame arriving at a queue is either accepted into the queue or is dropped immediately because the queue has no available space for the frame. [Ingress (egress) ports typically have storage for one arriving (departing) frame, but this is more a matter of detailed implementation than a significant feature of the queuing system.]

4. Multiplexers merge multiple frame flows into one united flow. Multiplexers have some associated policy that determines which source will be selected next. The variants of multiplexer selection policies are discussed later in the chapter, but for the moment we can think of a multiplexer that accepts frame flows at each of several frame priorities, and always picks the highest-priority available frame.

5. Demultiplexers accept one frame flow and separate it into multiple output flows. Demultiplexers typically examine some aspect of their incoming frames to determine which egress path to follow. For instance, a demultiplexer might accept frames from port 1 and direct them to any of N egress ports, depending on the address in the frame header.

6. Egress ports accept one frame at a time and then emit it onto the egress link. Other than the one frame that is in the process of being emitted, they offer no storage capability.

7. Implicit in each queuing system is a set of control signals that effectively push frames from their ingress port sources into queues, and pull frames from these queues toward ready egress ports. Control signals are not represented in the queuing models presented below, but are an important part of any actual implementation.

Each queuing system can be consistently labeled with the flow and queue notations described above. Without being formal, we introduce queuing system diagrams in Figure 11.2, which illustrates an $N = 2$ ingress–egress port system with a single queue and a single priority of traffic. The mux combines flows $F_{1,1,*}$ and $F_{2,1,*}$ to form $F_{*,1,*}$, which then fills queue $Q_{*,1,*}$. The demux accepts $F_{*,1,1}$ from the queue and separates it into $F_{*,1,1}$ and $F_{*,1,2}$. The policy of the mux with regard to possible priorities between the two ports is unspecified. The

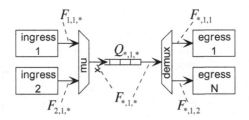

FIGURE 11.2   Example of a queuing system diagram, with flow and queue labels.

details of how flow control and speedup avoids collisions of packets on $F_{*,1,*}$ are not described. The demux must always deliver the frame at the head of the queue (if any) to the correct egress port. The details of this interaction are also unspecified. While the resolution of these details will be necessary before we have a implementable switching structure, as we shall see, the abstract queuing diagram presented in the figure already tells us a lot about any possible switch that can be constructed based on this schema. This notation allows us to explore systematically the range of logical queuing solutions that form the basis for most frame switches. Although practical switches may depart from the queuing solutions we are about to explore, these departures are generally only minor optimizations on the basic structures presented below.

11.4.4 Speedup and Blocking

Three types of problems can be caused by the design of a queuing systems: need for speedup, inability to support priorities, and head of line blocking. These three problems are examined below.

Speedup Requirements Whenever multiple ingress ports fill a queue through a multiplexer, as illustrated in Figure 11.3(a), the multiplexer and queue structure must operate sufficiently quickly to absorb simultaneous traffic from all of the input ports. Thus, if there are N inputs, a speedup factor of N is required in the multiplexer and the ingress side of the queue. A comparable situation occurs when a queue feeds egress ports through a multiplexer, as illustrated in Figure 11.3(b): the output stages of the queue and the demultiplexer must be capable of operating at N times the basic flow, where the demultiplexer feeds N sinks. In both situations, the speedup must extend from/to some buffer with comparable speedup. This commonly forces the ingress and egress ports to provide this faster buffer. With this sort of design, the mux and demux remain a sequentially switched data path, but operate N times faster, as do the port buffers with which they exchange data. This cost will commonly be hidden in our diagrams, or ignored in order to reduce clutter, but it is important to recall that this requirement exists.

Support for Priorities The next fundamental issue in queuing systems in frame switches involves priorities. The normal interpretation of priorities is

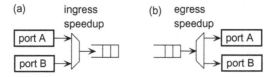

FIGURE 11.3 Structural speedup requirements.

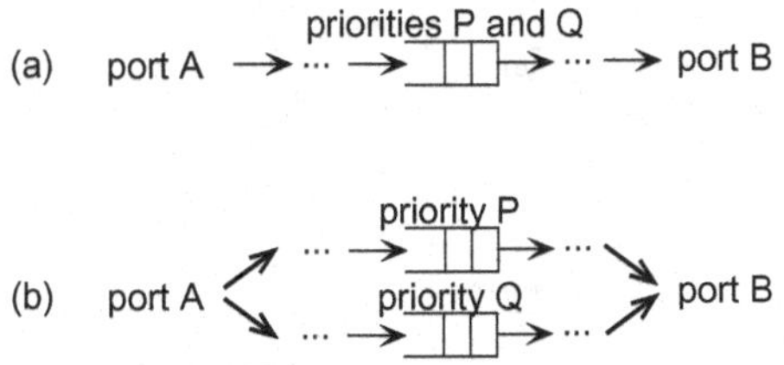

(a)   port A → ⋯ → ▢▢▢ → ⋯ → port B
                  priorities P and Q

(b)   port A ⟨ ⋯ → ▢▢▢ → ⋯ ⟩ port B
                priority P
                priority Q

FIGURE 11.4   Separate paths are required to support multiple priorities.

that higher-priority frames should be emitted first, when both higher- and lower-priority frames are present within a switch. Within an ingress-to-egress flow, this means that it must be possible for higher-priority frames to pass lower-priority frames. Thus, when a lower-priority frame has been in a switch for some time and a higher-priority cell arrives before the lower-priority cell is emitted, the newly arrived higher-priority cell should be emitted first. Recall that our queues are strictly FIFO. This property prevents higher-priority cells from passing lower-priority frames within any single queue. Similarly, all transfers on the connecting wires of our designs are inherently FIFO. Thus, it is not possible for higher-priority traffic to pass lower-priority traffic when all traffic from some ingress port A to some egress port B passes through one common queue anywhere in its ingress-to-egress path, regardless of its priority. This situation is illustrated in Figure 11.4(a). If, on the other hand, the ingress-to-egress datapath is sufficiently rich that there are multiple paths, and these multiple paths are used to direct different priority frames to different queues, as illustrated in Figure 11.4(b), it is possible for priorities to be respected, as the lower-priority queues can move slower than the higher-priority queues. At some point in the overall data paths, the differentiated priority flows must merge at a multiplexer that enforces the priority policy.

Head-of-Line Blocking and Corruption of Crossing Traffic   The final issue regarding queue structures arises when there are multiple fanout paths from a demultiplexer that follows a queue. Two related subproblems arise in this configuration. Figure 11.5(a) shows an instance of head-of-line (HOL) blocking. A queue feeds two ports: Port A is busy emitting a frame that it has already accepted from the queue; port B is idle. The queue contains, in order of decreasing age, a frame for A and a frame for B. The queued frame for port A cannot be sent until port A completes the emission of its current frame. The frame for B cannot be forwarded to port B because it is not at the head of the queue. Thus, the head of the queue is blocking useful work: The packet for B must wait, and port B must waste bandwidth by emitting idles.

Some switches have built additional features into the heads of queues that allow any of the first few items in a queue to be emitted out of FIFO order. Such a solution can reduce the probability of HOL blocking, but cannot remove it entirely, as the same phenomena will then occur farther back in the

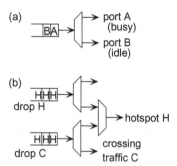

FIGURE 11.5 Sources of (a) HOL blocking and (b) disruption of crossing traffic.

queue. For our purposes we consider only straightforward queues with one accessible head; it turns out that such queues, used in the appropriate queuing system, lead to superior performance.

A related problem is the blocking of crossing traffic. Examine the situation illustrated in Figure 11.5(b). Egress port H is a hotspot in the sense that more traffic is arriving for that port than can be emitted. Naturally, traffic destined for H builds up in queues, awaiting a chance to get out port H. If one considers all ingress ports that may supply traffic to egress H, and all queues that lie between these multiple sources and the egress at port H, we see that this entire set of queues may become filled with traffic for the hotspot H. This is commonly referred to as a *cone* of congestion *rooted* at egress port H. In the figure, two queues are filling with traffic for H, as both receive substantial flows for H, and the sum of these flows exceeds the egress capability of port H. Eventually, these queues will fill. They may contain only traffic for H, or some mix of traffic for H with other traffic, but it is the traffic for H that is the source of the difficulty. If the hotspot persists long enough, all queues that carry traffic to H will fill. When the queues are full, arriving traffic must be discarded. Most of the discarded traffic will be destined to port H. This is unfortunate, but not surprising, as port H is simply overloaded, and some higher-level (protocol) response will be necessary to manage the congestion. But other, crossing traffic flows may be affected. Consider frames destined for egress port C. These cells are normally routed through either of these queues, but in particular we consider one such frame for C arriving at the bottom of the two queues. This arriving frame for C has a high probability of being discarded because that queue is full with traffic for egress port H. Such a frame is accepted only if space in the queue is available. Since the queue is largely full of hotspot traffic, and this hotspot traffic must move through the queue slowly as it competes with other hotspot traffic in other queues, opportunities for traffic to C to be accepted are (arbitrarily) rare. Thus, this crossing flow to a presumably uncongested egress port is damaged by frame discard, due only to the congestion at egress H. In general, many such crossing traffic flows can be damaged by congestion due to a single hotspot. Worse, the area of congestion can grow as traffic to non-hotspot egresses such as H backs up into other queues and causes damage to other crossing traffic flows.

11.4.5 Possible Queuing Systems

We are now prepared to begin our tour of the variety of queuing systems appropriate for most frame switching problems. We pursue this tour in a systematic fashion. First, we note that it is possible to have one or more queues in a switch, and that if we have more than one it is most natural to have one per ingress port, one per class, one per egress port, or one for each of some combination of ingress port, class, and/or egress port. This leads to a simple notation for the natural feasible queuing systems. A queuing system is denoted by a sequence of three letters, one each for the queuing treatments of ingress ports, classes, and egress ports: ICE. Each subscript can be the letter I, C, or E (in its respective position) or an asterisk. The presence of the letter means that the queuing system includes a separate queue for each element of that type. The presence of an asterisk means that the queuing system combines frames of all values of that category in each queue. For instance, I** denotes a queuing system with one queue for each ingress port (I), where each such queue holds frames of all classes (*) and destined for all egresses (*). The example *CE denotes a queuing system that provides a queue for each combination of class and egress port, where each queue can contain frames from all ingress ports. Since each of three subscript positions can have either of two values, there are eight possible queuing systems, as represented in Figure 11.6, where the queues are ordered from the least specific at the bottom (one queue

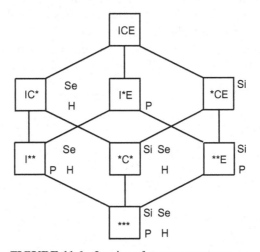

FIGURE 11.6 Lattice of queue structures.

for all traffic, ***) to the most specific (one queue for each combination of ingress port, class, and egress port, ICE).

Recall the queuing system problems discussed above. These problems are a direct consequence of the queuing system, as denoted by our subscripted-Q notation. Each type of queuing system is annotated (in Figure 11.6) with its characteristic problems: *Si* denotes a need for ingress speedup; *Se* denotes a need for egress speedup; *P* denotes an inability to support priorities; and *H* denotes HOL blocking and the related problem with crossing traffic. Notice that problem *Si* occurs whenever the queues accept traffic from all ingresses (I = "*") and hence must be prepared to accept from all ingress ports at once; problem *P* occurs when there are not separate queues for each priority (C = "*"); and problems *Se* and *H* occur when queues serve multiple outputs (E = "*"). Unsurprisingly, the queuing systems that are more specific (which have more, differentiated queues) have fewer problems. Queuing system ***, which has only one shared queue, exhibits all four problem types; ICE, with *ICE* separate queues, has none of the characteristic problems. Queuing system *CE behaves as well as queuing system ICE once the cost of ingress speedup is met. The remainder of this section examines a few representative queuing systems in more detail.

Queuing System ***   In Figure 11.7, we see the simplest queuing structure: one queue for all flows. The *N* ingress ports each supply their flow ($F_{1,*,*}$ through $F_{N,*,*}$) to a multiplexer that accepts all of these ingress flows and supplies the single queue, $Q_{*,*,*}$. This single queue supplies a demux that directs traffic to egresses by flow ($F_{*,*,1}$ through $F_{*,*,N}$). This solution has instructive limitations that motivate our continuing exploration of possible queuing solutions.

- *Speedup.* All *N* ports contribute (and consume) frame flows into (and from) the single queue; thus, the queue needs a speedup factor of *N* for ingress (and egress) traffic.
- *Priority Blocking.* There being only one queue, it is impossible for high-priority traffic to pass low priority traffic, so priorities cannot be recognized.

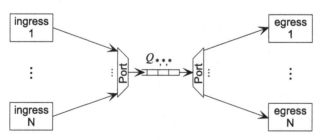

FIGURE 11.7   Queuing system ***.

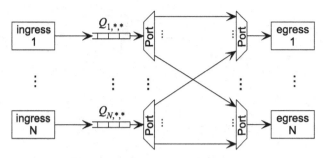

FIGURE 11.8 Queuing system I**.

- *Head-of-Line Blocking.* The single queue must supply all N egresses. If a particular egress is busy, and the HOL frame is destined for that busy egress, all following frames for other idle ports must wait. In the simplest implementation, only one of the N egress ports can be busy at any time. With egress speedup, this can be improved, but HOL blocking remains. The more serious problem of damage to crossing traffic can also occur: the single queue may fill with excess traffic to some hotspot, thereby causing packet discards to other noncongested flows.

Queuing System I Our next queuing system is I** (Figure 11.8), which has one queue per ingress port. Following the ingress queues, we find a switching structure that allows each ingress queue to forward a packet to some otherwise idle egress port. This switching structure is an N-to-N frame crossbar. Note the following properties:

- *Speedup.* This structure supports N simultaneous transfers from ingress ports to ingress queues. Consequently, there is no requirement of speedup into the queues. On the other hand, traffic leaving one queue may have to serve all egress ports in a burst, and consequently, requires a factor-of-N speedup.
- *Priority Blocking.* This queuing model does not have differential queues for differential priorities, so cannot support proper priority semantics.
- *Head-of-Line Blocking.* This system is subject to HOL blocking (e.g., egress port N is busy emitting a frame from $Q_{N,*,*}$; the oldest frame in $Q_{1,*,*}$ needs egress N; the next oldest frame in $Q_{1,*,*}$ needs idle egress 1 but is HOL blocked). This problem extends to damage to crossing traffic.

Queuing System *C* Our next logical queuing system, *C* (Figure 11.9), implements one queue for each priority class. The ingress switching structure is a frame crossbar from N ingress ports to C classes; the egress switching structure is a frame crossbar from those C classes to N egress ports. The key properties of this logical queuing system are:

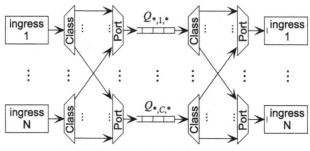

FIGURE 11.9   Queuing system *C*.

1. *Speedup.* Assuming that $N > C$, we need N/C average speedup to get the N ingress–egress flows through the C queues. But an instantaneous traffic flow may concentrate all frames into one priority class, so each queue needs a full N speedup.

2. *Priority Blocking.* Frames are organized by priority when moving from ingress ports to priority queues, and each egress has an independent path to each priority queue. This augurs well for respecting priorities, but trouble is caused by HOL blocking: An egress may violate priority (by moving a lower-priority frame before a higher-priority frame) due to HOL blocking on the higher-priority queue. If the higher-priority queue is busy and a lower-priority queue is not blocked, depending on the egress port's frame scheduling policy, priority may thus be violated.

3. *Head-of-Line Blocking.* Each priority queue is subject to HOL blocking. Any queue can be unable to move its next frame out because the frame goes to an otherwise busy egress.

Queuing System *CE*   The logical queuing structure shown in Figure 11.10 has separate queues for each combination of class and egress; thus, there are CN queues. The supply-side data path is a frame crossbar from N to CN; the demand-side data path is a set of N frame crossbars from C to 1. The properties of this logical structure are:

- *Speedup.* Each queue may have to accept frames from up to N ports at a time, so needs an ingress speedup factor of N.
- *Priority Blocking.* The use of differentiated queues for each priority at the egress ports allows the priority to be respected.
- *Head-of-Line Blocking.* The supply to the egress ports cannot exhibit HOL blocking. Damage to crossing traffic cannot develop because there is no demultiplexer farther downstream.

Queuing System *ICE*   Our final logical queuing structure (phew!) has a separate queue for each combination of ingress port, priority, and egress port

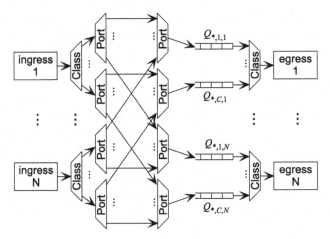

FIGURE 11.10 Queuing system *CE.

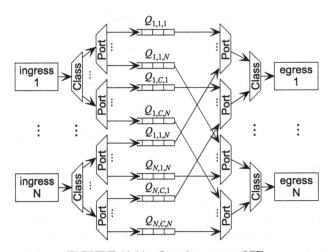

FIGURE 11.11 Queuing system ICE.

(Figure 11.11). Thus, there are NCN queues. The supply-side data path is a set of N frame crossbars from one ingress port to NC queues; the demand-side data path is a set of N frame crossbars from NC queues to one egress port. This properties of this final structure are:

1. *Speedup.* Each queue is fed by only one source and drained by only one destination, so speedup is not required.
2. *Priority Blocking.* Each priority has its own set of queues, so priorities can be respected.

3. *Head-of-Line Blocking.* The egress paths cannot HOL-block, as each queue supplies only one egress port. Damage to crossing traffic is not possible.

This final structure thus appears to be ideal in the sense of avoiding blocking, respecting priorities, not needing (per path) speedup, and not being subject to hotspot effects on crossing traffic. Thus, it appears that the ICE solution is superior to all the other solutions. Although this is certainly true in theory, there are often practical reasons to build switches with less powerful queuing systems, such as cost. In the transport network, however, we are generally forced to employ solutions, such as ICE or *CE, which do not affect crossing traffic because it is unacceptable to compromise independent services due to congestion in one service.

11.5   QUEUE LOCATIONS AND BUFFER SHARING

Several critical features of logical queues significantly affect switch architecture and performance:

1. Where the queues are located (ingress, central, or egress)
2. Whether co-located queues have independent physical memories or whether they share a common physical memory (and access to that memory)
3. Whether co-located queues in a common physical memory have distinct pools of frame buffers, or whether they share some common pool of frame buffers

There are three conventional places to locate queues, based on their relative position with respect to the core switching structure: on the ingress paths, before crossing the switching matrix; in the center of the switch, embedded within the switching matrix; or on the egress paths, after crossing the switching matrix. As we shall see, this choice has important consequences. Figures 11.12 to 11.14 illustrate these three positions. In each case we use the ICE queuing model from in Section 11.4. In these figures, the dashed outline boxes indicate the location of these queues. Those queues within the same dashed outline box are co-located.

In Figure 11.12 all the queues filled by ingress port I are co-located with port I. The heads of the various queues associated with ingress port I cross to the N egress ports via the heart of the switching data paths. In Figure 11.13, all of the queues are located in the same place. There are data paths from all ingress ports that reach this collection of queues, and from the collection of queues to all egress ports. These queues are centrally located. In Figure 11.14 all the queues drained by egress port E are co-located with egress port E. The

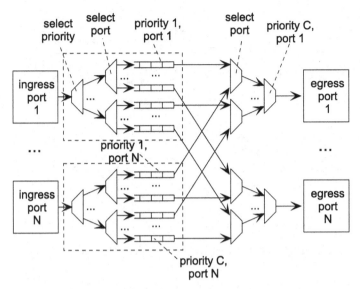

FIGURE 11.12 Ingress co-located ICE queues.

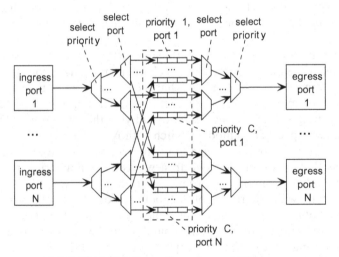

FIGURE 11.13 Centrally co-located ICE queues.

tails of the various queues associated with egress port E are fed by data path wires crossing from the N ingress ports via the heart of the switching data paths.

Each of the three possible queue locations have the same abstract properties. But two factors make the location of queues of critical importance: the

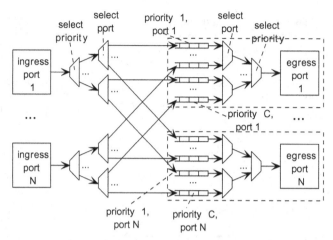

FIGURE 11.14   Egress co-located ICE queues.

possibility of implementing multiple queues in the same memory, and the consequent possibility of sharing buffers between logically separate queues. As mentioned before, co-implementation (or not) of queues in memories has important consequences for any implementation. We consider the the possibility of sharing buffers among co-located queues in the remainder of this section.

When we are forced to have a separate queue for each of N^2C combinations of ingresses, classes, and egresses, and each queue must have some reasonable number of frame buffers, W (an issue that is dealt with later), the cost in terms of total bits of memory can be impractically high. Suppose that our frame buffers are designed for our maxigram of M bytes, then we require $N^2CWM \times$ 8 bits of queue space. For a modest switch with $N = 32$, $C = 8$, $W = 8$, and $M = 2048$, we find that we require 1 gigabit of storage! Yet we have seen the negative effects that occur when we go to a weaker queuing structure than our preferred ICE. A common compromise is to design with the ICE system in mind, but to allow the sharing of buffers between different queues. Thus, there is some pool of buffers that can be dynamically allocated to individual queues as frames arrive (and returned to the common pool when frames depart).

Buffer sharing policies can be complex. It is beyond the scope of this book to study them in detail. Instead, we focus on the fundamental issues involved in sharing. There are seven obvious ways to share buffers:

1. Share among ingresses at each unique class-X-egress.
2. Share among classes at each unique ingress-X-egress.
3. Share among egresses at each unique ingress-X-class.
4. Share among ingresses and classes at each unique egress.
5. Share among ingress and egress at each unique class.

6. Share among class and egress at each unique ingress.
7. Share among ingress, class, and egress over the entire queue structure.

Central location of all queues permits any of the sharing modes above. If we adopt an ingress buffer location, then only sharing modes 2, 3, and 6 are possible. If we adopt an egress buffer location, only sharing modes 1, 2, and 4 are possible. Sharing among queues can be arranged in many ways. Some interesting approaches that have been used in various switch designs, include:

1. Equal sharing, in which all sharing queues have equal rights to any of the buffers, on a first-come first-served basis
2. Sharing with limits, in which each sharing queue has some assigned maximum number of allowed buffers, where the sum of these maximums is greater than the number of shared buffers
3. Sharing, in which each sharing queue has some assigned minimum number of allowed buffers, where the sum of these minimums is less than the number of shared buffers, so some pool remains to allow queues to go above their minimums and usually not above some maximum
4. Sharing with eviction, in which some queues can steal buffers from other (usually, lower priority) queues

Whenever buffers are shared, a new form of blocking becomes possible. If the sharing scheme permits any particular queue to have few or no buffers when they are required, that queue can be effectively blocked unless it has some buffer preemption rights. There are many mechanisms by which buffer sharing can cause one congested traffic stream to affect other streams which "cross" through the shared buffer pool. The details of these mechanisms are highly dependent on the queue model and the buffer sharing scheme and go beyond the scope of this book.

Ideally, we might prefer to allow sharing between any queues, perhaps with one large buffer pool with which each queue borrows buffers on demand. Such a scheme requires that all the buffers and queues be co-located. Thus, only the centrally buffered switch architecture is suitable for this scheme, as it is not possible to share if buffers are not co-located. Central buffering is ideal for this reason, but in general it imposes unrealizable demands on our implementation technologies. The problem is that central buffering demands that all the frame buffers and all the switching data paths and logic be co-located. For nontrivial switch sizes, this is not feasible. So we are forced to move something off the central switching chip. The most rewarding and feasible thing to move off the central switch chip(s) are the buffers. That leads us to ingress or egress-buffered designs. Notice that Figure 11.12 co-locates the queues at the ingress side of the core switching paths. These queues could be placed on the switching chip itself, but they can also be placed with the ingress ports, on separate chips, of which we have one per port. Since we are forced to avoid the centrally buffered design (for all but the smallest switching requirements), we must chose between the ingress-buffered (co-located) model and the egress-buffered (co-located) model.

11.6   FILLING AND DRAINING QUEUES

Filling queues seems to be a simple-minded process: A frame comes along, so put it at the tail of its queue, or drop it if there is no space. But it is rarely that simple. In fact, the topics of classification of frames into flows, queue filling, and draining policies could probably fill a book of size equivalent to the one you are reading now. Although we do not wish to minimize the importance of these queue mechanisms, we must restrict the size of this book and leave material for other authors. Consequently, we present only a brief summary of these important topics.

11.6.1   Filling

Token Bucket Policing   It is sometimes necessary to "police" incoming flows; that is, an identifiable flow runs under an agreement or conditions that limit the amount of bandwidth to be consumed by that flow. The flow may be identified by ingress, class, and egress; the flow may be further identified by MPLS tag, by protocol carried, by tags or fields within the frame, or by other features. Agreements may be service-level agreements, which are commercial contracts negotiated between service providers and their customers. Or agreements can be a promise of high-quality service for certain traffic types, such as voice. Conditions that limit the bandwidth may be network provisioning considerations imposed by network operators.

In any of these cases, an attempt is made to measure and control the rate of traffic in identifiable flows. The standard mechanism used to police flows is the *token bucket*. A token bucket allows a certain smoothed average number of bytes per second to pass for a given flow but also allows limited short-term bursts of traffic. The mechanism is simple. The bucket is essentially a counter. The counter increments a constant rate, up to some maximum value, at which point any attempt to increment the counter results in the counter overflowing and simply staying at its maximum. The count corresponds to the number of tokens available to pay for the transmission of data on the controlled flow. Each token corresponds to a fixed number of bytes of transmission (e.g., 1 byte). Each time a frame comes to the token bucket policing station, the number of tokens required to pay for that frame are taken from the bucket (counter). If there are sufficient tokens, the frame is accepted normally into the queue. If there are insufficient tokens, the frame is either dropped, put into a lower-priority queue, or marked in some way that prioritizes it for possible loss later.

The average rate of appearance of tokens controls the average flow rate. The limit on the bucket controls the maximal credit the flow can accumulate; this maximal credit can be used to legitimize periodic bursts of traffic.

Buffer-Sharing Policies   Where buffers are shared between queues, more policies need to be considered. Each queue may have some reserved number of buffers available only to it, as well as shared access to some other pool of buffers. When a new frame arrives, any such sharing scheme must be interrogated to see if a buffer is available to that queue. Additionally, it is sometimes possible for higher-priority flows to steal buffers from lower-priority flows.

Random Early Discard   A simpleminded frame drop policy is to have queues begin to drop packets exactly when there is no more space for frames. However, where multiple TCP/IP flows compete for a common pool of buffers, all such flows experience frame drop at the same time. When TCP/IP flows experience frame drop, they (may) back off; as a result, the network is starved of traffic as all active flows go idle at once. Random early discard (RED) remedies this synchronization problem. A set of queues sharing a common buffer pool and managed by RED has zero probability of dropping a new frame when the queues are nearly empty (or the free pool is sufficiently large). Once a threshold of fullness has been reached, however, RED begins to drop frames randomly. The probability of dropping rises monotonically until it becomes one when the queues are full. The advantage of RED is that it avoids synchronization of frame drop backoffs and thereby keeps the link full of traffic. Some flows have suffered frame drop prematurely, but the overall use of the expensive resource (the egress link) is maintained. The increasing probability of frame drop provides a gently increasing throttling of the overall flow, right up to the point at which the queue systems must drop all new frames. Of course, RED incurs no frame drop when the loads are light and the queues are nearly empty.

11.6.2   Draining

Again, taking frames from the head of a queue seems like a simple proposition but it is complicated by many possible network flow control features.

Leaky Bucket Mechanism   First, we consider the *leaky bucket*. The idea of this flow control mechanism is to smooth potentially bursty traffic flows. The goal is to reduce burstiness in the network, thereby bring links and queues to a more efficient average condition. In this case, the bucket holds incoming frames. Of course, the frames are actually in a queue—the leaky bucket is a control mechanism for scheduling frames from that queue for egress. The bucket can emit only a fixed rate of flow (from a hole in the bottom of the bucket). Arriving packets flow into the bucket from the top. Should the input flow exceed the allowed egress flow, the bucket will tend to fill up. Should this continue, the bucket will fill, and frames will spill over the top of the bucket and be lost (i.e., packet drops). Should the ingress flow be sufficiently low, the bucket will tend to drain. When any accumulated frames are gone and the ingress flow is below the maximal egress flow, the actual egress flow will drop to the current ingress flow. This leaky bucket mechanism delays bursts, letting them flow downstream in the network only at some preset maximal rate. The logic of the leaky bucket implies that it is part of the scheduling of a packet from the head of a queue.

Round-Robin Mechanism   *Round robin* is a method to share egress link capacity evenly among multiple queues or flows competing for service at a mux that feeds the output link. Round robin is the simplest such sharing mechanism: Each queue is permitted to emit one frame, should it have an available frame, in strict rotation with the other queues. In its pure form, round robin pays no attention to the length of the packets, so the vagaries of packet-length distributions may result in unfair use of link capacity, but over time such unfairness is generally assumed to even out.

Weighted Round-Robin Mechanism   The weights of weighted round robin allow competing flows or queues to have different access to the shared egress path. WRR assumes an average packet length, then computes a normalized, weighted number of packets to be emitted by each queue in turn, based on the weight assigned to each queue.

11.7   CENTRAL MEMORY PACKET–CELL SWITCHES

The first frame switch architecture we examine is our old friend, the shared central memory switch. Since all the queues are supported in a single common memory, it is reasonable to support the superior ICE queuing model using any sensible buffer-sharing scheme. All N ingress ports deposit R bits per second into this central memory; all N egress ports remove R bits per second from this central memory; thus, the central memory must support $2NR$ bits/second. As with the central memory switches examined in Chapter 10, this architecture runs into technology limits having to do with the maximal speed of memories. Thus, if we limit ourselves to 500-MHz central memories and, for example, 64-bit-wide access on one read/write port, we find that we can support only

$$N = \frac{500\,\text{MHz} \times 64\,\text{bits}}{2 \times 2.48832\,\text{Gb/s}} = 6$$

OC-48C ports (carrying frames in the SPEs). But this is only the beginning of the story about the limits of central memory frame switches. In the rest of this section we examine a series of related limits on the central memory model for frame switching.

Notice that we have exactly balanced memory bandwidth against port bandwidth. Also notice that the preceding example assumed that the central memory was $W = 64$ bits in width, allowing the memory to produce a net bandwidth of $500\,\text{MHz} \times 64$ bits, allowing $N = 6$ ports. We must wonder whether $W = 128$ supports $N = 12$ ports, and so on. For the ports to use a central memory of width W, there must be a process in each ingress port that chops frames into W-wide slices for storage in the common central memory, and there must be a corresponding process in each egress port that assembles frames from these W bit memory slices. Choosing W to be the same as the maxigram size would remove these problems entirely, but is absurdly expensive.

If a frame is some even multiple of W bits in length, these W-slicing processes will work without a problem. But most protocols allow a wide range of frame lengths, usually varying in bytes. So when a frame that is not an even multiple of W in length has to be stored/retrieved, the last memory access must use only a subset of the W bits available. Some number of bits is wasted in this final slice for each frame that is not some integral number of W-bit segments.

Yet we computed the number of ports to be supported by a given memory W and speed without considering wastage. Clearly, if some of the memory bandwidth is wasted systematically, the number of ports to be supported by a given memory must be reduced, or there must be some speedup to/from the memory system. We now turn our attention to characterizing the performance impact of this memory segmentation issue.

Our simple example above assumed that the bit rate of the N ports of OC-48C produced $2N \times 2.5\,\text{Gb/s}$ of load on the central memory (ignoring SONET/SDH overheads) and that the central memory provided this amount of bandwidth. Thus, the losses produced by partial use of memory transfer sizes must be accommodated in some form, or the memory will fall behind the N ports and eventually lose frames. This problem occurs in two distinct forms: one is referred to as the $W + 1$ *problem*, the other, as the *minigram problem*.

The $W + 1$ problem occurs when a frame (or worse, an unbounded series of frames) are of length $W + 1$. Such a frame requires two writes to central memory and two reads from central memory. The first operation carries W bits of the frame; the second carries only 1 bit. With most protocols, we must think of W in terms of bytes, and the "+1" is an additional byte, but the same principle applies. We get only $W + 1$ units of frame size moved to/from memory in a unit of time when the memory is providing $2W$ of available bandwidth. To accommodate for this inefficiency, the memory must run $2W/W + 1$ times faster than the aggregate port bandwidth. This is almost a factor of 2.0. It effectively reduces the number of ports that can be accommodated in our earlier example by half, to $N = 3$, assuming that we keep W and the memory speed constant. The same conceptual problem occurs for $W + 2$ frames, but the frame arrival rate is slightly slower, so the required speedup is less. Again, the same conceptual problem occurs for $nW + 1$, for $n > 1$, but the required speedup is $(n + 1)W/nW + 1$, which is strictly less than $2W/W + 1$ for $n > 1$. Thus, a speedup of 2.0 is sufficient to cover the worst impact of the $W + 1$ problem.

The minigram problem is due to the fact that the shortest frame, the minigram, must also be moved to/from the central memory without overrunning the memory's capability. If the length of the minigram is greater than $W/2$, the speedup used to solve the $W + 1$ problem takes care of any inefficiency caused by having to transfer minigrams. But as the size of the switch grows with network demands, W tends to be pushed up to extend the life of the central memory architecture while the size of the minigram remains constant. When

$$\frac{W}{len(minigram)} > \frac{2W}{W + 1}$$

or usually, $W/len(minigram) > 2$, additional speedup of $W/len(minigram)$ is required to avoid memory starvation in the presence of a sequence of minigrams (and nothing in our packet protocols prevents a long sequence of minigrams).

However, there is an independent constraint. We must allow each port to source and sink an unlimited sequence of minigrams, and each such ingress or egress minigram must consume at least one central memory operation during the transfer period of the minigram. No matter how large we make W to improve the memory bandwidth, we must support enough minigram read/write events per second to support all ports. Thus,

$$2N \frac{R}{length(minigram)} \leq memory\ speed \cong 500\,MHz$$

Let's consider an example switch as it encounters the foregoing limits. Suppose that our application switches IP frames carried over OC-48 links. Assume that our minigram is 48 bytes, and that 2.5 Gb/s is available to carry frames on each port. Our memory limit is 500 MHz. We first attempt to build a $N = 40$ port switch.

1. First, we check the minigram arrival/departure rate. Minigrams must arrive at

$$\frac{2.5\,Gb/s}{8 \times 48\ frame} \cong 6.5 \times 10^6\ frames/s$$

and must depart at the same rate. With $N = 40$ duplex ports, we find a minimal memory rate of 520 MHz, which slightly exceeds our memory technology. We must retreat to a smaller design. $N = 32$ is permissible in terms of the minigram arrival rate, as only 416 MHz of memory operations is required.

2. Given the $W + 1$ problem, what value of W is appropriate for $N = 32$ and the other parameters of this problem? We expect a $W + 1$ speedup of 2.0. This leads us to a minimal W of

$$\frac{2 \times 2NR}{500\,\text{MHz}} = 640\,\text{bits} = 80\,\text{bytes}$$

We use this minimal value to reduce hardware costs.

3. We must check this design for the *minigram inefficiency problem*. Our minigram is 48 bytes; our W is 80 bytes. Thus, our minigram is greater than half of our memory transfer size, so the minigram inefficiency problem is less severe than the $W + 1$ problem, and our speedup factor of 2.0 covers both issues, given that we reduced the number of ports to $N = 32$ to meet the minigram frequency limit.

Thus, it appears that we could build the frame data path of this $N = 32$, $R = 2.5$ Gb/s, len(*minigram*) = 48 byte switching product with a 500-MHz memory rate, with $W = 640$ bits. It is tempting to think that we could do much better (in size) by using two central memories and alternating between them, as we did with TDM switches in Chapter 10. But that trick does not work here because we have no control over when packets arrive and depart. If we put a particular packet in one of our two memories, we must schedule its departure for when that memory is being read from by the egress ports. This is, in general, impossible (or at least very difficult) given the mix of packet lengths and the varying times spent in the different queues.

All cell protocols have one fixed length for each frame. Thus, there can be no $W + 1$ problem or minigram problem. Making W equal to the fixed cell length is generally a good design choice. The number of ports that can be supported by a single central memory is thus

$$N = \frac{W \times 500\,\text{MHz}}{2R}$$

This seems to be an enormous advantage for cell protocols, allowing at least twice as many ports as a segmenting frame protocol. However, this advantage is mitigated by the fact that the cell flow often originates in a segmented packet flow, and that the segmentation process—however remote— will produce a cell stream that needs the same $W + 1$ speedup. So the cell stream is subject to having to carry twice the bandwidth of the original packet stream. In usual practice, cell streams of segmented packets are not given this doubled bandwidth, the reason being that such cell streams usually carry many separate packet streams, and it is felt that it is unlikely that all of them would exhibit the $W + 1$ problem at the same time.

Queue Descriptor Access   So far, we have looked only at the raw frame data path. Another issue arises when we consider that the queues (FIFOs) that contain the packets are also data structures which must be implemented economically in some fashion such that they can be accessed sufficiently frequently to keep up with the raw data flow. Each insertion/removal of a frame into/from a queue requires that that queue data structure be modified to reflect the change. The standard methods of representing queues in memories all require two memory operations to complete these enqueue/dequeue operations: The first operation reads the current status of the queue, the second operation modifies the queue to reflect the frame addition or removal. We know that the worst frame arrival–departure rate is determined by the length of the minigram. Thus, any central queue descriptor memory must support $(2 \times 2NR)/len(minigram)$ reads or writes per second. For our earlier example this is 833×10^6 operations per second. Clearly, this heavy memory access load cannot be added to the central memory that holds the frames themselves without seriously slowing or shrinking the switch. Instead, a separate data structure, usually implemented in flip-flops instead of RAMs, is used for this purpose.

11.8   INGRESS-BUFFERED PACKET–CELL SWITCHES

Today's and tomorrow's transport networks cannot accept the frame switch size limitations uncovered in Section 11.7. Just as we saw with physical circuit switches and logical circuit switches, it is necessary to continue to architectures that support larger aggregate traffic loads. In the next two sections we examine some possible solutions.

11.8.1   Blocking Ingress-Buffered Frame Switches

Our first attempt to increase the scalability of our frame switches will be to divide the central memory into one separate memory per ingress port. These much smaller memories can be faster and must support only $(1/N)$ th the load of the central memory design, if we are careful. The use of ingress buffering permits a range of queuing system models. To produce a switch capable of good performance under heavy or hotspotted traffic, we work with the ICE model, giving a separate queue to each combination of ingress port, class, and egress port. All the queues for each ingress will be implemented in the memory

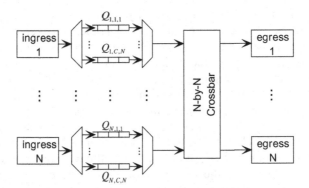

FIGURE 11.15 Blocking ingress-buffered frame switch.

module found with each ingress port. Thus, each ingress port will have a queue for each combination of class and egress port. This situation is illustrated in Figure 11.15.

The ingress process is straightforward. Each frame is analyzed for class and egress port and is added to the appropriate queue. Given that the queues are in a common memory, buffer-sharing policies can be implemented if desired. But since we are writing incoming variable-length frames into memories, the $W + 1$ problem and the two minigram problems must be considered in determining the width (W) of the memory (both for writing new frames into memory and for reading frames to be emitted from the other side of the queues). Examine the data flow paths that follow the ingress queues in Figure 11.15. All the queues for each ingress feed into a multiplexer, which in turn feeds into an $N \times N$ crossbar (of course, the crossbar is just a set of N fanout structures followed by a set of N fanin structures). The implications of this data path structure are that each ingress port can select one packet from one of its NC queues and route it through its mux to the crossbar and hence to the desired egress port. Each ingress port has this capability; thus, the overall switch supports a permutation of frame connections, with each ingress supplying one frame to one egress, and each egress accepting one frame from one ingress. Or, of course, lighter loads with idle ingress and egress ports are possible. We have not yet considered how frames are scheduled through this switch (i.e., via an appropriate control structure). This issue is deferred until Section 11.9.

This is a fairly robust switch architecture. It completely respects priorities, and hotspots cannot affect other egresses or higher-priority classes. But this switch does exhibit a form of internal blocking. Suppose that ingress port A has frames queued for egress ports B and C and that no other ingress port has any frames queued for B or C. We would like to use ingress A as a source for packets to both B and C, but as shown in Figure 11.14, our data path does not permit this option. Only one path exists by which A can emit a frame to an egress port, so the control system must make a choice between ports B and C. The losing egress port must sit idle, even though a frame sits ready to be emitted in the queues associated with ingress port A.

Notice that each set of ingress queues can feed any of the output ports. The highest-performance implementation of this architecture would give a speedup of N to the heads of these queues and the crossbars feeding the outputs, thereby allowing one ingress port to supply up to all egress ports at once (avoiding internal blocking). But this imposes high costs in terms of the very fast memory reads and the width and speed of the crossbar connections. For this architecture we avoid the use and cost of this speedup, and accept that each ingress port can supply only one egress port at a time. This makes the architecture dramatically more practical and scalable, but decreases its overall switching capability.

It is common to apply an extra 10 to 20% speedup to the internal paths of a switch like this, in order to provide compensation for the internal blocking phenomena that we have just uncovered. It is also necessary to add egress buffers to the architecture to absorb the speedup of the switch core and to allow the frames to be emitted at line speed. In practice, this solution is generally effective, but it does not allow hard guarantees of switching performance—just general statistical statements.

11.8.2 Nonblocking Ingress-Buffered Frame Switches

It is possible to remedy the internal blocking that exists in the ingress-buffered architecture examined in Section 11.8.1. Figure 11.16 illustrates an architecture that achieves this goal and avoids the temptation to use expensive speedup. The architecture in Figure 11.16 also uses the ICE queuing model and is identi-

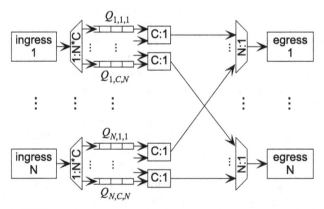

FIGURE 11.16 Nonblocking ingress-buffered frame switch.

cal to that in Figure 11.15 up to and including the ingress buffers. But instead of having a selection mux that restricts each ingress to emitting only one cell to only one of the egress ports, this architecture routes all ingress queues directly to selection muxes at the appropriate egress port. Thus, ingress port 1 has C queues that support egress port 2; each of these queues is connected directly to the selection mux for egress port 2; ingress port 1 also sends C queues directly to each of the other $N - 1$ egress ports; and all $N - 1$ other ingress ports send C queues directly to egress port 2's selection mux. With this architecture, each ingress can emit from zero to N frames at any time. Thus, blocking due to otherwise committed ingress mechanisms has disappeared. Each egress port can be supplied by any ingress port, regardless of other commitments made on behalf of the ingress port desired. This architecture does not require additional internal speedup, as there is no source of internal blocking.

The control system required to schedule packets from ingress to egress ports is quite simple in this architecture. Each egress port is given visibility and control of the head of each queue. Thus, each egress can see what frames are queued for it and can implement any policy it choses to select the next frame. In particular, each egress need not worry about the choices being made by the other egress ports, because the ingress queues can freely support any combination of egress-dictated traffic. Control of the (ingress) queues used in this configuration is interesting. Additions to the tail of the queue occur at the ingress port, where the queue's buffers are located. But removals of frames from the head of the queue are controlled by the egress port. Thus, the data structures that represent the queue must be accessible from two distinct locations in the switch, and they must permit (and control) concurrent access by the ingress insertion process and the egress removal process.

The combination of ingress buffering and visibility of queue heads at the egress ports associated with each queue is an attractive solution for frame switches. The combination is generally known as "ingress buffered, virtual output queued." We avoided discussing the control system that schedules frames from ingresses to egresses in this section because the problem is significantly more difficult in that architecture: The egresses must agree on which egress gets access to each particular ingress, because each ingress can serve only one egress. This problem is fully addressed when we consider the next frame switch architecture.

11.9 REQUEST–GRANT CELL SWITCHES

The last two switches we have examined both locate their queues with their ingress ports. This solution has advantages, but there is further gain to be had by fully exploiting this architectural approach. Thus far, we have implicitly assumed that the ingress buffers are implemented on the single VLSI substrate that implements the switch. This has two disadvantages: It places the entire buffering implementation cost on one VLSI chip and thereby limits the scalability of the architecture, and by forcing the entire switch onto the one chip it prevents the solutions that are explored in the next section. Thus, it would be advantageous to move these ingress buffers off the single core switch chip. The natural alternative position is to place these buffers on the port cards, associated with the ingress data paths. In this location we are required to place only $(1/N)$th of the switch's buffers on any one port; this is highly advantageous, as it avoids the architectural bottleneck of trying to pack too much memory on one chip or in one part of the system. With this approach, the switch core itself becomes a simple frame crossbar (allowing all possible connections of ingress and egress ports, in permutation patterns), and the control system determines when the various ingress–egress connections are to be made and broken.

The control system becomes the interesting issue. How can it know the depths of the various queues in order to make useful connections between ingress and egress ports? And how can the control system communicate with the various ingress and egress ports to start and stop connections? Three key ideas are required to make this system work: the use of cells instead of packets, the use of a *request–grant protocol* to communicate between ingress/egress ports and the switch core, and the use of an appropriate arbiter on the switch core to determine which cells and which connections should proceed at any time. We address these three issues in turn.

11.9.1 Use of Cells

The frame crossbar at the center of this switching architecture allows permutation connections between ingress and egress ports. The scheduling of the connections in this frame crossbar must be determined and controlled by a distributed algorithm, as the switch core and each of the ports must be involved in some sense. Distributed algorithms are difficult to design and implement, as, in general, events occur across the system in some unknown or uncontrollable fashion. A special case of a distributed system is a system in which all timing is controlled by a central clock and there is a fixed, known schedule of events based on this clock. The result remains a distributed system, but its control is synchronous, making it much simpler and easier to design and verify.

The most straightforward way to place this problem under a synchronous control system is to have all transfers through the frame crossbar be of fixed size and to have all such transfers start and end on common temporal boundaries. Thus, the emission of a set of fixed-length cells through a cell crossbar becomes the common, coordinating schedule. This solution has the added advantage that practical scheduling algorithms can pack switching work more efficiently into a cell crossbar than into a frame crossbar, with the frame crossbar's less disciplined packing of transfers into available time. The most serious disadvantages of this approach are that:

- Frames in the ingress ports must be segmented into cells.
- Cell headers must be added to each segment, thereby decreasing the efficiency of the switching operation.
- Frames in the egress ports must be reassembled from cells.
- Cell streams speed up, due to packet segmentation in the ports.

However, these disadvantages are not sufficiently serious to compromise the benefits of this architecture.

11.9.2 Request–Grant Protocol

The frame queues reside on the port cards; the cell scheduling control system resides in the switch core. Clearly, some control protocol is required to make this system work. To make sensible scheduling decisions, the central control system must know the state of all ingress queues. If it does not have this information, it cannot make useful decisions. The solution is to maintain a count of cells in each queue in the switch core. For decision-making purposes, a count of the depth of a queue is every bit as useful as the queue itself.

A request–grant protocol maintains the counts on the switch core. When the switch is initialized, all of the queue-depth counts are zeroed. Each time an ingress port creates another cell (out of some received packet) and places it in one of its ingress queues, it sends a request message to the switch core. The request message contains the number of the queue to which the frame was just added. On receiving the request message, the switch core increments a counter associated with that queue. For each cell transfer time, the switch core will examine the set of all queue counters and pick a set of ingress–egress connections to schedule for the next cell transfer time. We discuss below how this process proceeds, but for now we just notice that the winning queues should have their depth counts decremented to reflect their true depth after the newly scheduled set of cell transfers.

The winning ingress ports must be informed of their recent victories and must know which of their multiple queues was their winning candidate. This is accomplished by a grant message from the switch core to the ingress ports. Once the ingress port receives the grant, it immediately forwards the winning cell to the switch core. By this time, the switch core has set the cell crossbar to the correct settings to cause each incoming cell to be forwarded to the appropriate egress port. Also, the egress ports know from the global cell schedule when they should expect the next cell to arrive (if any has been scheduled for it during the current cell transfer cycle).

The request and grant messages are very small. They are defined as part of the cell headers created when the segments of the packets are formed. The details of these formats depend on a variety of issues, such as the number of ports, N, and classes, C. We need not be concerned with these details at this time.

11.9.3 Permutation Arbitration by Wavefront Computation

During each cell transfer period, each ingress port can provide one cell and each egress port can accept one cell. This leads us to our familiar permutation pattern. (Well, technically, this leads us to any set of 1-to-many mappings without reuse of egress ports, but we deal with this more general form when we discuss multicast in Section 11.12.) We require a control mechanism that can examine the counts representing the depths of the various ingress queues and find a subset of those queues that represent as nearly as possible a complete permutation load for the core cell crossbar. We would like each such schedule to be a complete permutation to maximize use of the cell crossbar, but under lightly loaded conditions we must expect not to be able to fill out this schedule completely.

Figure 11.17 represents a set of ingress queue depth counts. There are $N = 4$ ports. For the moment, there is only $C = 1$ class. The figure is organized such that the ingress ports are represented by the rows, and the egress ports are represented by the columns. Thus, for $I = 2$ and $E = 1$ we find a count of 3, which indicates that $Q_{2,1,1}$, located on the port card for port $I = E = 2$, contains three cells. Generalizing, we see that there are N^2 counts in our example. If there were more than one priority class, there would be N^2C counts.

A permutation load has a total count of one per row and one per column, for a total count of N. Figure 11.18 illustrates a (complete) permutation in which cells are scheduled from ingress 1 to egress 2, from 2 to 4, from 3 to 1, and from 4 to 3. These 1's occur at nonzero points in Figure 11.17. Also, we see that the sum of each row and each column is 1 (or zero). This is a suitable set of queues to schedule for the next cell transfer period, as it will keep all ingress ports and all egress ports engaged productively, without any form of

	Egresses			
	1	2	3	N=4
1	1	2	0	0
2	3	2	0	1
3	4	0	1	0
N=4	0	2	7	2

FIGURE 11.17 Request counts for $N = 4$ and $C = 1$.

Ingresses \ Egresses	1	2	3	N=4
1		1		
2				1
3	1			
N=4			1	

FIGURE 11.18 Permutation solution for $N = 4$.

Ingresses \ Egresses	1	2	3	N=4
1	1	1	0	0
2	3	2	0	0
3	3	0	1	0
N=4	0	2	6	2

FIGURE 11.19 Residual request count for $N = 4$.

blocking. If we schedule the queues indicated in Figure 11.18, it is necessary to reduce the counts recorded in Figure 11.17 to reflect this progress. Figure 11.19 shows the residual counts after a simple matrix subtraction of the permutation (Figure 11.18) from the original counts (Figure 11.17). Unless newly arriving requests increase any of the counts, these counts will be used to find the next permutation load for the next cell transfer period.

The overall process of accepting requests, finding maximal permutations, and issuing grants is represented in Figure 11.20. The process begins by initializing all request counts to zero. Then two separate processes are launched. The process to the right accepts every request incoming from port I for an egress E and increments the appropriate cell in *Request[*, *]*. The process to the left finds a maximal input-to-output permutation in *Request[*, *]*, subtracts that permutation out of *Request[*, *]*, and issues grants for each transfer in the permutation. This process is completed once in each cell time.

The question becomes: How do we identify suitable maximal permutation loads? The problem is clearly related to finding open paths for physical or logical circuits that we explored in the two preceding chapters. But now we have a particular urgency to our problem. While we have not yet discussed the appropriate length for the cells that are transferred by the switch, we can anticipate that they must be reasonably small, certainly not much larger than the minigrams of any frame traffic that is being carried. In fact, we can anticipate from the discussion of minigram speedup issues in central memory switches that it may be very sensible to have the cell size be exactly twice the

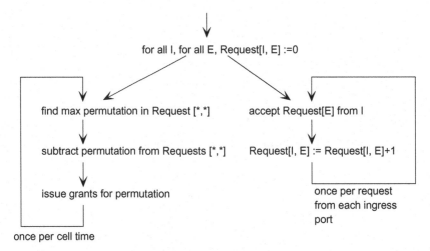

FIGURE 11.20   Process view of request accumulation and permutation extraction.

minigram size. For a pure TCP/IP application, this would lead to a cell size of 80 bytes (including the cell header). Suppose that our ports and internal switch links are running at 1 Gb/s. Then our cell time is only 256 ns. Our control system must compute a new maximal permutation each cell time. As N increases and as realistic C values of 4 or 8 are considered, there are many counts (queues) to be considered in finding each permutation. The constraints of this problem dictate that it be done in parallel in hardware, as opposed to the sequential software solutions we considered in earlier chapters.

We look to the ideas of systolic or wavefront arrays to find practical solutions to this problem. These architectural approaches use arrays of small processing elements to implement a parallel algorithm. Figure 11.21 represents an abstract view of a wavefront array for our maximal permutation problem. The array has one row for each ingress port and one column for each egress port. Each array element contains the queue count for the appropriate ingress–egress pair. Computation is initiated on the left and the top boundaries by inserting initial values. Computational progress proceeds down and from left to right across the array. On the left, initial values are inserted indicating that the ingress port corresponding to each row is still "hunting" for an idle egress with which it shares a nonempty queue to form a connection for the next cell transfer time. On the top, initial values are inserted indicating that the egress port corresponding to each column is still "available" for any ingress that is still hunting, as long as the ingress–egress pair has a nonempty shared queue. In both cases, these initial values are represented by arrows entering the wavefront array.

Each cell in the wavefront array attempts to match a hunting ingress with an available egress when the two ports share a nonempty queue. When such

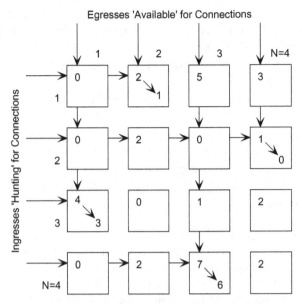

FIGURE 11.21 Conceptual wavefront arbiter.

a match is made, the cell does not pass the hunting signal farther to the right, to reflect the fact that the ingress corresponding to that row is no longer hunting. Similarly, when a match is made, the cell does not pass the available signal farther down the column, to reflect the fact that the egress corresponding to that row is no longer available. However, the cell does decrement the request count and signals the fact that it has scheduled the corresponding cell transfer to the rest of the control system.

Whenever the conditions for a successful match (ingress hunting, egress available, and nonempty shared queue) are not met, any received hunting and available signals are passed to the right and down, respectively. Naturally, the count in the cell is left unchanged. Let's examine this process as illustrated in Figure 11.21 (we denote the cells of the array by their address as <ingress row, egress column>):

1. The process begins in cell <1, 1>, which receives a hunting signal and an available signal. However, the queue count at <1, 1> is zero, so no cell can be scheduled. Cell <1, 1> passes its received hunting and available signals to <1, 2> and <2, 1>, respectively.

2. Cell <1, 2> receives a hunting signal from <1, 1> and an available signal from the initial values given to the top row. Cell <1, 2> also has a nonempty queue, so the conditions for scheduling a cell transfer have been met. As a consequence, <1, 2> does not emit either a hunting or an

available signal, does decrement its count, and does signal the rest of the control signal that it has scheduled a transfer from ingress 1 to egress 2 (this step is not shown).

3. Cells <1, 3> and <1, 4> do not receive a valid hunting signal, so have no chance to schedule a cell transfer.
4. Cell <2, 1> receives valid hunting and available signals, but has a count of zero.
5. Cell <2, 2> receives a hunting signal and has a count, but does not receive an available signal because cell <1, 2> consumed egress 2.
6. Cell <2, 3> has hunting and available signals but an empty queue.
7. Cell <2, 4> has hunting and available signals and a nonempty queue, so schedules another cell.

The remaining eight cells of this wavefront array follow the examples we have seen in the upper half of the array. When the computation has progressed from cell <1,1> to cell <4,4>, the computation is complete. In this example, four cells have self-scheduled cell transfers (<1,2>, <2,4>, <3,1>, and <4,3>). These four cells decrement their request counts. In logic not shown in Figure 11.20, these four cells are communicated to the appropriate ingress ports as grants in the request–grant protocol.

The logic required of each cell in the wavefront array is quite simple. Figure 11.22 specifies this logic. The hunting and available signals are treated as simple active-high Boolean values (e.g., *True* means still hunting). $Connect_{I,E}$ is part of the system not shown in Figure 11.21 with which winning cells identify themselves. The individual cells of the wavefront array are quite small in terms of gates. From the array architecture shown in Figure 11.21 it is easy to see that the cost of the array scales as $O(N^2)$. Each cell of the array cannot compute its result until both of its inputs ("hunting" and "available") are presented to it. Thus, the worst-case latency of this systolic computation comes from the available signal propagating down the first column, and then allow the hunting signal of the bottom row to propagate across to the bottom right

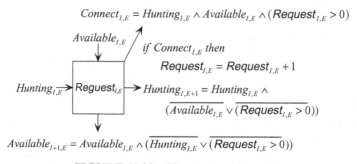

FIGURE 11.22 Wavefront arbiter cell.

cell, at which point the computation is complete. Thus, $2N - 1$ cells must compute their results, so the time cost of this solutions is $O(N)$. Understanding the logic of Figure 11.22 completes our initial view of wavefront arbitration. We have arrived at an effective and neat solution. However, five serious issues remain: latency, systematic arbitration bias, treatment of multiple classes, problem partitioning, and multicast scheduling.

Latency For a cell to transit the request–grant architecture that has been described above, it is necessary for the packet to be received in the input port, the first (and succeeding) request to be issued to the switch core, that request(s) to win in arbitration, the grant(s) to be returned to the ingress port, the cell(s) to be transmitted through the cell crossbar to the egress port, and the full packet to be reassembled in the egress port. For each cell, a three-phase operation is required: request–grant–cell. This is triple the minimal latency consumed by just sending the cell (and hoping). Thus, request–grant-based switches do increase the latency packets experience. In WAN applications, the time consumed in these switches in negligible compared to the latency incurred by the fiber links. In any congested network, the time that packets reside in queues can exceed either of these latencies. Nevertheless, there are applications for which the added latency of request–grant semantics is a problem.

Arbitration Bias As the wavefront array has been defined, with ingress 1 in the first row and egress 1 in the first column, we have built serious bias into the algorithm. Ingress 1 always has the first chance to grab any egress. Egress 1 is the first path that will be found by any ingress that is still hunting. Thus, in terms of both input and output, the array is biased to give the best service to port 1. Although this is useful and can be exploited in some special switching applications, it is a serious flaw for most uses in a transport network, which should generally avoid any form of port bias. The problem with the wavefront array is that each ingress and egress is locked in a particular position in the rows and columns, which determines its relative priority. Yet the wavefront array concept seems to require this geometric commitment. We are very reluctant to abandon the wavefront array concept, as it is not clear that there are other avenues to solutions that can run fast enough.

In Figure 11.21 the row–cell processors are wired together in a fixed pattern (from lower to higher port numbers). Instead of using fixed wiring to connect these cells, we use a switch! Each cell receives one hunting signal and emits one hunting signal. If all such hunting signals are connected by a switch (a crossbar), it becomes possible to change the order of the wavefront array cells. If we use one such switch for each column and one for each row (with all the column switches using the same connection pattern at any time, and all row switches being ganged similarly), we can arbitrarily permute the row and column ordering of the wavefront array from time to time. By changing the orderings for each arbitration cycle, we can smooth out the fixed bias we found in our earlier solution by varying the relative positions of the rows and columns over time.

It might seem ideal to build additional hardware that could generate all possible permutations of the N ports and feed these permutations to the wavefront array to smooth the inherent bias perfectly over time. But no one seems to know how to build a (completely fair) permutation generator in digital hardware that can produce its next permutation in fixed time (there's a problem for your future Ph.D. research, but be careful, as the problem probably has no complete solution). An alternative might be to precompute all the permutations and store them in a memory in the control system. Before you proceed too far down this path, consider the number of bits you would need to represent all permutations for N ports as N grows to practical values such as $N = 64$. These sequences are so long that they cannot be stored an any practical control system. Even if they could, they are so long that a switch running under their control could remain significantly biased for long periods of time, as the permutation sequence happened to be passing through a part of the sequence that was biased against a particular port. Before you wrack your brains over this issue, you probably should accept that it is one of those insolvable problems that occur frequently in algorithms.

Rather than continue to search for an unachievable perfection, it is appropriate to seek a useful compromise. A useful solution is to precompute a subset of permutations such that every pair of ports A and B occur half of the time with A first and half of the time with B first. This gives a sort of first-order fairness. What it does not give us is the property that the other ports $(C, D, \ldots)$ are equivalently fairly treated when A precedes B as when B precedes A. However, it leads to an acceptably small sequence with acceptable fairness. Figure 11.23 illustrates the use of programmable column and row switches to

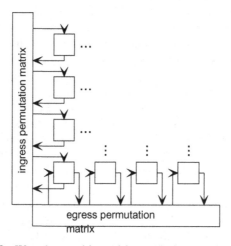

FIGURE 11.23 Wavefront arbiter with reordering permutation switches.

permit the controlled shuffling of the order of the rows and columns in a wavefront array. There are many ways to build permutation-based fair wavefront arrays. The style presented in this section was selected primarily for clarity. Depending on the underlying technology, other styles may be more efficient.

Arbitration computations based on stored, partial permutation sets are feasible to implement and can be made sufficiently fair for all practical purposes. Although this architecture is a sophisticated design, it remains subject to internal blocking. Consider two situations in which ingress port I has cells destined for egress ports E_1 and E_2. In the first situation, no other ingress port has traffic for E_1 or E_2. The arbiter binds ingress port I to egress port E_1. E_2 is then left without an ingress partner even though traffic for that port does exist—the problem is that ingress port I cannot supply both egresses, so must deny one or the other of the egresses. The data path is not capable of dual transfers from one ingress port. In the second situation, ingress port J has traffic only for egress port E_1. The arbiter considers ingress port I before port J and binds ingress I to egress E_1. When the arbiter considers ingress J, egress E_1 is already consumed, so ingress J must remain idle (as it has no other traffic). If the arbiter could reconsider its action in binding ingress I to egress E_1, it could allocate ingress I to egress E_2, then allocate ingress J to egress E_1. But this would require backtracking, which is too expensive to include in practical implementations. So we are left with an "eager" arbiter which makes a series of irrevocable binding decisions and may work itself into a less than optimal corner. At the speed that these arbiters must operate, there appears to be no choice. Practical switch fabrics must employ some degree of internal speedup to overcome such inefficiencies.

Multiple Classes Thus far, we have considered the arbitration problem only for single-priority-class situations. It is necessary to generalize the problem to accept multiple classes. This generalization is logically straightforward. One $N \times N$ wavefront array is used for each priority class. The highest-priority class is initialized with all ingress ports hunting and all egress ports available. That first wavefront array contains and uses the ingress-to-egress queue counts for the top-priority queues. This first wavefront arbiter then makes any possible assignments for the top priority. When it is done, some ingress may still be hunting, and some egresses may still be available. The status of these ingress/egress ports is passed to the next wavefront arbiter for the next priority class to use and in N inputs for hunting and the N inputs for available. Thus, this second wavefront arbiter accepts the residual opportunities from the first arbiter and tries to make further useful assignments. The residual opportunities from this second arbiter can be passed downward to lower priorities in a continuation of this chain of arbiters. This approach is illustrated in Figure 11.24. While this solution for multiple classes is effective, it has three drawbacks:

1. It can only support rigid priorities. "Softer" priority schemes that attempt some balance between priority classes are not supported.
2. The time required for the process to complete [$O(NC)$] can become too long. Recall that the arbitration process must be complete before the end of a single cell time. Increasing the depth of the arbitration logic chain by a factor of C can lead to timing problems.
3. The area (gates) required by the circuit, which is C times larger than that of a single-class arbiter [$O(N^2C)$], can be a problem.

Problem Partitioning   We have seen that the area or gate cost of large multiclass arbiters may be excessive. As long as there is enough time to run the full N^2C arbitration problem, there are simple techniques to control the size of the required circuit. Figure 11.24 shows the replication of a wavefront arbiter to process multiple classes. Rather than replicate the actual wavefront circuit, it would be possible to have one wavefront arbiter and use it iteratively to cover the C classes. This solution requires that registers (flip-flops) be used to hold the results found for one class, to be used as inputs to the next class.

Similarly, if the $N \times N$ wavefront array is too large, it can be divided into several equal parts, each of which is run on a small wavefront array, with registers to connect the subproblems in the appropriate way. A simple solution would be to divide the array into four logical parts and to build a wavefront array processor for one of these quarters. It is necessary that the order of the subproblems correspond to the forward (or causal) directions in the overall wavefront array problem (i.e., the upper left quadrant must be run first, it then feeds both the upper right and lower left quadrants, which can be run in either order; finally, the lower right quadrant must be run last, as it requires inputs from the lower left and the upper right quadrants).

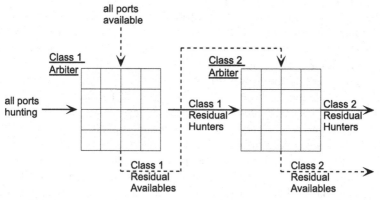

FIGURE 11.24   Wavefront permutation for multiple priority classes.

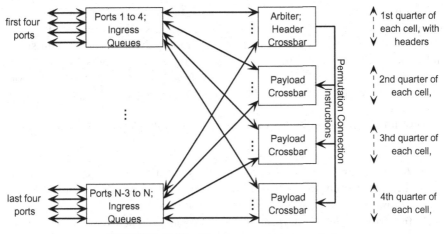

FIGURE 11.25 Sliced request–grant switch.

11.10 SLICED REQUEST–GRANT SWITCHES

The frame switch architecture presented in Section 11.10 has many advantages, but of course it has limits. The total amount of traffic that can be put through any crossbar chip is limited by power and pinout considerations. However, it is possible to slice this architecture, thereby allowing multiple chips to support the aggregate traffic flow. Suppose that we slice the architecture into $S = 4$ layers, with each layer supported by its own chip (Figure 11.25). Each port segments its cells into S subcells, with all the first subcells being sent to the first layer, and so on to the last layer. Each subcell is $1/S$ as long as the original cells. The first subcell carries the complete cell header to the top switch chip, which contains the request counts and the wavefront arbiter. The arbiter works as before but communicates the crossbar settings it determines to all the other $S - 1$ chips in the core switch stack. By the time the subcells corresponding to a set of grants arrive at the S cores, all the crossbars are set for that group of subcells, and synchronous timing of the subcell time has been maintained.

This solution allows the input/output load to be divided, which is a move in the right direction. However, it reduces the time allowed for each arbitration to $1/S$ as long. It also introduces the problems of coordinating the stack temporally and communicating the switch settings for each subcell time period. However, this approach can be used to increase the aggregate bandwidth of truly large frame switches.

11.11 MULTISTAGE FRAME NETWORKS

We have seen multistage architectures for physical and logical circuit-switching applications. The same possibility applies to frame switching. Many proposals for multistage frame switches have been made. Some of these proposals have fascinated many researchers and students for several decades. One of the present authors spent years designing and studying one such multistage network. A few such ideas have been built as practical products. But the track record of practical application of these lovely ideas has been poor—so poor, in fact, that we do not feel that these networks belong in this practically oriented book.

The fundamental problems of most of these multistage frame switch proposals have been:

1. The number of chips and internal links within these networks increase their implementation costs dramatically. If the goal is to reduce costs, it is always better to have a single-stage network.
2. In terms of blocking, practical multistage switches have much worse switching performance than that of single-stage switches.

For these two reasons, the motivation to avoid multistage frame switches is so strong that, in general, practical products have built up to the limit of what single-stage networks can provide and not attempted larger switches. This single-stage limit couples nicely with the stack of switching solutions from frame to logical circuit to physical circuit. It has always been possible to build larger switches as the layer moves toward the physical circuit switch. So when an upper layer of switching runs out of steam, networks tend to move switching to the next lower layer rather than build "hero" switches at the upper layers. Hence, industry has largely avoided multistage ideas. This is a bit sad, as some of the multistage ideas are quite lovely. But as designers (or students) of network elements, we are constrained to be practical.

11.12 MULTICAST

We have seen the provision of multicast in the logical circuit switches of Chapter 10. The SNB single-stage switches had no problem with any degree of multicast. However, the multistage fabrics consistently had difficulties with multicast. In this section we consider how multicast can be provided for frames based on the switches examined in earlier sections of the chapter. In general, providing true multicast in frame switches is a very challenging problem. Multicasting incorporates all forms of 1-to-many switching. Two subsets of the general problem are relatively straightforward and will be dispensed with quickly.

1. Broadcast is the one-to-all problem. It is relatively straightforward to add special case control logic to any switch that causes a single frame to be copied to all egress ports. The principal difficulties are to determine when broadcast should be employed and how much bandwidth should be devoted

to it, and to coordinate transmission to all egress ports. It is possible to wait for all egress ports to become idle and then use the normal unicast switching data paths to copy the frame to all egresses. It is also possible to implement a special-purpose broadcast data path to a broadcast frame buffer in every egress port. When all egress broadcast frame buffers are empty, the central broadcaster can copy new frames to all such buffers. Then each egress port inserts this broadcast frame into its outgoing stream at the next packet boundary.

2. Egress port duplication is a form of 1-to-2-casting in which the two egress ports are intended to carry exactly the same stream of frames (and normally, the same lower-layer protocol streams). This feature is used for protection in SONET/SDH systems—at any time one emitted copy is the working stream and the other copy serves as a backup protection copy. A frame switch may absorb this function when it is part of an integrated network element that acts as both a frame switch and a SONET/SDH element. It is not difficult to add special logic and/or data paths to implement this form of bicasting to the frame switches we have examined.

The general multicast problem is to copy a frame received on any ingress port to any subset of the N egress ports of the switch. There are 2^N possible combinations of output ports in an N-ported switch. Logically, each such output pattern represents a queue. If we consider that each ingress and each class may require its own set of multicast queues, we are faced with as many as $2^N NC$ multicast queues to add to a system that had no more than $N^2 C$ queues before multicast was added. Thus we have gone from a number of queues quadratic in N to a number of queues exponential in N. It should be no wonder that multicast adds enormously to the degree of difficulty of frame switching.

In practice, no approach based on 2^N queues has been used. A common solution is to have one multicast queue, or perhaps one multicast queue for each ingress. A multicast server takes one multicast cell at a time to copy to all required outputs. The multicast server must compete with all the other ports and queues in the arbitration process. Sometimes a limited flow of multicast is given priority over unicast on a priority class-by-priority class basis.

The destination of each frame (or of all the frames in a given multicast queue) is indicated by a bit vector of length N with a 1 in each position that corresponds to an egress port to which the frame should be copied. A sensible approach to completing the emissions required by any multicast frame is to begin by initializing an N-bit vector to describe all the egress ports that should receive a copy, then to copy the frame to the various ports as soon as possible given other events in the switch, turning off the bits for the completed copies as they occur. When the bit vector is all zeros, the multicasting of that frame is complete.

We must consider the origin of this N-bit multicast flow vector. In a switch that is richly endowed with multicast capabilities, the ingress ports may send a complete N-bit multicast flow vector with each frame (or at least with special multicast frame). This is expensive in bandwidth, especially in cell switches, where each cell must carry the overhead of this flow vector. An alternative is to have a subset of the normal routing codes which indicate multicast (e.g., the numbers $1 \ldots N$ indicate unicast to the numbered port, but numbers $>N$ are taken as multicast codes). The switch contains a multicast code table that translates from multicast code to a full N-bit multicast vector. Thus, no frames are burdened with large multicast vectors. The shortcoming is that only a limited number of multicast flows can be in use at any time, as the code space and lookup table space limit the number.

General multicast is at its easiest in the central memory frame switch. One copy of the frame to be multicast can be stored in the common memory, where each required egress port can find it. The queue representation and egress choice/arbitration logic must represent and take account of the special needs of multicast. As we react to the limit of scalability of central memory switches and build any of the distributed memory switches, the separation of the memories and the various data paths makes multicast more difficult. We consider briefly how multicast might be added to the blocking ingress-buffered design, the nonblocking ingress-buffered design, and the request–grant design.

11.12.1   Multicast in the Blocking Ingress-Buffered Architecture

In the blocking ingress-buffered architecture, we find that reasonable solutions for multicast are available. Each ingress can add a single multicast queue (or one at each priority), as many of the egress commitments for each multicast cell are scheduled at the same time. For each such scheduling, the cell is copied into the crossbar and is read from the crossbar by those egresses which are scheduled for the time period. It is necessary to coordinate the egress ports that are to participate.

11.12.2   Multicast in the Nonblocking Ingress-Buffered Architecture

This architecture is better for unicast than is the blocking version. Thus, it is surprising to find that the architecture cannot perform multicast as it is drawn. The problem is that each ingress queue is going to only one egress port, so there is no suitable place to store the multicast cell while incrementally fulfilling its fanout requirements.

11.12.3   Multicast in the Request–Grant Architecture

In this architecture, the ingress queues have been moved off the switch core, back to the ingress line cards. This ingress queuing system has only a single data path into the switch core. Consequently, we have a choice of either doing a limited amount of multicast by sequential repetition of the multicast cells from the ingress port to the various egress ports, or to have the core of the switch carry out the multicast. Whenever multicast loads are at all significant, they must be carried out in the switch core, to avoid congesting the ingress port (as it tries to inject more bandwidth into the switch fabric than it receives on its telecommunications links). However, multicast within the switch core is a difficult problem. The key issue is that the wavefront arbiter presented earlier in this chapter is not capable of allocating multicast traffic. It is necessary to design a separate multicast arbiter. Also, in order to have respectable multicast performance, it is necessary that multicast take precedence over unicast (at least at otherwise equal-priority levels). Finally, it is not reasonable to expect that all multicasts can be accomplished in one cell time. If the multicast core imposed a restriction that all multicasts must be accomplished at once, overall multicast bandwidth would be limited by the inability of the multicast arbiter to find sufficient multicast requests that did not conflict in their needs for some particular output port. Consequently, it seems that the only reasonable multicast arbiter must be capable of partially fulfilling a multicast request in one cell cycle and then keeping the residue of the request for fulfillment in a later cell time. These various difficult demands of core-based multicast can be solved, but the solution is beyond the scope of this book.

KEY POINTS

In this chapter we covered a great amount of ground. We have:

- Defined the frame switching problem
- Examined the difficulties raised by fundamentally uncontrolled IP packet flows
- Considered the basic ideas of queues and the various forms of blocking
- Examined a complete set of queuing systems for frame switches, based on ports and classes
- Concluded that the one queuing system that differentiates flows by ingress port, egress port, and priority class (all other queuing systems possible lead to unpleasant consequences)
- Discussed the implications of the location of the queues on the possibilities for buffer sharing among queues and for the overall architecture of switching systems
- Looked briefly at policies for filling and draining queues
- Presented and analyzed a simple central memory frame switch and discovered several reasons for speedup of the central memory subsystem
- Proposed and analyzed a blocking ingress-buffered frame switch
- Improved the blocking ingress-buffered switch to produce a nonblocking ingress-buffered frame switch

- Described how to move ingress queues away from the central switch implementation and onto the ingress port cards using a request–grant protocol
- Described how to find cell transfer permutation within the context of the request–grant protocol on our ingress-buffered virtual output queued switch architecture
- Considered briefly how to build yet-larger frame switches by slicing their operation across a stack of core switching elements
- Considered why multistage frame switches have rarely been used in practice
- Discussed the difficulty of providing a general frame multicast capability
- Discussed how a frame multicast capability could be added to the various architectures presented throughout the chapter

12

NETWORK ELEMENTS

12.1 INTRODUCTION

Whether they are corporate intranets, wireline carrier networks, or wireless networks, networks are comprised basically of nodes and transmission channels. Nodes, also known as network elements (NEs), support a variety of communication functions. Several dozen different NEs exist for the aforementioned networks. Many of these NEs reside at a particular layer of the Open Systems Interconnection Reference Model (OSIRM); others support several layers of the protocol model. (Generally, it is more cost-effective to have an NE cover multiple layers than discrete devices; this way, there is no need for multiple

---

Network Infrastructure and Architecture: Designing High-Availability Networks,
By Krzysztof Iniewski, Carl McCrosky, and Daniel Minoli
Copyright © 2008 John Wiley & Sons, Inc.

chassis, multiple power supplies, multiple racks with interconnecting cables, multiple network monitoring systems, and so on.) Regardless of the functional bundling, NEs are critical to proper functioning of any network.

In this chapter we look at some of the basic NEs that are commonly used to construct communication networks. Specifically, the purpose of this chapter is to review basic networking functions and contrast switching to grooming and switching to routing. NEs such as synchronous optical network (SONET) add–drop multiplexers (ADMs), switches, and routers are discussed in terms of both their functionality and their prototypical internal construction. A number of these NEs have actually been discussed (indirectly) in earlier chapters; however, the discussion here takes a more systematized view.

12.2 NETWORKING FUNCTIONS

As we have seen in Chapter 1, fundamentally, a network exists to move information (voice, video, data, instant messages, etc.) from a source point A to a remote sink point B (this being the case for unicast communication environments; in multicast environments, the sinks will be B1, B2, . . . , Bn; in broadcast environments, there will be, in principle, infinitely many sinks; see Figure 12.1). The transmission channel can operate as a guided medium, where the signal is contained within the physical medium itself (e.g., cable, fiber-optic link), or as an unguided medium, where the signal is sent without a physical medium but only as electromagnetic energy in the free-space field (e.g., radio communication). Furthermore, the signal can be carried in baseband mode, where the signal changes are coupled directly onto the channel (e.g., by direct connection to the medium itself), or in a broadband mode, where the information is modulated (added, superimposed) onto an adequate underlying carrying signal called the *carrier*. For example, local area networks operate at baseband; fiber-optic links also operate at baseband; cable modem communication, commercial radio, and traditional over-the-air TV, among others, use carrier for transmission.

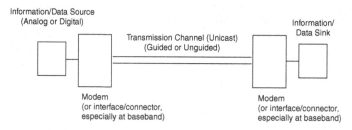

FIGURE 12.1 Basic communication framework.

Almost invariably, additional requirements beyond this basic transmission function are imposed on a commercial network, such as, but not limited to, the fact that information will be transmitted reliably, rapidly, with low latency/jitter/loss, cost-effectively, in an integrated fashion, and in a secure manner. Some of these requirements are met by selection of the proper transmission channel; other requirements are met by employing appropriate NEs in the network, such as NEs supporting multiplexing, grooming, switching, and routing (many other functions are supported in a network, but our discussion here is focused on these key functions).

Networks can be classified in a number of ways. The classification can be in terms of (but not limited to):

- *Information-handling approaches:* circuit mode, packet mode, circuit emulation over a packet mode, reservation mode (packet-mode reservation of a circuit-mode channel), hybrid (some have called this pacuit mode), where the network combines packet switching and time-division circuit switching in a common system
- *Information media:* † voice, data, video, multimedia/converged
- *Transmission class:* wireline, wireless, other
- *Transmission mode:* physical or virtual (tunneled/encapsulated/overlay)
- *Transmission channel technology:* optical, cable, twisted pair, short-hop radio frequency, long-hop radio frequency, free-space optics, other
- *Information-handling protocols* (particularly at a given OSIRM layer)
- *Administration type:* private, public, hybrid
- *Geographic technology:* personal area network (PAN), local area network (LAN), metropolitan area network (MAN), regional area network (RAN), wide area network (WAN)
- *Logical scope:* intranet, extranet, internet
- *Function/scope:* commercial, public safety, military, other

Clearly, it follows that a comprehensive discussion of networking requires one to take a multidimensional view; in this chapter, however, we look at only a small subset of these dimensions. Many of these factors drive network design considerations as discussed in more detail in Chapter 13.

One fundamental classification is related to information-handling approaches, as identified above. Communication started out to be *circuit mode–based* (also called *connection-oriented communication*), where a session (agreement to exchange information) is established before the information is transmitted between the source and the sink. To accomplish this, the network

†Notice that the term *media* is used in two distinct ways: to describe the transmission medium (channel) and to describe the information stream—the intended meaning should be clear from the context.

must maintain "state" information about the session. In connection-oriented systems, the content is almost always transmitted monolithically in its entirety between the source and the sink. This mode has been used in voice telephony since its inception in the nineteenth century and it is reasonably well suited for certain types of communication applications, such as high-throughput, long-duration, fixed-bandwidth environments. However, this mode has limitations in terms of efficiency and cost-effectiveness for a number of media, particularly data.

More recently (since the late 1960s), designers have deployed *packet mode communication* (also called *connectionless communication*). Here, a session (agreement to exchange information) is *not* required to be established before information is transmitted between the source and the sink; instead, information is sent without formal preestablishment of "terms" and the network does not maintain "state" information[†]. The content is always segmented into (relatively) small packets [also called protocol data units (PDUs) or datagrams] and the packets are transmitted between the source and the sink and reassembled sequentially at the sink. This mode has been used for data communications and the Internet; it is well suited for data communication applications particularly for medium-throughput, short-duration, bursty-bandwidth environments. Recently, voice and video applications have also seen migration to the packet mode.[‡] In particular, packet mode is ideal for multimedia environments. Lately, carriers have seen a price erosion of their traditional services, and to compensate for this, have being pursuing a strategy called *triple play*, where they would aim at providing voice, data (Internet access), and entertainment-level video. To do this effectively, packet-mode communications are indispensable.

Circuit emulation over a packet mode, reservation mode (packet-mode reservation of a circuit-mode channel), and hybrid have seen much less penetration than the circuit- and packet-modes systems just described. Hence, we focus here on the circuit vs. packet classification. We also allude to the information media mode (voice, data, video, multimedia/converged). The other classifications will be cited only parenthetically.

In the subsections that follow we examine five basic functions used to meet the above-named goals: regenerating, multiplexing, grooming, switching, and routing; information transfer (transmission) will be discussed only briefly. All three of these functions can occur at several layers of the OSIRM model, particularly the lower layers. See Table 12.1, which also includes information transfer for completeness. It should be noted, however, that some forms of

---

[†]This is particularly true in the case of user datagram protocol (UDP) transmissions. Although one could argue that transmission control protocol (TCP) transmissions entail some sort of end-to-end handshake, it is still considered to be a connectionless service since the *network* does not maintain any state (although the endpoints maintain some state information).

[‡]However, as of press time only about 3 to 4% of the global voice traffic (as measured by revenue) was carried in packet mode. This percentage will grow over time, but the rate of growth is unknown at this time.

TABLE 12.1   Communication Functions Across OSIRM Layers

	Information Transfer	Multiplexing	Grooming	Switching	
Transport layer: end-to-end communication	L4	L4	L4	L4	Other
Network layer: communication across a set of links	L3	L3	L3	L3	Other
Data link layer: communication across a single link	L2	L2	L2	L2	Other
Physical layer: communication over a physical channel	L1	L1	L1	L1	Other

multiplexing, grooming, or switching are more important and/or well-known and/or supported by commercially available NEs than others; Table 12.2 depicts some of the best known NEs. Figure 12.2 provides a generalized view of the functions of these key NEs. As noted in the introduction, a number of these NEs have already been discusses in the text, but from a technology-specific perspective. We define the following terms:

- *Grooming:* combining a relatively large set of low-utilization traffic streams to a higher-utilization stream or streams. Generally, done at layer 1 of the OSIRM, but can also be done at other layers. Differs from multiplexing in the sense that in the latter case the incoming streams are usually considered to be fully loaded and their entire content needs to be aggregated time slot by time slot onto the outgoing channel; in grooming, the incoming streams are usually lightly utilized, and not every incoming time slot needs to be preserved.
- *Information transmission:* physical bit transmission (transfer) across a channel. Transmission techniques are specific to the medium (e.g., radio channel, fiber channel, twisted-pair copper channel). Often entails transducing, modulation, timing, noise/impairments management, signal-level management. Transmission is generally considered a layer 1 function in the OSIRM.
- *Multiplexing:* basic communication mechanism to enable sharing of the communication channel, based on cost-effectiveness considerations. Multiplexing allows multiple users to gain access to the channel and to do so

TABLE 12.2 Well-Known Communication Functions Across OSIRM Layers[a]

	Information Transfer	Multiplexing	Grooming	Switching	Other
Transport layer: end-to-end communication	—	L4 multiplexing	L4 grooming	L4 switching	Other
Network layer: communication across a set of links	IP routing	L3 multiplexing	L3 grooming	IP switching, MPLS, VoIP (soft switch)	Other
Data link layer: communication across a single link	Relaying	ATM, frame relay, Ethernet	L2 grooming	ATM switching, frame relay switching, Ethernet switching	Other
Physical layer: communication over a physical channel	Bit transmission; bit retransmission (repeater) (a.k.a. reamplification); 3R: reamplify retime, reshape (optical systems)	T1 multiplexing, SONET/SDH multiplexing, SONET/SDH ADM, wavelength-division multiplexing	Digital cross-connect systems, SONET/ SDH ADM, all-optical cross-connect systems	Matrix switching, voice/ISDN switching, optical switching, digital cross-connect systems, ASON, video router (switch)	Other

[a]ADM, add–drop multiplexer; ASON, automatic switched optimal network; ATM, asynchronous transfer mode; IP, Internet protocol; ISDN, integrated services digital network; MPLS, multiprotocol label switching; SDH, synchronous digital hierarchy; VoIP, voice over Internet protocol.

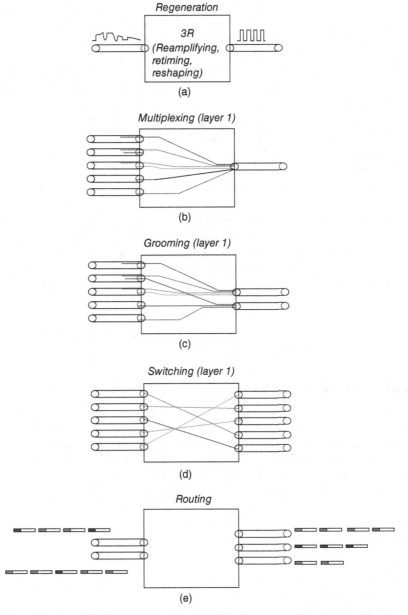

FIGURE 12.2 Key NE functions: examples.

in a reliable manner that does not impede or frustrate communication. Multiplexing is achieved in a number of ways, depending on technology, channel characteristics, and so on. Generally done at layer 1 of the OSIRM, but can also be done at other layers.

- *Regeneration:* restoring the bit stream to its original shape and power level. Regeneration techniques are specific to the medium (e.g., radio channel, fiber channel, twisted-pair copper channel). Typical transmission problems include signal attenuation, signal dispersion, crosstalk; regeneration correctively addresses these issues via signal reamplification, retiming, and reshaping. Regeneration is generally considered a layer 1 function in the OSIRM.

- *Routing:* the forwarding of packets based on the header information included in the protocol data unit. Routing is a layer 3 function in the OSIRM. The forwarding is based on topology information and other information, such as priority and link status. Topology information is collected (often in real time) through the use of an ancillary routing protocol.

- *Switching:* the mechanism that allows information arriving on any inlet (port) to be forwarded/relayed/interconnected with any outlet. The function is technology dependent. Furthermore, switching can be undertaken at layer 1 of the OSIRM (e.g., voice switching), layer 2 (e.g., cell switching), layer 3 (e.g., multiprotocol label switching), or even at higher layers.

Next, we discuss the four key functions we highlighted earlier in some detail. Figure 12.2 depicts the concepts pictorially.

12.2.1 Information Transfer and Regeneration Function

This function entails bit transmission across a channel (medium). Since there are a variety of media, many of the transmission techniques are specific to the medium at hand. Functions such as, but not limited to, modulation, timing, noise/impairments management, and signal-level management are typical in this context; these are not, however, discussed here. Typical transmission problems (e.g., in optical communication) include the following:

- Signal attenuation
- Signal dispersion
- Signal nonlinearities
- Noise, especially due to simple repeaters and optical amplifiers used along the way
- Crosstalk and intersymbol interference, especially occurring in optical switches and optical cross-connects

Some of these impairments can be dealt with by a regenerator. Figure 12.2(a) depicts this signal regeneration function pictorially. Regeneration is an important function, especially in fiber-optic systems. Figure 12.3 depicts three versions of the repeating function, with increasing levels of sophistication.

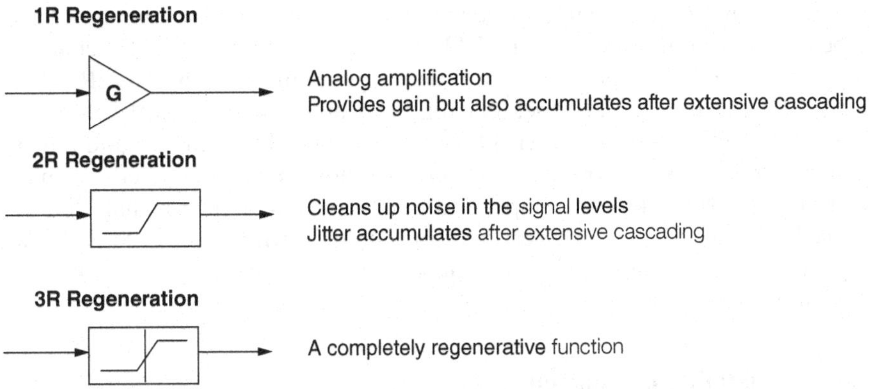

1R Regeneration

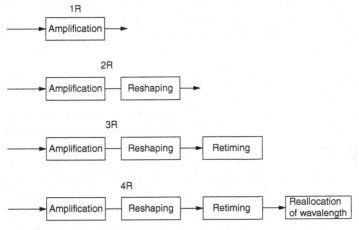

Analog amplification
Provides gain but also accumulates after extensive cascading

2R Regeneration

Cleans up noise in the signal levels
Jitter accumulates after extensive cascading

3R Regeneration

A completely regenerative function

FIGURE 12.3   Examples of repeaters.

1R

→ Amplification →

2R

→ Amplification — Reshaping →

3R

→ Amplification — Reshaping — Retiming

4R

→ Amplification — Reshaping — Retiming — Reallocation of wavalength

FIGURE 12.4   Basic functionality of various regenerators.

Figure 12.4 depicts the basic building blocks of various regenerators. A high-end regenerator includes the reamplification, retiming, and reshaping (3R) functionality; a low-end regenerator includes only the reamplification (1R) function. 1R NEs are generically called *repeaters*. The functions of a 3R NE are:

- *Reamplification:* increases power levels above the system sensitivity
- *Retiming:* suppresses timing jitter by optical clock recovery
- *Reshaping:* suppresses noise and amplitude fluctuations by decision stage

Regenerators are invariably technology specific. Hence, one has LAN repeaters (even if rarely used); WiFi repeaters; copper-line (T1 channel) repeaters; cable TV repeaters; and optical regenerators (of the 1R, 2R, or 3R form). Originally, optical repeaters required optical-to-electrical (O/E) followed by electrical-to-optical (E/O) conversion, which added cost and performance limitations. All-optical regenerators now being deployed in many carrier networks retime, reshape, and retransmit an optical signal (the all-optical scheme achieves the regeneration function without O/E conversion). Again, some of these NEs were discussed in earlier chapters from a technological point of view.

12.2.2 Multiplexing Function

Figure 12.1 depicted a basic communications model where a source makes use of a channel to transmit information to a remote location. It turns out that in most situations, the channel is relatively expensive. This is because the channel may require a dedicated cable, wire pair, or fiber-optic link to be strung between the two locations; this entails raw materials (e.g., copper, cable), rights of way, maintenance, protection, and so on. Clearly, in the case of multicast or broadcast communication, this becomes even more expensive. (One can see why there is so much interest in wireless communication.)

It follows that one of the most basic communication requirements is to use some mechanism to enable the sharing of the communication channel, based on cost-effectiveness considerations. This requirement is met using the mechanism of multiplexing. Figure 12.2(b) depicts this function pictorially. Multiplexing, which can be achieved in a number of ways, depending on technology, goals, channel, and so on, allows multiple users to gain access to the channel, and to do so in a reliable manner that does not impede or frustrate communication (Figure 12.5). Generally, there will be a large set of incoming links operating at lower speed; the NE will collect traffic from these links and aggregate the traffic into a single higher-speed bundle in such a manner that the traffic can be redistributed at the remote end such that source–destination

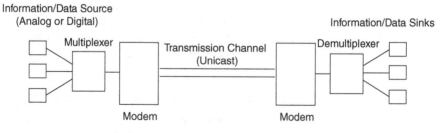

FIGURE 12.5 Concept of multiplexing.

pairs are unmistakably identifiable. Again, some of these NEs were discussed in earlier chapters.

Examples include a multiplexer that takes 24 DS0 channels and muxes them to a T1 line; or a multiplexer that takes 28 DS1/T1 channels and muxes them to a DS3/T3 line; or a multiplexer that takes four OC-3 channels and muxes them to an OC-12 line. What drives the need for muxing is the fact that there are consistent economies of scale in transmission systems: for example, for the cost of half-a-dozen DS0 channels, one can already use a T1 line that can, in fact, carry 24 DS0. Hence, for example, if one needed to transmit remotely eight DS0s, one would be much better off purchasing a T1 mux and using a T1 line (the cost of the T1 mux is usually recoverable very quickly). These economies of scale are applicable at all speed levels; again, for the cost of two OC-3 channels, one can use an OC-12 line that in fact can carry four OC-3s. SONET ADMs can be seen as multiplexers or as grooming devices (see Section 12.2.3).

The multiplexing technique depends on a number of factors, including but not limited to, underlying channel (e.g., radio, fiber-optic); technology (e.g., space-division multiplexing, frequency-division multiplexing, wavelength-division multiplexing, time-division multiplexing, code-division multiplexing, demand-assignment multiple access, random access); discipline (e.g., deterministic multiplexing, statistical multiplexing); the protocol layer (e.g., physical layer, data link layer, packet layer); and the purpose of the function.

Multiplexing is a fundamental network function that is indispensable to modern networking. Space-division multiplexing is typical of satellite technology, cellular telephony, WiFi, WiMax, and over-the-air commercial radio and TV broadcasting, to name a few. Frequency-division multiplexing is very typical of traditional radio and wireless transmission systems, such as satellite technology, cellular telephony, WiFi, WiMax, and over-the-air commercial radio and TV broadcasting (in all of these cases it is used in addition to space-division multiplexing and optical systems). Wavelength-division multiplexing (a form of frequency-division multiplexing) is now very common in optical transmission systems. During the past quarter of a century, the trend has been in favor of time-division multiplexing as an initial first step and statistical multiplexing as a follow-on next step. Code-division multiplexing has been used in military application and in some cellular telephony applications. Demand-assignment multiple access has seen limited applicability in satellite communication. Random access techniques were very common in traditional LANs, before the shift to switched LANs that has occurred since the mid-1990s.

In deterministic multiplexing the sender and receiver of the information have some kind of well-established time-invariant mechanism to identify the data sent; in statistical multiplexing data ownership is generally achieved through the use of a label process (the label mechanism depends on the OSIRM layer); the mix of traffic is based on the arriving rate from different

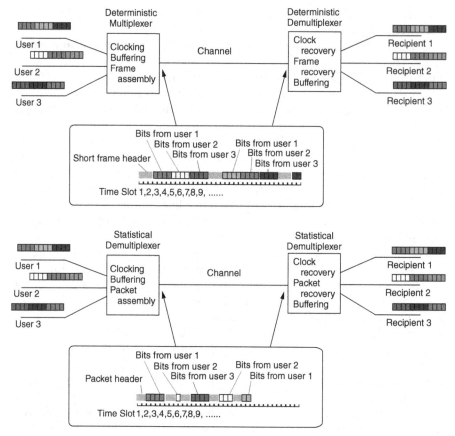

FIGURE 12.6 Deterministic multiplexing function vs. statistical multiplexing function.

users, not fixed a priori (Figure 12.6). Deterministic multiplexing is also called time-division multiplexing (TDM).

It is interesting to note that at the commercial level, physical layer multiplexing, such as that achieved with T1/T3 digital transmission lines, was important in the 1980s[†]; data link layer multiplexing, such as that achieved in frame relay service and in cell relay service, was important in the 1990s; while in the 2000s, network-layer multiplexing (packet technology in IP and MPLS) has become ubiquitous.

Not only is multiplexing done in the network, but it is also done routinely at the end-system level. As an example we discuss briefly a type of multiplexing from the Moving Picture Expert Group (MPEG) 2 and/or 4 video environ-

[†]It continues to be important at this time in the context of fiber-optic systems and traditional telephony.

ment which finds applications in Internet protocol TV (IPTV) environments, in support of carriers' triple play services. The MPEG-2 standard consists of three layers: audio, video, and systems. The systems layer supports synchronization and interleaving of multiple compressed streams, buffer initialization and management, and time identification. The audio and video layers define the syntax and semantics of the corresponding elementary streams (ESs). An ES is the output of an MPEG encoder and typically contains compressed digital video, compressed digital audio, digital data, and digital control data. The information corresponds to an access unit (a fundamental unit of encoding) such as a video frame. Each ES is in turn an input to an MPEG-2 processor that accumulates the data into a stream of packetized elementary stream (PES) packets (Figure 12.7). A PES can be as long as 65,536 octets and includes a six-octet header; it typically contains an integral number of ESs.

PESs are then mapped to transport stream (TS) unit(s). Each MPEG-2 TS packet carries 184 octets of payload data prefixed by a four-octet (32-bit) header. The TS header contains the packet identifier (PID); the PID is a 13-bit field that is used to uniquely identify the stream to which the packet belongs (e.g., PES packets corresponding to an ES) generated by the multiplexer. The PID allows the receiver to differentiate the stream to which each received packet belongs; effectively, it allows the receiver to accept or reject PES packets at a high level without burdening the receiver with extensive processing. Often, one sends only one PES (or a part of single PES) in a TS packet (in some cases, however, a given PES packet may span several TS packets, so that the majority of TS packets contain continuation data in their payloads). MPEG transport streams are typically encapsulated in the user

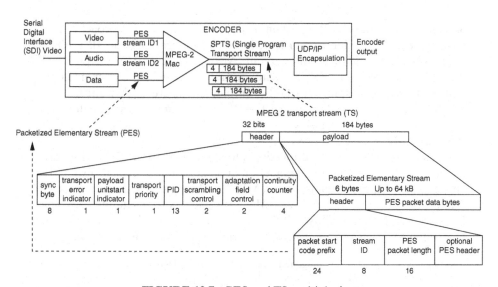

FIGURE 12.7 PES and TS multiplexing.

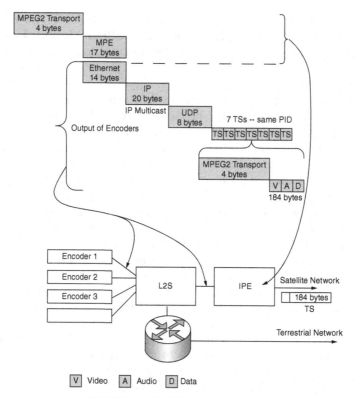

FIGURE 12.8   IPE protocol stack.

datagram protocol (UDP) and then in IP. *Note:* Traditional approaches make use of the PID to identify content; in IPTV applications, the IP multicast address is used to identify the content. Also, the latest IPTV systems make use of MPEG-4-coded PESs. For satellite transmission, and to remain consistent with already existing MPEG-2 technology,[†] TSs are further encapsulated in multiprotocol encapsulation (MPE—RFC 3016) and then segmented again and placed into TS streams via a device called the IP encapsulator (IPE) (Figure 12.8).

IPEs handle statistical multiplexing and facilitate coexistence. IPE receives IP packets from an Ethernet connection and encapsulates packets using MPE and then maps these streams into an MPEG-2 TS. Once the device has encapsulated the data, the IPE forwards the data packets to a satellite link. *Note:* IPEs are usually not employed if the output of the layer 2 switch is connected to a router for transmission over a terrestrial network; in this case the headend

---

[†]Existing receivers [specifically, integrated receiver decoders (IRDs)] are based on hardware that works by deenveloping MPEG-2 TSs; hence, the MPEG-4-encoded PESs are mapped to TSs at the source.

is responsible for proper downstream enveloping and distribution of the traffic to the ultimate consumer.

12.2.3 Grooming Function

Network grooming is a somewhat loosely defined industry term that describes a variety of traffic optimization functions. A basic description of grooming is the *repacking* of information from a (large) number of incoming links that are not fully utilized to a (smaller) set of outgoing links that are much better utilized. Grooming is the optimization of network traffic handling; Figure 12.2(c) depicts this function pictorially. One real-life example would be repacking (say, grooming) of airline passengers arriving at an airport from various regional locations to be reloaded/repacked onto another airplane (of the same or larger size) to achieve higher utilization (seat occupancy). Clearly, the grooming of a network to optimize utilization of its traffic-carrying capacity has a significant impact on the network's cost-effectiveness and availability. The availability figure of merit is improved by grooming the network in such a manner that traffic can easily be rerouted in case of a link (or other equipment) failure; this is often achieved by grooming capacity so that an available pool of alternative resources is guaranteed to be available in case of failure. Table 12.3 identifies functionality of NEs at a more detailed level.

Grooming makes use of the fact that there typically are multiple layers of transport within a carrier-class (or, at least, fairly complex) network. Generally, there will be a large set of incoming links operating at a given speed; the NE collects traffic from these links and aggregates it into a set (maybe as small as one, but could be more than one) of outgoing links (usually of the same speed as the incoming lines, but could also be a higher-speed line). The aggregation is done in such a manner that traffic can be redistributed at the remote end such that source–destination pairs are unmistakably identifiable. The term *concentrator* is sometimes also used to describe this function. [However, grooming is usually done in a nonblocking manner while pure concentration could be done in a blocking (read overbooked) manner.]

As implied above, grooming has come to encompass a variety of meanings within the telecommunications industry and literature. Grooming spans multiple distinct transmission channels or methods and can occur within multiple layers of the same technology or between technologies: it can be performed when signals are bundled for extended-distance transmission and when cross-connection equipment converts signals between different wavelengths, channels, or time slots (Barr et al., 2005). Some grooming terms are defined here.

- *Core grooming:* traffic grooming achieved inside a network; combining low-utilization traffic streams to a higher-utilization stream over a carrier's high-speed core/backbone network; typically, done by the carrier

TABLE 12.3   Grooming Functionality

Function	Definition	Example
Packing	Grouping lower-speed signal units into higher-speed transport units; operates on a digital signal hierarchy such as the traditional asynchronous digital signal hierarchy (embedded in each DS3 are 28 DS1 circuits; in turn, each DS1 carries 24 DS0 circuits) or synchronous SONET/SDH hierarchy	Traditional telco cross-connect system
Assigning	Binding flows to transmission channels (e.g., time slots, frequencies, wavelengths) within a given transport layer	Assigning bit streams to SONET time slots; assigning wavelength-division-multiplexing light paths to specific wavelengths on each span of a given mesh or ring network
Converting	Altering signals between channels in the same transport layer	Reshuffling time slots of transiting traffic with time-slot interchange within a SONET ADM; optical cross-connection to convert light path wavelengths (also called *wavelength cross-connects*)
Extracting/inserting	Taking lower-speed bit streams to/from higher-speed units	Using an ADM to terminate a lower-rate SONET stream
Physical-level routing	Routing speed bit streams between their origins and destinations	Determining the path that each OC-3 needs to follow and creating a set of light paths in an optical network

Source: Adapted from Barr et al. (2005).

- *End-to-end grooming:* traffic grooming achieved outside the network (i.e., beyond the edge of the network); typically, done directly by the end user; also sometimes called *bypass*
- *Grooming:* combining low-utilization traffic streams to a higher-utilization stream; a procedure of efficiently multiplexing/demultiplexing and switching low-speed traffic streams onto/from high-capacity band-

width trunks in order to improve bandwidth utilization, optimize network throughput, and minimize network cost (Zhu et al., 2003); term used to describe the optimization of capacity utilization in transport systems by means of cross-connections of conversions between different transport systems or layers within the same system (Barr and Patterson, 2001); entails complex routing and often implicitly assumes bundling or multiple capacities or multiple layers of transmission (Barr et al., 2005)

- *Grooming architectures:* a strategy for the placement of intermediate grooming sites, routing of traffic, and rules for how often traffic is groomed as it traverses the network (Weston-Dawkes and Baroni, 2002)
- *Hierarchical grooming:* combination of end-to-end (subrate) and intermediate (core) grooming (Cinkler, 2003)
- *Next-generation optical grooming:* traffic grooming in the context of next-generation optical WDM networks to cost-effectively perform end-to-end automatic provisioning (Zhu et al., 2003)
- *Optical grooming:* grooming done on optical links, usually by utilizing optical switches or optical cross-connects
- *WDM grooming:* techniques used to combine low-speed traffic streams onto high-speed wavelengths to minimize network-wide cost in terms of line-terminating equipment and/or electronic switching (Dutta and Rouskas, 2002); bundling of low-speed traffic streams onto high-capacity optical channels (Zhu and Mukherjee, 2003)

Grooming differs from multiplexing in the sense that in the latter case the incoming streams are usually considered to be fully loaded and their entire content needs to be aggregated time slot by time slot onto the outgoing channel; in grooming the incoming streams are usually lightly utilized, and not every incoming time slot needs to be preserved. The example given earlier about air travel is useful: In the airline grooming function, not every seat of arriving planes needs to have a corresponding seat in the outgoing plane, because many of those seats (on the regional planes) were empty (in this example[†]). It should be noted that sometimes (but not commonly) grooming entails the simple but complete reassignment of slots, one by one, but for an entire transmission facility.

The best known example of a grooming NE is a digital cross-connect system (DCS); see Figure 12.9 for an example. SONET ADMs can be seen as grooming devices or as TDM multiplexers. ADMs demultiplex a given signal (say, an OC-3) off the backbone for local distribution at a given point in the network

---

[†]On the other hand, consider, for example, a fully loaded plane that was ready to leave a terminal but was found to have a mechanical problem; if a new aircraft is brought to the terminal to replace the impaired aircraft, there would have to be an exact seat-by-seat transfer from the old plane to the new plane.

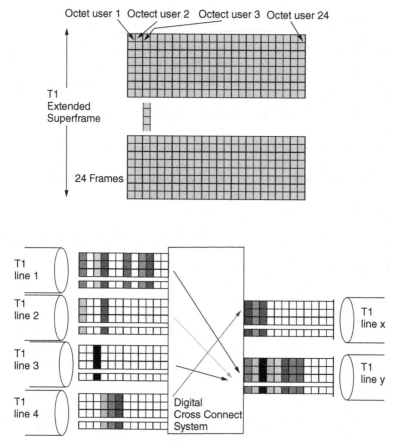

FIGURE 12.9   Example of T1-level grooming using a DCS.

(e.g., a metropolitan ring), and at the same time reinject (remultiplex) a different signal onto the ring, which originates at that local point and needs to be transported over the backbone to some remote location. Network grooming methods are particular to the (multiplexing) technology in use, such as TDM links, SONET/SDH rings, WDM links, and WDM mesh networks. Again, some of these NEs were discussed in earlier chapters.

12.2.4   Switching Function

As seen in Table 12.2, switching is a commonly supported function; 12 types of switches are identified there. The basic goal of switching is to allow information arriving on any inlet (port) to be forwarded/relayed/interconnected with any outlet. The function is technology dependent. Furthermore, switching can be undertaken at layer 1 of the OSIRM (e.g., voice switching), layer 2 (e.g.,

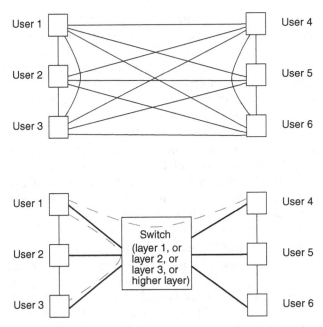

FIGURE 12.10   Connectivity advantages of switching.

cell switching), layer 3 (e.g., multiprotocol label switching), or even at higher layers. Figure 12.2(d) illustrates the basic concept.

Figure 12.10 shows the advantages of switching. Fundamentally, switching greatly reduces costs by allowing any-to-any connectivity without requiring channels to be deployed between any two pairs of entities that are required to communicate: without switching or relaying one would need $n(n-1)/2$ links to interconnect n users. Definition of some of the key switching NEs follow.

- *Any-layer switch, at layer x:* a generic switch offered for the first time by the authors of this book to define such a switch as an NE that (1) has protocol stacks on the line interface engine that support layer x protocols with an equivalent layer x protocol peer at a remote end, and (2) allows any incoming protocol data unit (PDU) at layer x from a line interface engine to be forwarded to an outgoing line/trunk interface engine by using/acting-on information contained at the layer x header. A line interface is an unmultiplexed interface that supports information from/to a single transmitting user; a trunk interface is a multiplexed interface that supports information aggregated and proxied from/to the interface from a multitude of transmitting users.
- *Any-layer switch, at layer x, protocol y:* an any-layer switch where the protocol stack supports protocol y. An example is a VLAN switch: a layer 2 switch that supports the IEEE 802.1p/1q protocol.

- *Matrix switch:* a layer 1 switching system that transparently maps an incoming bit stream on a given port to a specified output port. The switching is typically done via a bus and/or internal memory infrastructure. It enables one to select an input source and connect it to one or more outputs. A matrix switch enables rapid reconfiguration under software control of many connections. A digital cross-connect system is an example. Usually, such a switch does not have a sophisticated session (call) control mechanism. Circuit-switched technology.

- *Voice switch:* a layer 1 traditional NE that allows incoming voice lines (or trunks) to be connected with low or no blocking to outgoing lines (or trunks). Originally, these switches handled analog voice streams, but since the 1970s the technology has moved to digital pulse-code-modulated voice streams. Usually, such a switch has a sophisticated call control mechanism. Circuit-switched technology.

- *Integrated services digital network* (ISDN) *switch:* a layer 1 switch that allows incoming ISDN voice lines (or trunks) to be connected with low or no blocking to outgoing lines (or trunks). Switches support an out-of-band signaling protocol stack (specifically, link access control D) in its control plane, as well as an information-transfer stack in its user plane. Circuit-switched technology.

- *Optical switch:* a layer 1 switch that handles optical streams. Typically, it terminates an optical-level interface such as SONET and either goes through an O/E/O conversion, or is completely O/O-based. More advanced systems switch optical-level wavelengths. Circuit-switched technology.

- *Digital crossconnect system:* a Layer 1 switching matrix optimized (standardized) for DS0, DS1, and/or DS3 facilities switching or grooming. Granularity of switchable entities varies, but can be all of the above (DS0, DS1, or DS3). Circuit-switched technology. Typically does not have a sophisticated real-time control plane for session control, but the switching is accomplished by non- or near-real-time provisioning.

- *Automatically switched optical network* (ASON) *switch:* a layer 1 optical switch typically able to support incoming trunks supporting the ITU-T Rec.8080/Y.1304, *Architecture for the Automatically Switched Optical Network (ASON)*, November 2001. ASON has the objective of supporting a scalable, fast, efficient, and simple transport network architecture. ASON introduces flexibility to the optical transport network by means of a control plane. It describes a reference architecture for this control plane and its interfaces specifying the basic relation between the three different defined planes: control plane, user (transport) plane, and management plane (Escalona et al., 2005). The control plan allows for real-time control of the connections through the switch. Circuit-switched technology. Switching typically occurs at the OC-12, OC-48, OC-192, or optical lambda level.

- *Video router switch:* a layer 1 switching system that allows one to switch video signals, typically in synchronous digital interface format (or, in earlier days, analog video signals). For example, it allows more than one camera, digital video recorder, video server, and similar, to more than one monitor, video printer, and so on. An audiovideo matrix switcher typically has several video and stereo/audio inputs, which can be directed by the user in any combination to various audiovideo output devices connected to the switchers. Circuit-switched technology.

- *Asynchronous transfer mode* (ATM) *switch:* a layer 2 switch designed to support (high-speed) switching of ATM cells. ATM cells are comprised of 48 octets of payload and five octets of header; the header contains a pointer to the destination (but not the destination itself). The statefull switch supports end-to-end virtual connections. ATM is defined by an extensive set of ITU-T standards. The fixed-length nature of the cells allows optimized high-throughput inlet-to-outlet (input port-to-output port) forwarding (read switching) of cells. Rigorous queue management allows the delivery of very well-defined QoS connectivity services. In theory, three planes are supported: control plane, user (transport) plane, and management plane, enabling the support of switched virtual connection; in reality, only permanent virtual connections were actually supported by the majority of commercially available ATM switches. Typical line interface rates are at the OC-3 and OC-12 rates, but in some cases other rates are also supported. Packet (cell)-switched technology.

- *Frame relay switch:* a layer 2 switch designed to support megabit-per-second-level switching of frame relay frames. The statefull switch supports end-to-end virtual connections. Frame relay is defined by an extensive set of ITU-T standards. Limited queue management supports a limited set of QoS-based services. In theory, three planes are supported: control plane, user (transport) plane, and management plane, enabling the support of switched virtual connections; in reality, only permanent virtual connections were actually supported by the majority of commercially available switches. Typical line interface rates are at the DS0 and DS1 rates. Packet (frame)-switched technology.

- *Ethernet* (*local area network*) *switch:* a layer 2 switch designed to support 10-, 100- 1000-, and 10,000-Mbps switching of Ethernet connections (Ethernet frames). Ethernet is defined by an extensive set of IEEE standards. Packet (frame)-switched technology.

- *Virtual local area network* (*Ethernet*) *switch:* a layer 2 switch designed to support 10-, 100- 1000-, and 10,000-Mbps switching of Ethernet connections (Ethernet frames) when the Ethernet frames support IEEE 802.1q/1p headers defining VLAN domains. Most Ethernet switches are, in fact, VLAN switches. Packet (frame)-switched technology.

- *IP switch:* a layer 3 switch that aims at treating (layer 3) packets as if they were (layer 2) frames. The goal is to expedite the packet forwarding time and simplify the forwarding process. Relies on switching concepts by using a label (at a lower layer) instead of a full layer 3 routing function. Attempts to create a flat(er) network where destinations are logically one hop away rather than being logically multiple hops away. Supports the concept of "routing at the edges and switching at the core." Seeks to capture the best of each "ATM switching and IP routing." Started out as vendor-proprietary technology; now mostly based on multiprotocol label switching. An "any-layer" switch operating at layer 3. True packet-switched technology.
- *Multiprotocol label switching* (MPLS) *switch:* a layer 3 IP switching technology based on the IETF RFCs defining MPLS. True packet-switched technology.
- *VoIP switch* (*soft switch*): a layer 3 IP switching technology optimized for handling voice over IP (VoIP). In this environment voce is digitized and compressed by an algorithm such as the one defined in ITU-T G.723.1; voice is then encapsulated in the real-time protocol, user datagram protocol, and then IP. Replaces a traditional TDM (layer 1) voice switch. Often based on software running on a general-purpose processor rather than being based on dedicated hardware. True packet-switched technology.
- *Layer 4 switch:* an any-layer switch operating at layer 4. True packet-switched technology.

Switches have line interfaces and trunk interfaces. A *line interface* is an unmultiplexed interface that supports information from/to a single transmitting user; a *trunk interface* is a multiplexed interface that supports information aggregated and proxied from/to the interface from a multitude of transmitting users. Switches typically incorporate call/session control mechanism that allow rapid and dynamic call/session cross-connection to the outgoing line, trunk, or ultimate destination.

12.2.5 Routing Function

Routing is the forwarding of *datagrams* (*packets*) based on the header information included in the protocol data unit. Routing is a layer 3 function in the OSIRM. Routed networks use statistically multiplexed packets to transfer data in a connectionless fashion.[†] A *router* is a device or, in some cases, software in a computer, that determines the next network node to which a packet should be forwarded in order to advance the packet to its intended destina-

---

[†]Some older systems, such as ITU-T X.25, also support a connection-oriented switched virtual connection mode.

tion. Routing is usually done on a "best-effort" manner, where there is no absolute guarantee (at the network layer) that the packets will, in fact, arrive at the destination. The forwarding is based on topology and other information, such as priority and link status. Topology information is collected (often in real time) with the use of an ancillary routing protocol. The routing (forwarding) information is maintained in the routing table; this table contains a list of known routers, the addresses they can reach, and a cost metric associated with the path to each router so that the best available route is chosen. Routers are generally more complex (on a comparable interface basis) than other NEs.

Routing protocols can be of the internal gateway protocol (IGP) type or of the external gateway protocol (EGP) type. IGP is a protocol for exchanging routing information between gateways (hosts with routers) within an autonomous network. Two commonly used IGPs are the routing information protocol (RIP) and the open shortest path first (OSPF) protocol. OSPF is a link-state, routing protocol defined in IETF RFC 1583 and RFC 1793; the multicast version, multicast OSPF (MOSPF), is defined in RFC 1584 (some routing protocols are distance-vector type of protocols). Enhanced interior gateway routing protocol (EIGRP) is a well-known vendor-specific network protocol (an IGP) that lets routers exchange information more efficiently than with earlier network protocols.

Some basic routing terminology follow.

- *Border gateway protocol* (BGP): a routing protocol used for exchanging routing information between gateway hosts in an autonomous system (AS) network; it is an interautonomous system routing protocol. An AS is a network or group of networks under a common administration and with common routing policies. BGP is used for exchanging routing information between gateway hosts (each with its own router) in a network of ASs. BGP is used to exchange routing information for the Internet and is the protocol used between Internet service providers (ISPs); that is, BGP is often the protocol used between gateway hosts on the Internet. Intranets used by corporations and institutions generally employ an IGP such as OSPF for the exchange of routing information within their networks. Customers connect to ISPs, and ISPs use BGP to exchange customer and ISP routes. When BGP is used between ASs, the protocol is referred to as external BGP. If a service provider is using BGP to exchange routes within an AS, the protocol is referred to as interior BGP.
- *Enhanced interior gateway routing protocol* (EIGRP): a vendor-specific network protocol (an IGP) that lets routers exchange information more efficiently than with earlier network protocols. EIGRP evolved from.
- *Exterior gateway protocol* (EGP): a protocol for distribution of routing information to the routers that connect autonomous systems (here *gateway* means *router*).

- *External border gateway protocol* (eBGP): an exterior gateway protocol (EGP), used to perform interdomain routing in transmission control protocol/Internet protocol (TCP/IP) networks. A BGP router needs to establish a connection to each of its BGP peers before BGP updates can be exchanged. The BGP session between two BGP peers is said to be an external BGP (eBGP) session if the BGP peers are in different autonomous systems.

- *Interior gateway protocol* (IGP): a protocol for exchanging routing information between gateways (hosts with routers) within an autonomous network. Two commonly used IGPs: the RIP and OSPF protocol.

- *Interior gateway routing protocol* (IGRP): a routing protocol (an IGP) developed in the mid-1980s by Cisco Systems.

- *Internal border gateway protocol* (iBGP): an EGP used to perform interdomain routing in TCP/IP networks. A BGP router needs to establish a connection to each of its BGP peers before BGP updates can be exchanged. A BGP session between two BGP peers is said to be an internal BGP session if the BGP peers are in the same autonomous systems.

- *Open shortest-path first* (OSPF): a link-state routing protocol; an IGP defined in RFC 1583 and RFC 1793. The multicast version, multicast OSPF, is defined in RFC 1584 (some routing protocols are distance-vector type protocols).

- *Route:* in general, the *n*-tuple <prefix, nexthop, [other routing or nonrouting protocol attributes]>. A route is not end to end, but is defined with respect to a specific next hop that should take packets on the next step toward their destination as defined by the prefix. In this use, a route is the basic unit of information about a target destination distilled from routing protocols. This term refers to the concept of a route common to all routing protocols. With reference to the definition above, typical nonrouting-protocol attributes would be associated with diffserv or traffic engineering (Berkowitz et al., 2005).

- *Route change event:* route change: implicit, replacing it with another route, or explicit, withdrawal followed by the introduction of a new route. In either case, the change may be an actual change, no change, or a duplicate (Berkowitz et al., 2005).

- *Route flap:* a change of state (withdrawal, announcement, attribute change) for a route. Route flapping can be considered a special and pathological case of update trains (Berkowitz et al., 2005).

- *Route mixture:* the demographics of a set of routes.

- *Route packing:* the number of route prefixes accommodated in a single routing protocol UPDATE message, either as updates (additions or modifications) or as withdrawals.

- *Route reflector:* a network element owned by a service provider (SP) used to distribute in BGP routes to an SP's BGP-enabled router (Nagarajan, 2004; Andersson and Madsen, 2005).
- *Router:* (*gateway* in original parlance—this term now has limited use) a relaying device that operates at the network layer of the protocol model. Node that can forward datagrams not specifically addressed to it. An interconnection device that is similar to a bridge but serves packets or frames containing certain protocols. Routers interconnect logical subnets (e.g., implemented at a local area network level) at the network layer. A computer that is a gateway between two networks at layer 3 of the OSIRM and that relays and directs data packets through that internetwork. The most common form of router operates on IP packets (Shirey, 2000). Internet use: In the context of the Internet protocol suite, a networked computer that forwards Internet protocol packets that are not addressed to the computer itself (IPv6, 2006).
- *Router advertisement:* a neighbor discovery message sent by a router in a pseudoperiodic way or as a router solicitation message response. In IPv6 the advertisement includes, at least, information about a prefix that will be used later by the host to calculate its own unicast IPv6 address following the stateless mechanism (IPv6, 2006).
- *Router discovery:* the process by which a host discovers the local routers on an attached link and configures a default router automatically. In IPv4, this is equivalent to using Internet control message protocol v4 (ICMPv4) router discovery to configure a default gateway (Narten et al., 1998).
- *Routers discovery:* a neighbor's discovery process that allows the discovery of routers connected to a particular link (IPv6, 2006).

EGP is a protocol for distribution of routing information to the routers that connect autonomous systems (here *gateway* means *router*). Intranets used by corporations and institutions generally employ an IGP such as OSPF for the exchange of routing information within their networks. Customers connect to ISPs, and ISPs using an EGP such as border gateway protocol to exchange customer and ISP routes.

In general terms, the functions of a router include:

- To interface physically to the inbound and outbound network channel by supporting the appropriate layer 1 and layer 2 protocols (e.g., SONET, ATM, Gigabit Ethernet)
- receive packets
- buffer packets
- process IP-level packet headers
- forward packets to appropriate destination, as specified in IP-level packet header over appropriate outbound interface

- manage queues so that packets can be forwarded based on the priority specified in the IP-level packet header (when priorities are specified)
- process topology information via the appropriate internal/external gateway protocol
- maintain routing tables
- To perform firewall function (optional)

Most, if not all, of the NEs described in previous sections handle incoming traffic in a fairly direct and transparent manner. While routers (also) do not deal with the content of the information, they do perform a significant amount of processing related to either protocol management (several dozen protocols are typically supported by a high-end router), queue management, and media management, or combinations of all three.

At a broad level, routers include the following components: network interface engines (specifically, line cards) attached to the incoming and outgoing telecommunication links (each link can be of a different type/technology); processing module(s); buffering module(s); and an internal interconnection module (or switch fabric). Datagrams are received at inbound network interface cards; they are processed by the processing module and, even if transiently, stored in the buffering module. The type of processing is based on the protocol envelope they carry: The header is examined and understood in the context of the protocol syntax; the protocol type is derived from a field contained in the lower layer (typically the Ethernet frame). [In Ethernet-encapsulated datagrams the field is the EtherType (see Figure 12.11); Hex 0800 is obviously the most common protocol ID at this juncture.] Queue management is critical to the operation of a router, particularly if it is intended to support multimedia and QoS. Many queueing algorithms have been developed over the years. One of the most common such models is the weighted fair queueng algorithm; other algorithms include FIFO and class-based queueing. Datagrams are then forwarded through the internal interconnection unit to the outbound interface engines (line cards), which transmits them on the next hop. The destination is reached by traversing a series of hops; the sequence of hops selected is based on the routing tables which are maintained through the routing protocols. The aggregate packet stream of all incoming interfaces needs to be processed, buffered, and relayed; this implies that the processing and memory modules need sufficient power. Often, this is accomplished by replicating processing functions either fully or partially on the interfaces to allow for parallel operation.

Router vendors are now beginning to look into the possibility of adding job functions to a router, such as network-based computing. With what some vendors call *application-oriented networking* or *application-aware networking*, functions such as, but not limited to, proxing, content-level inspection, and advanced voice–data convergence are being advocated. In-network processing and data analysis have been proposed for large sensor networks; however,

EtherType Codes

decimal	Hex	decimal	octal	
0000	0000-05DC	-	-	IEEE802.3 Length Field
0257	0101-01FF	-	-	Experimental
0512	0200	512	1000	XEROX PUP (see 0A00)
0513	0201	-	-	PUP Addr Trans
	0400			Nixdorf
1536	0600	1536	3000	XEROX NS IDP
	0660			DLOG
	0661			DLOG
2048	0800	513	1001	Internet IP (IPv4)
2049	0801	-	-	X.75 Internet
2050	0802	-	-	NBS Internet
2051	0803	-	-	ECMA Internet
2052	0804	-	-	Chaosnet
2053	0805	-	-	X.25 Level 3
2054	0806	-	-	ARP
2055	0807	-	-	XNS Compatability
2056	0808	-	-	Frame Relay ARP
2076	081C	-	-	Symbolics Private
2184	0888-088A	-	-	Xyplex
2304	0900	-	-	Ungermann-Bass net debugr
2560	0A00	-	-	Xerox IEEE802.3 PUP
2561	0A01	-	-	PUP Addr Trans
2989	0BAD	-	-	Banyan VINES
2990	0BAE	-	-	VINES Loopback
2991	0BAF	-	-	VINES Echo
4096	1000	-	-	Berkeley Trailer nego
4097	1001-100F	-	-	Berkeley Trailer encap/IP
5632	1600	-	-	Valid Systems
16962	4242	-	-	PCS Basic Block Protocol
21000	5208	-	-	BBN Simnet
24576	6000	-	-	DEC Unassigned (Exp.)
24577	6001	-	-	DEC MOP Dump/Load
24578	6002	-	-	DEC MOP Remote Console
24579	6003	-	-	DEC DECNET Phase IV Route
24580	6004	-	-	DEC LAT
24581	6005	-	-	DEC Diagnostic Protocol
24582	6006	-	-	DEC Customer Protocol
24583	6007	-	-	DEC LAVC, SCA
24584	6008-6009	-	-	DEC Unassigned
24586	6010-6014	-	-	3Com Corporation
25944	6558	-	-	Trans Ether Bridging
25945	6559	-	-	Raw Frame Relay
28672	7000	-	-	Ungermann-Bass download
28674	7002	-	-	Ungermann-Bass diagnostic/loopback
28704	7020-7029	-	-	LRT
28720	7030	-	-	Proteon
28724	7034	-	-	Cabletron
32771	8003	-	-	Cronus VLN
32772	8004	-	-	Cronus Direct
32773	8005	-	-	HP Probe
32774	8006	-	-	Nestar
32776	8008	-	-	AT&T
32784	8010	-	-	Excelan
32787	8013	-	-	SGI diagnostics
32788	8014	-	-	SGI network games
32789	8015	-	-	SGI reserved

FIGURE 12.11   EtherType (partial list).

32790	8016	–	–	SGI bounce server
32793	8019	–	–	Apollo Domain
32815	802E	–	–	Tymshare
32816	802F	–	–	Tigan, Inc.
32821	8035	–	–	Reverse ARP
32822	8036	–	–	Aeonic Systems
32824	8038	–	–	DEC LANBridge
32825	8039-803C	–	–	DEC Unassigned
32829	803D	–	–	DEC Ethernet Encryption
32830	803E	–	–	DEC Unassigned
32831	803F	–	–	DEC LAN Traffic Monitor
32832	8040-8042	–	–	DEC Unassigned
32836	8044	–	–	Planning Research Corp.
32838	8046	–	–	AT&T
32839	8047	–	–	AT&T
32841	8049	–	–	ExperData
32859	805B	–	–	Stanford V Kernel exp.
32860	805C	–	–	Stanford V Kernel prod.
32861	805D	–	–	Evans & Sutherland
32864	8060	–	–	Little Machines
32866	8062	–	–	Counterpoint Computers
32869	8065	–	–	Univ. of Mass. @ Amherst
32870	8066	–	–	Univ. of Mass. @ Amherst
32871	8067	–	–	Veeco Integrated Auto.
32872	8068	–	–	General Dynamics
32873	8069	–	–	AT&T
32874	806A	–	–	Autophon
32876	806C	–	–	ComDesign
32877	806D	–	–	Computgraphic Corp.
32878	806E-8077	–	–	Landmark Graphics Corp.
32890	807A	–	–	Matra
32891	807B	–	–	Dansk Data Elektronik
32892	807C	–	–	Merit Internodal
32893	807D-807F	–	–	Vitalink Communications
32896	8080	–	–	Vitalink TransLAN III
32897	8081-8083	–	–	Counterpoint Computers
32923	809B	–	–	Appletalk
32924	809C-809E	–	–	Datability
32927	809F	–	–	Spider Systems Ltd.
32931	80A3	–	–	Nixdorf Computers
32932	80A4-80B3	–	–	Siemens Gammasonics Inc.
32960	80C0-80C3	–	–	DCA Data Exchange Cluster
32964	80C4	–	–	Banyan Systems
32965	80C5	–	–	Banyan Systems
32966	80C6	–	–	Pacer Software
32967	80C7	–	–	Applitek Corporation
32968	80C8-80CC	–	–	Intergraph Corporation
32973	80CD-80CE	–	–	Harris Corporation
32975	80CF-80D2	–	–	Taylor Instrument
32979	80D3-80D4	–	–	Rosemount Corporation
32981	80D5	–	–	IBM SNA Service on Ether
32989	80DD	–	–	Varian Associates
32990	80DE-80DF	–	–	Integrated Solutions TRFS
32992	80E0-80E3	–	–	Allen-Bradley
32996	80E4-80F0	–	–	Datability
33010	80F2	–	–	Retix
33011	80F3	–	–	AppleTalk AARP (Kinetics)
33012	80F4-80F5	–	–	Kinetics
33015	80F7	–	–	Apollo Computer
33023	80FF-8103	–	–	Wellfleet Communications
33031	8107-8109	–	–	Symbolics Private
33072	8130	–	–	Hayes Microcomputers
33073	8131	–	–	VG Laboratory Systems
33074	8132-8136			Bridge Communications

FIGURE 12.11   (*Continued*)

33079	8137-8138	-	-	Novell, Inc.
33081	8139-813D	-	-	KTI
	8148			Logicraft
	8149			Network Computing Devices
	814A			Alpha Micro
33100	814C	-	-	SNMP
	814D			BIIN
	814E			BIIN
	814F			Technically Elite Concept
	8150			Rational Corp
	8151-8153			Qualcomm
	815C-815E			Computer Protocol Pty Ltd
	8164-8166			Charles River Data System
	817D			XTP
	817E			SGI/Time Warner prop.
	8180			HIPPI-FP encapsulation
	8181			STP, HIPPI-ST
	8182			Reserved for HIPPI-6400
	8183			Reserved for HIPPI-6400
	8184-818C			Silicon Graphics prop.
	818D			Motorola Computer
	819A-81A3			Qualcomm
	81A4			ARAI Bunkichi
	81A5-81AE			RAD Network Devices
	81B7-81B9			Xyplex
	81CC-81D5			Apricot Computers
	81D6-81DD			Artisoft
	81E6-81EF			Polygon
	81F0-81F2			Comsat Labs
	81F3-81F5			SAIC
	81F6-81F8			VG Analytical
	8203-8205			Quantum Software
	8221-8222			Ascom Banking Systems
	823E-8240			Advanced Encryption Syste
	827F-8282			Athena Programming
	8263-826A			Charles River Data System
	829A-829B			Inst Ind Info Tech
	829C-82AB			Taurus Controls
	82AC-8693			Walker Richer & Quinn
	8694-869D			Idea Courier
	869E-86A1			Computer Network Tech
	86A3-86AC			Gateway Communications
	86DB			SECTRA
	86DE			Delta Controls
	86DD			IPv6
34543	86DF	-	-	ATOMIC
	86E0-86EF			Landis & Gyr Powers
	8700-8710			Motorola
34667	876B	-	-	TCP/IP Compression
34668	876C	-	-	IP Autonomous Systems
34669	876D	-	-	Secure Data
	880B			PPP
	8847			MPLS Unicast
	8848			MPLS Multicast
	8A96-8A97			Invisible Software
34915	8863	-	-	PPPoE Discovery Stage
34915	8864	-	-	PPPoE Session Stage
36864	9000	-	-	Loopback
36865	9001	-	-	3Com(Bridge) XNS Sys Mgmt
36866	9002	-	-	3Com(Bridge) TCP-IP Sys
36867	9003	-	-	3Com(Bridge) loop detect
65280	FF00	-	-	BBN VITAL-LanBridge cache
	FF00-FF0F			ISC Bunker Ramo
65535	FFFF	-	-	Reserved

FIGURE 12.11   (*Continued*)

commercial deployments of sensor networks have by and large not yet implemented these in-network features. Proponents view this trend for in-network computing as the ultimate form of convergence: here the network function is more holistically integrated with the application, particularly where quality-of-service requirements exist. Others see this trend as a mechanism to move to a service-oriented architecture (SOA) where not only the application is expressed in terms of functional catalogs, but also the networking functions are seen as functional blocks that are addressable and executable; Web services support this view.

These recent forays into the network-computing hybridization are reminiscent of the efforts by the telephone carriers in the late 1980s to build an advanced intelligent network (AIN) that would indeed provide value-added capabilities in the transport fabric. They also remind us of the tug of war that has existed between two camps: the traditional carriers that would like to maximize the intelligence and value proposition of the network and the IP data people that would like the network to have very thin functionally, and to be general, simple, and able to carry any type of traffic. In trying to make the network intelligent, some inherent generality is lost as functions specific to targeted tasks and applications are overlaid. Nonetheless, it is interesting to note that the very same people who once argued against network intelligence, now that box-level revenues may experience erosion, advocate this once-discarded idea. It should be noted that the AIN efforts of the late 1980s and 1990s have generally failed to gain traction. The idea of a network-resident service management system has not materialized.

12.3   NETWORKING EQUIPMENT

In this section we briefly discuss the internal architecture of some NEs. As we have seen, there is an abundance of network devices. Hence, only a very small subset is discussed here; even then, some of the devices are specific to the information-handling approach (e.g., circuit mode), the information media (e.g., voice), the transmission class (e.g., wireline), the transmission channel technology (e.g., optical), and the information-handling protocols, to list a few. It follows that only some specific instances are examined. Finally, there generally is not a canonical form or architecture, even for the simplest NEs; vendor-based approaches are fairly common. Just focusing at the simplest NEs, such as DCSs or traditional voice switches, one finds that there are multiple internal and/or logical architectures, as discussed in other chapters of this book; this plethora of architectures is generally even more pronounced for the other NEs.

12.3.1   Regeneration Equipment

As implied earlier in the chapter, regeneration equipment is available for various types of transmission channels; it is also available for wired and wireless LANs (here covering the PHY layer and portion of the data link layer). Figure

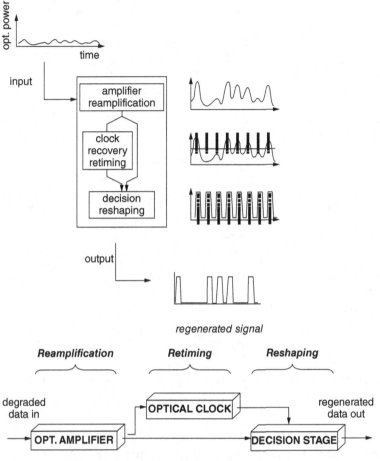

FIGURE 12.12 Basic architecture of a 3R regenerator.

12.12 depicts the internal architecture of an optical regenerator that includes the reamplification, retiming, and reshaping (3R) functionality (as noted earlier, a regenerator can also be of the 1R and 2R types). Figure 12.13 depicts an interesting type or regenerator: a satellite transponder. In fact, a typical satellite will have 24 transponders: 12 to regenerate the consecutive 12 36-MHz blocks that comprise the bandwidth block assigned for operation at the C-band, Ku-band, or Ka-band (500 MHz total) and that entails the *vertical signal polarization* mode; and 12 for the *horizontal signal polarization* mode. Commercial communication satellites perform the following functions:

- Receive signals from the ground station (uplink beam)
- Separate, amplify, and recombine the signals
- Transmit the signals back to (another) Earth station (downlink beam)

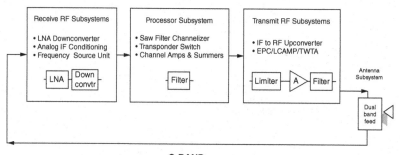

C-BAND
TRANSPONDER BLOCK DIAGRAM

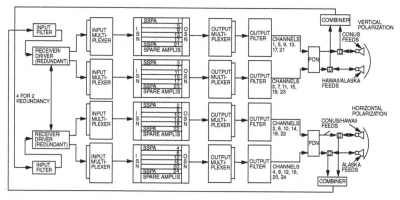

FIGURE 12.13 Architecture satellite regenerator, a transponder. (Courtesy of SES Americom.)

Some advanced functions include digital signal processing and are called *regenerative and nonregenerative on-board processors*.

12.3.2 Grooming Equipment

As we discussed earlier, grooming entails optimizing physical-level information-routing and traffic-engineering functions in a network. As in the case of the regenerator, there are various types of grooming NEs based on the transmission channel and other factors. Figures 12.14 and 12.15 provide a view into the architecture and function of a simple and more complex DCS, which is the most basic and typical grooming NE in a traditional telco network.

12.3.3 Multiplexing Equipment

As we implied earlier, multiplexing entails the combining of multiple streams into a composite stream that operate at higher speeds and with higher transport capacities. We noted that a number of approaches exist, based on the

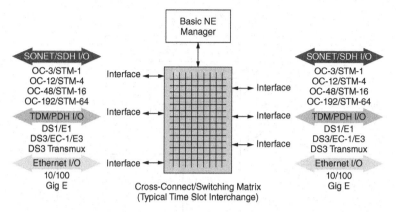

FIGURE 12.14 Basic digital cross-connect system.

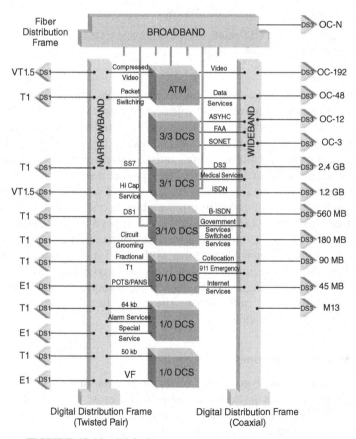

FIGURE 12.15 Digital crossconnect system arrangements.

M13MD Block Diagram

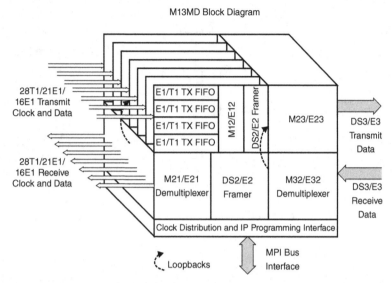

FIGURE 12.16 Basic TDM multiplexer (multiplexes DS1s to DS3s, a.k.a. M13). (Courtesy of Silicomotive Solutions Inc.)

underlying channel (e.g., twisted-pair/coaxial cable, radio, fiber-optic); the technology (e.g., space-division multiplexing, frequency-division multiplexing, wavelength-division multiplexing, time-division multiplexing, code-division multiplexing, demand-assignment multiple access, random access); the discipline (e.g., deterministic multiplexing, statistical multiplexing); and the protocol layer (e.g., physical layer, data link layer, packet layer). If one were to take just these factors, one would already have $3 \times 7 \times 2 \times 3 = 126$ different multiplexers, each with one or more internal architectures. Figures 12.16 to 12.18 depict three illustrative examples. Figure 12.16 shows a basic TDM multiplexer that is very commonly used in a traditional telephone plant: multiplexes DS1s to DS3s; this multiplexer is also known as the M13 multiplexer. Figure 12.17 shows a basic SONET multiplexer that is also very commonly used in a traditional telephone plant. Figure 12.18 depicts an example of an advanced multifunctional/multiservice multiplexer. These are just illustrative examples; many more could be provided.

12.3.4 Switching Equipment

As we saw earlier, there is quite a plethora of switching equipment. This is related to the fact that switches are one of the most fundamental types of network NEs, being that they drive the economics of a network. Without switches, every source and sink would have to be connected either directly or via a series of links (the latter is undesirable in many instances because of the accumulation of delay). The most common types of switches are:

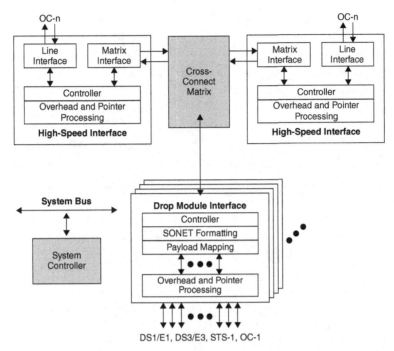

FIGURE 12.17 Basic SONET multiplexer. (Courtesy of Freeseale Semiconductase.)

- Traditional layer 1 voice switch (originally, analog switching, but now exclusively digital)
- Ethernet/VLAN layer 2 switches
- ATM/frame relay layer 2 switches
- Layer 3 IP switches

Traditional voice switches can be of the single-stage or multistage type. Single-stage switches typically are found in smaller and non-blocking applications (say, <10,000 lines); multistage switches are found in large-population environments (say, 50,000 lines)—here there may be a level of concentration and ensuing blocking. In nonblocking environments any inlet can always be connected to any specific unused outlet; in blocking environments an inlet may or may not be able to be connected to a specific outlet at a given instance even if the outlet is unused, because of the internal architecture of the switch. Blocking is, however, usually designed to be at a very minimum, say <1% of the cases (call attempts). Both the multistage feature and the possibly resulting blocking are utilized to decrease the cost of the switch fabric. The basic switching mechanism for traditional digital voice switches is the time slot interchange (TSI).

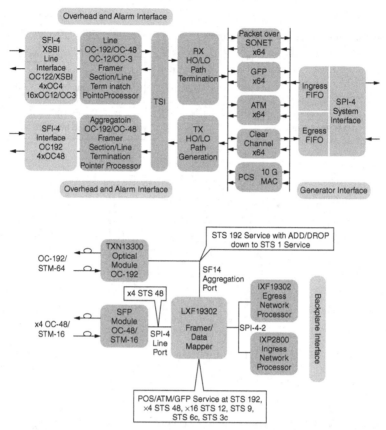

The Intel® IXF19302 10Gbps Bandwidth Aggregation and Channelizer Device service framer is a highly integrated interface solution for the transport of multi-service traffic. The device supports a single STS-192/STM-64, 4 STS-48/STM-16, 16 STS-12/STM-4, 16 STS-3/STM-1 or combination of these via 4 distinct ports per interface. Intel IXF19302 also provides an aggregation/protection interface for single STS-192/STM-64 or quad STS-48/STM-16. The integrated 10 Gigabit Ethernet Media Access Controller (MAC) allows Wide Area Network (WAN) and Local Area Network (LAN) operations. It supports concatenated and non-concatenated payloads STS-1, STS-Xc (where X = 3, 6, 9, ..., 48) and STS-192c. The multiplexer supports the mapping/de-mapping of Packet Over SONET (POS), General Framing Procedure (GFP), X.85, X.86, Link Access Procedure SDH (LAPS) and Asynchronous Transfer Mode (ATM) data for up to 64 SONET/SDH virtually or contiguously concatenated payloads/containers. The packets and cells that are mapped/demapped to/from these containers are transferred to/from the next packet device using an SPI-4 Phase 2 system interface.

FIGURE 12.18 Advanced multiplexer: Intel example. (Courtesy of Intel.)

Figure 12.19 depicts the architecture of a basic switch. The switch has incoming line cards (to terminate the incoming loop, which could be a discrete loop or a T1 access line—either in copper form or even in fiber form); outgoing line cards or trunks to terminate the outgoing loop or trunk, which could be a discrete loop or a T1/T3/SONET line—either in copper form or even in fiber form; and a call control mechanism to set up and terminate end-to-end connections. Note that outgoing line cards are just loop interfaces if the call

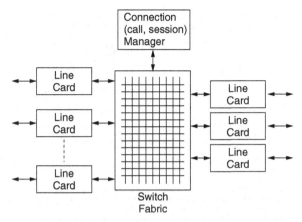

FIGURE 12.19 Basic voice switch architecture.

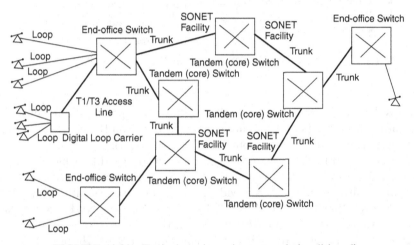

FIGURE 12.20 Typical carrier voice network (traditional).

needed to reach a local subscriber, or multiplexed trunks if the switch were to connect to a remote switch, as shown in Figure 12.20. Figure 12.21 depicts an example of a TSI-based architecture.

Next we look at asynchronous transfer mode (ATM). ATM is a connection-oriented high-speed service that was standardized by the ITU in the late 1980s and early 1990s that is based on cell forwarding. A cell is a small (53-octet) layer 2 PDU. With the use of interface cards, an ATM switch accepts multiple streams of data in many different formats, segments them into cells, and multiplexes them statistically for delivery at the remote end, where they are reassembled into the original higher-layer datagram. For example, ATM uses

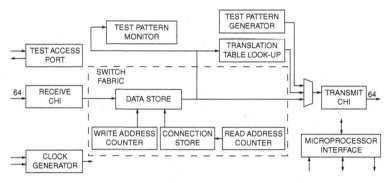

The TSI-16 time-slot interchanger (TSI) is a time/space switch with DS0 granularity. The fabric is a nonblocking structure with 16,384 input channels that may be interchanged to any of 16,384 output channels. The input and output channels are arranged on time-division multiplexed serial highways. The timing and structure of these highways complies with Agere Systems Inc. concentration highway interface (CHI) standard. Each CHI is independently programmed, and the output CHIs support multidriver busing. The CHIs have a programmable data rate (up to 16.384 Mbits/s) and frame offset. Transmit and receive configurations are also independently programmable. The TSI-16 is configured via a 16-bit synchronous microprocessor interface, which is used to control the connection data and to access the device's registers.

The switch fabric performs the nonblocking switching function. It can switch any of the 16,384 possible incoming time slots to any of the 16,384 possible outgoing time slots. It uses the classic configuration of two memories, one containing the traffic data and the second containing the switching configuration. The switch fabric performs this switching function without regard to the physical link from which the time slot was taken; hence, the TSI-16 TSI is a time-space switch. Time slots are rearranged in order within a frame (time) and among physical ports (space).

FIGURE 12.21 Example of TSI architecture. (Courtesy of Agere Systems.)

encapsulation of input IP packets before they are segmented into cells; after transiting the network, the cells are reassembled into an IP packet for delivery (typically, to a router or traditional-or-IP voice switch).

Sophisticated traffic management techniques are used to achieve a tightly defined QoS level (usually, a small set of QoS levels are supported). ATM can support multiple media (voice, video, and data). Access links range from 1.544 Mbps (DS1/T1) to 45 Mbps (DS3/T3), 155 Mbps (SONET), as well as, in some cases, OC-12 and OC-48 links (although these rates are much less common). ATM cells are typically transported via SONET- or SDH-based facilities. Mapping and delineation of cells into a SONET/SDH frame is done by the PHY layer's telecommunication convergence (TC) sublayer. The streams of information are managed using virtual circuits (VCs). VCs are permanent or semipermanent logical ("soft") connections are set up from source to destination that are utilized as long as they are needed and then torn down.

Cell forwarding is based on *label switching*. Here cells with a common destination are assigned a label that the ATM switch uses to index a routing table to determine the outgoing port on which the cells needs to be transferred. Note that the label is not the address of the destination; it is just a shorter pointer into a statefull table. The switch maintains "state" information on each active connection that indicates which outgoing port is to be used in order to

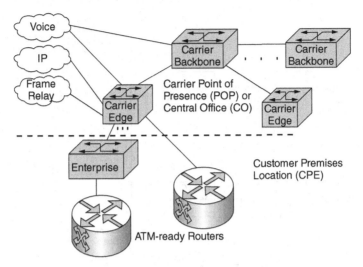

FIGURE 12.22   Typical ATM network environment.

reach another switch along the way to the destination or the destination itself. The switching label has only local (not end-to-end) significance. As part of the switching process the ATM switch may assigns new labels to new cells that are going to other switches, a technique referred to as *label swapping*.

Types of ATM switches typically include (see Figure 12.22): carrier backbone (the tier 1 core of a public ATM Network); carrier edge, located in carrier locations closer to the actual information sources or sinks (typically, a local central office); and enterprise backbone switches, although of late, enterprise networks have transitioned to high-speed core IP/MPLS routers for this application. Some ATM switch key attributes include:

- Blocking behavior (e.g., blocking, virtually nonblocking, or nonblocking)
- Switch fabric architecture (e.g., single bus, multiple bus, self-routing, augmented self-routing)
- Buffering method (e.g., internal queuing, input queuing, output queuing, output queuing/shared buffer)

Figure 12.23 depicts one example of an ATM switch; this design is by no means unique. Ethernet layer 2 switching has become very prevalent in enterprise networking applications. Basically, an Ethernet switch supports multiple Ethernet ports on the access side (e.g., 24, 96, 386) and a number of trunks on the uplink side. In addition, many Ethernet switches support VLAN management (subswitching based on the VLAN ID contained in IEEE 802.1p/1q-conformant frames.) Figure 12.24 depicts a typical enterprise environment

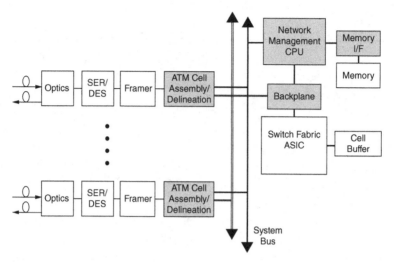

FIGURE 12.23   Typical ATM switch.

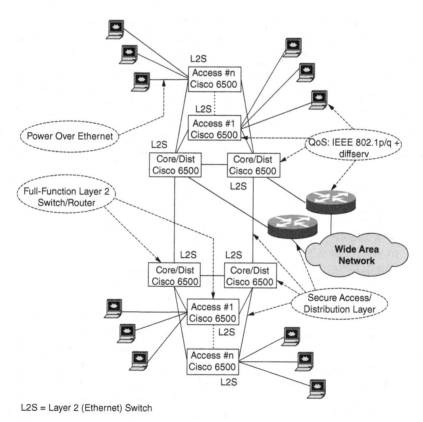

L2S = Layer 2 (Ethernet) Switch

FIGURE 12.24   Typical switched Ethernet environment.

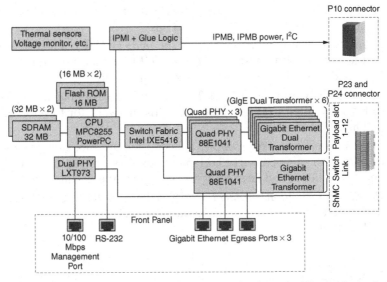

FIGURE 12.25 Example of Ethernet switch. (Courtesy of Zynx Networks.)

and Figure 12.25 shows one example of an Ethernet switch. Following are typical functions of an Ethernet switch; of course, frame forwarding is its basic function.

- *802.1D spanning tree protocol* (STP): a layer 2 feature that enables switches to find the shortest path between two points and eliminates loops from the topology.
- *802.1p packet prioritization:* a layer 2 feature that supports QoS queuing to reserve bandwidth for delay-sensitive applications (e.g., VoIP).
- *802.1q and port-based VLAN:* a layer 2 feature that isolates traffic and enables communications to flow more efficiently within specified groups.
- *802.3 frame forwarding:* a layer 2 feature that enables Ethernet frames to be forwarded on the uplink of the switch.
- *802.3ad link aggregation:* a layer 2 feature that allows multiple network links to be combined forming a single high-speed channel.
- *802.3x flow control:* a layer 2 feature that implemented in hardware to eliminate broadcast storms.
- *Distance vector multicast routing protocol* (DVMRP): a layer 3 feature for routing multicast datagrams.
- *GARP multicast registration protocol* (GMRP): a layer 2 feature that determines which VLAN ports are listening to which multicast addresses to reduce unnecessary traffic through the switch.

- *GARP VLAN registration protocol* (GVRP): a layer 2 mechanism for dynamically managing port memberships.
- *Internet group management protocol snooping* (IGMP): a layer 2 feature that allows switch to "listen in" on the IGMP communications between hosts and routers to add port entries automatically.
- *Jumbo frames support:* a layer 2 feature that enables packets up to 9K to be transmitted unfragmented across the network for lower overhead and higher throughput.
- *Open shortest path first* (OSPF): a layer 3 feature that supports the calculation a shortest path tree and maintains a routing table to reduce the amount of hops (and latency) it takes to reach the destination.
- *Routing information protocol* (RIP): a layer 3 feature that supports determination of a route based on the smallest hop count between source and destination.

12.3.5 Routing Equipment

Routers are very common NEs in IP networks such as Internet backbone/ access networks, VoIP networks, IPTV networks, and enterprise networks. Routers perform two main functions: control path functions and data path control (switching) functions. As noted earlier, these functions include:

- *Classification:* classification of packets for handling/queuing/filtering (specifically, comparing packets to classification lists and performing control via a variety of sophisticated queue management techniques)
- *Packet switching:* layer 3 switching based on routing information and QoS markings; includes generating outbound layer 2 encapsulation, performing layer 3 checksum, managing time-to-live (TTL)/hop count update
- *Packet transmission:* accessing outbound transmission channels
- *Packet processing (manipulation):* changing the contents of a packet (e.g., compression, encryption)
- *Packet consumption:* maintaining/manipulating routing information (tracking updates/update neighbors, e.g., absorbing routing protocol updates; issuing service advertisements/routing protocol packets
- *Support for management functions:* interface statistics, queue statistics, Telnet, SNMP alerts, ping, trace route

How some of these functions as supported is illustrated in the figures that follow. Figures 12.26 and 12.27 depict some basic router internals. Figure 12.28 shows the typical hardware modules of a router. Finally, Figure 12.29 illustrates some vendor- and model-specific architectures.

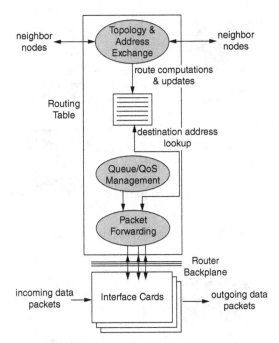

FIGURE 12.26 Basic router internals.

12.4 SUMMARY

In this chapter we provided a survey of NE functionality and hardware across several OSIRM layers. As we have seen, there is an abundance of NEs. Many of the NEs are specific to the information-handling approach (e.g., circuit mode), the information media (e.g., voice), the transmission class (e.g., wireline), the transmission channel technology (e.g., optical), and the information-handling protocols. Also, we have learned that generally there is no canonical form or architecture, even for the simplest NEs; vendor-based approaches are fairly common. Yet, NEs are fundamental to the realization of a reliable, cost-effective, well-designed network; therefore, it is important to have a good understanding of their role and general architecture.

KEY POINTS

- Various network elements (NEs) support variety of communication functions.
- Some NEs reside at the particular level of the open systems interconnection reference model, while others cover multiple levels.

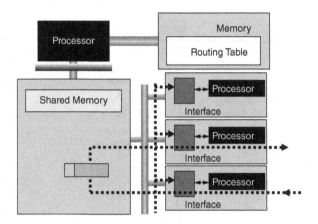

Shared Memory Distributed Processors Architecture

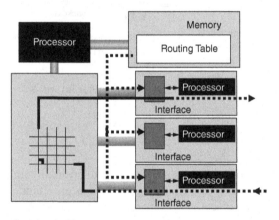

Crossbar Architecture

FIGURE 12.27   Basic router internals (another view).

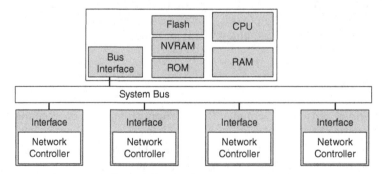

FIGURE 12.28   Typical router hardware.

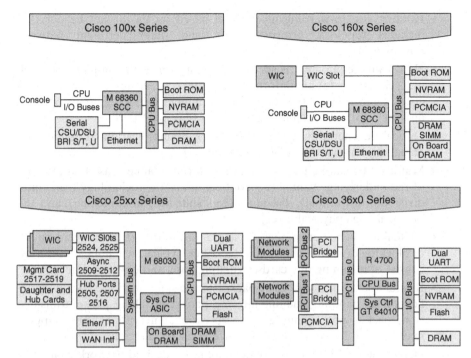

FIGURE 12.29 Examples of router architecture. (Courtesy of Cisco Systems.)

- Examples of most popular NEs in data networks are SONET add–drop multiplexers, Ethernet switches, and IP routers.
- Communication systems can be circuit-mode-based (where a session is established before the information is transmitted) and packet-mode-based (where the information is sent without establishment of the session).
- Basic networking functions include data transmission, regeneration, mutiplexing, grooming, switching, and routing.
- Transmission is a physical transfer of information across a communication channel (optical fiber, for example).
- Regeneration is a process of restoring the physical data stream to its original shape and power level.
- Grooming function relies on combining a large set of low-utilization streams to a higher-utilization stream.
- Multiplexing allows multiple users to gain access to a communication channel.
- Switching is a mechanism that allows information arriving on any input port to be forwarded to any output port.

- Routing is a forwarding mechanism based on the packet header that includes topology, priority, status, and other pertinent network information.

- Networking equipment is specific to the information-handling approach (e.g., circuit mode), the information media (e.g., voice), the transmission class (e.g., wireline), the transmission channel technology (e.g., optical), and the information-handling protocols (e.g., SONET).

- Optical regeneration typically includes reamplification, retiming, and reshaping functionality (3-R regeneration).

- Multiplexing can be performed in a different manner based on the type of the underlying channel (e.g., optical fiber), the technology (e.g., time-division multiplexing), discipline (e.g., statistical multiplexing), and the protocol (e.g., physical layer).

- Switching can be performed in a single- or multistage fashion and can be of blocking or nonblocking characteristics. Typical switching architecture consist of incoming line cards, switching core, and outgoing line cards.

- Routers perform a variety of packet-processing functions: packet classification for handling/queuing/filtering, packet switching, packet transmission, packet processing (e.g., compression, encryption), and support management functions (e.g., interface statistics, alerts, trace route).

- Deployment of NEs is highly dependent on desired network characteristics. NEs are fundamental to the realization of a reliable, cost-effective, and well-designed network.

REFERENCES

Andersson, L., and T. Madsen, *Provider Provisioned Virtual Private Network (VPN) Terminology*, IETF RFC 4026, March 2005.

Barr, R., and R. Patterson, Grooming telecommunications networks, *Optical Networks Magazine*, vol. 2, no. 3, pp. 20–23, 2001.

Barr, R. S., M. S. Kingsley, and R. A. Patterson, *Grooming Telecommunications Networks: Optimization Models and Methods*, Technical Report 05-emis-03, June 22, 2005.

Berkowitz, H., E. Davies, Ed., S. Hares, P. Krishnaswamy, and M. Lepp, *Terminology for Benchmarking BGP Device Convergence in the Control Plane*, RFC 4098, June 2005.

Cinkler, T., Traffic and λ grooming, *IEEE Network*, vol. 17, no. 2, pp. 16–21, 2003.

Dutta, R., and G. N. Rouskas, Traffic grooming in WDM networks: past and future, *IEEE Network*, vol. 16, no. 6, pp. 46–56, 2002.

Escalona, E., S. Figuerola, S. Spadaro, and G. Junyent, Implementation of a management system for the ASON/GMPLS CARISMA network, White Paper, CARISMA project, http://carisma.ccaba.upc.es, 2005.

IPv6 Portal, http://www.ipv6tf.org/meet/faqs.php, 2006.

Nagarajan, A., *Generic Requirements for Provider Provisioned Virtual Private Networks (PPVPN)*, RFC 3809, June 2004.

Narten, T., E. Nordmark, and W. Simpson, *Neighbor Discovery for IP Version 6 (IPv6)*, RFC 2461, December 1998.

Shirey, R., *Internet Security Glossary*, RFC 2828, May 2000.

Weston-Dawkes, J., and S. Baroni, Mesh network grooming and restoration optimized for optical bypass, *Technical Proceedings of the National Fiber Optic Engineers Conference (NDOEF 2002)*, pp. 1438–1449, 2002.

Zhu, K., H. Zang, and B. Mukherjee, A comprehensive study on next-generation optical grooming switches, *IEEE Journal on Selected Areas in Communications*, vol. 21, no. 7, pp. 1173–1186, 2003.

Zhu, K., and B. Mukherjee, A review of traffic grooming in WDM optical networks: architectures and challenges, *Optical Networks Magazine*, vol. 4, no. 2, pp. 55–64, 2003.

13

NETWORK DESIGN: EFFICIENT, SURVIVABLE NETWORKS

13.1 INTRODUCTION

Organizations have become highly dependent on their network infrastructure. As early as the mid-1990s it was recognized that "...we have reached a state where '*the corporation is the network*.' Hence, the entire business viability of a corporation may well depend on the network it has in place ..." (Minoli and Alles, 1996). One need only think of the importance of the network to a bank, a brokerage firm, an online business such as reservations or e-commerce, the aviation industry, or the military, to name just a few examples. It follows that

Network Infrastructure and Architecture: Designing High-Availability Networks,
By Krzysztof Iniewski, Carl McCrosky, and Daniel Minoli
Copyright © 2008 John Wiley & Sons, Inc.

organizations seek to have networks with high availability and self-healing capabilities. Requirements for availability in the 99.999% range are common for many institutional networks. Carriers and service providers need to deploy network infrastructures that meet these end-user needs. A properly designed carrier network will achieve high availability. Such a network will make full use of the network elements (NEs) and advanced technologies discussed in earlier chapters.

There typically are a number of network design objectives, such as cost-effectiveness, reliability, high-availability, bandwidth efficiency, bandwidth elasticity, scalability, and security. These objectives drive design principles that must take into account interrelated factors. In particular, some of these objectives demonstrate orthogonal impacts on each other; for example, a network that is maximally cost-effective will not generally be the one with the highest intrinsic availability. Therefore, the designer needs to perform an engineering analysis that deals with an overall compromise of the various parameters in question. In this chapter we focus on issues related to reliability and availability; we do not emphasize the cost implications of achieving these goals, which should be somewhat self-evident. The challenge is, and designers have accepted, to solve a min-max problem: Minimize the cost of maximizing availability. The techniques discussed in the chapter, in fact, embody these min-max considerations. Evolving networks will consist of NEs such as routers, switches, dense wave-division multiplexing (DWDM) systems, add–drop multiplexers (ADMs), photonic cross-connects (PXCs), optical cross-connects (OXCs), and other advanced NEs that will use signaling protocols to provision resources dynamically and provide network survivability using protection and restoration techniques (Mannie, 2004).

The purpose of this chapter is to provide a brief overview of networks as a complex synthesis of the NEs described in earlier chapters and of a variety of transmission channels, specifically with reliability in mind. The chapter focuses on protection schemes and traffic management. Protection schemes can be applied to any layer of the open systems interconnection reference model (OSIRM), but generally they are most effective when applied at the lowest possible layer. Most networks rely on protection being implemented at a few layers; however, if the lower-layer protection is in place, it will relieve the burden from higher layers, where protection may be slower and/or more expensive.

In this chapter we focus on physical layer protection switching, which has been standardized extensively by the International Telecommunications Union (ITU). Layer 2 switching exists for local area networks (LANs), asynchronous transfer mode (ATM), and frame relay (FR) technologies; layer 3 switching is supported in multiprotocol label switching (MPLS) and Internet protocol (IP) using either internal gateway protocols (IGPs) or exterior gateway protocols (EGPs). Service impairments range from complete link or node failure to channel suboptimality (e.g., high bit/block/packet loss), to traffic congestion (and ensuing delay). Schemes have been developed to deal with all of these issues. Physical layer protection switching typically operates at the 50-ms level; layer 2 switching operates in the second-to-multisecond level; layer 3 operates on the tens-of-seconds level, since it requires topology information from significant portions of the entire network.

In Chapter 12 we discussed the fact that the most important NEs are ADMs (which support multiplexing, grooming, and low end switching), switches, and routers. In this chapter we focus by and large on the multiplexing and grooming capabilities of ADMs in support of network goals for reliability; these same NEs come into play in other objectives, such as performance, but these issues will not be covered. One critical issue that was not highlighted directly in Chapter 12, but that is of critical importance to networking, relates to the use of open industry standards for the user plane, control plane, and management plane of *all* the various NEs. It would not be possible to achieve any of the restoration, performance, service, interworking, and transport goals of the NEs if they did not conform to standards such as those promulgated by the ITU or other recognized standards-making body. The *user plane* refers to data forwarding in the transmission path; these standards deal with transmission parameters such as electrical, optical, mechanical, frame and packet formats, and so on. The *control plane* refers to the session setup and teardown mechanism and is applicable to a variety of situations. Two switches (of any of the types described in Chapter 12) cannot interwork effectively to establish a session unless the same session control mechanisms are used. The *management plane* deals with procedures, mechanisms, and standards to deal with fault, configuration, accounting, and performance management. Because NEs have to interact with each other to achieve restoration and rerouting (and to provide consistent event definition and performance thresholds), it is indispensable that they make use of standards.

The rest of the chapter focuses on physical layer protection switching in the synchronous optical network/synchronous digital hierarchy (SONET/SDH) environment, in the optical transport network (OTN) environments and in the automatically switched optical network (ASON) environment (Minoli, 2003). The emphasis of this chapter is on standards and less on technology. This is for two reasons: (1) as noted, it would not be possible to achieve rapid automatic restoration and the performance, service, interworking, and transport goals of the NEs if they did not conform to standards; and (2) with the exception of SONET/SDH, the other topics discussed in this chapter (OTN and ASON) are still, in general, conceptual frameworks, designs, and road map carrier plans, with technology in the process of development.

13.2 SONET

13.2.1 Overview

Although there are a number of evolving technologies on the deployment horizon, SONET and SDH are the de facto infrastructure technologies for communications services at the present time and for the near-term future. It is estimated that there were about 400,000 SONET/SDH rings deployed worldwide at press time (Minoli, 2006). Effectively, all communication in North America, Europe, and Asia ranging between DS3 and OC-768 takes place over SONET or SDH systems. A portion of the DS1 traffic is also carried over virtual tributaries (VTs) in SONET. Vendors are also bringing out data-aware "next-generation" SONET platforms that support data interfaces (e.g., Gigabit Ethernet) directly on SONET NEs. SONET was discussed in previous chapters, but we review here major features and focus on network restorability.

SONET was developed in the United States through the ANSI T1X1.5 committee. ANSI's work started in the mid-1980s, and the Consultative Committee for International Telephony and Telegraphy (CCITT) (now the ITU) initiated a standardization effort at basically the same time. The U.S. government sought a data rate close to 50 Mbps in order to carry DS1 (1.544 Mbps) and DS3 (44.736 Mbps) signals. The Europeans needed a system to carry E1 (2.048 Mbps), E3 (34.368 Mbps), and 139.264 Mbps signals efficiently. The Europeans did not accept the 50 Mbps proposal from the United States and focused on a base signal rate close to 150 Mbps. Eventually, a compromise was reached that allowed the U.S. data rates to be a subset of the ITU specification for SDH. The SONET/SDH specifications of the late 1980s and early 1990s were designed for supporting the transmission of a single wavelength (also known as a beam, a light path, or λ), per fiber; typically, a send and a receive wavelength pair operate over two fibers.

Figure 13.1 depicts the SONET reference model. The ADMs allow traffic to enter or leave the system; the traffic that is injected or removed from the system can be at various levels in the SONET/SDH hierarchy. For example, a DS1 link may arrive at an incoming interface and an ADM; the ADM places the DS1 traffic into a VT, which is then placed into an STS-1 SPE (payload); in turn, this STS-1 may be multiplexed into a higher-level signal such as an

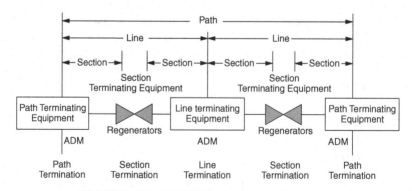

FIGURE 13.1   SONET reference model.

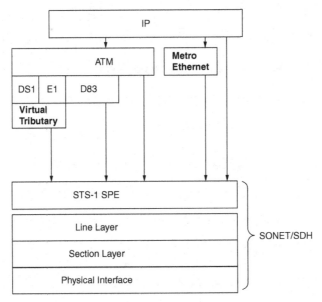

FIGURE 13.2 Mapping of higher-layer protocols over a SONET transport infrastructure.

STS-768. The DS1 signal is then transported to another ADM, where the traffic may be removed and routed to a local outgoing interface. In this example the VT carrying the DS1 traffic is a path. The only restriction in path switching is that both the entry and exit nodes for a path operate at the same level. Figure 13.2 depicts the mapping of higher-layer protocols over a SONET transport infrastructure; increasingly, IP is an important protocol that requires efficient support by the underlying transport mechanism. SONET's transport mechanisms and frame formats were discussed in earlier chapters, hence are not be covered here.

13.2.2 Protection Switching

Automatic protection switching (APS) is the mechanism used in SONET/SDH to restore service in the case of an optical fiber failure or a NE failure. SONET standards require that restoration of service occur within 50 ms, which is very desirable since this generally makes the failure transparent to the higher layer protocols. Fiber cuts are the most common failures in SONET/SDH rings; NEs usually have on-chassis and power redundancy and therefore tend to fail catastrophically less often (mean time between failures of 150,000 hours for carrier-class NEs—about 20 years—are often cited by manufacturers).

SONET/SDH networks can be configured as linear networks; here the SONET/SDH ADMs are connected serially, as shown in Figure 13.3. In this arrangement there may be two fiber connections between the ADMs (as is the

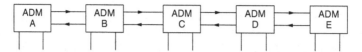

FIGURE 13.3   Linear SONET/SDH network.

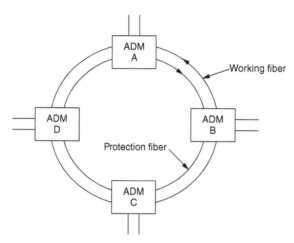

FIGURE 13.4   Two-fiber SONET/SDH ring.

case in Figure 13.3), or there may be four fiber connections, with one set serving as a backup, or *protection*, pair. Linear networks are typically used in low-end applications. Linear networks have limitations; for example, even if two sets of fiber were used between the nodes, it is conceivable that all the fibers be cut at the same time unless explicit design goals for diversified routing are adhered to.

The most common SONET topology for carrier networks is the ring. Rings are preferred because they provide an alternative path to support communication between any two nodes. Figure 13.4 depicts a unidirectional fiber ring where all the traffic can be carried on one fiber (note that a two-fiber ring can also operate as a bidirectional ring). Figure 13.5 shows a four-fiber (two-pair) SONET/SDH ring; note that this type of ring is always bidirectional. A two-fiber ring can operate as a unidirectional or bidirectional ring. In a unidirectional ring, traffic is limited to one fiber and always flows in the same direction around the ring; the second fiber is the protection path. With bidirectional designs, information is sent on both fibers; when data are sent between nodes, say nodes 1 and 2, it flows over the two fibers connecting the two nodes. To provide backup, however, each fiber in a bidirectional ring can be utilized to only half its capacity—the other half of the capacity is reserved for backup.

Four-fiber rings always operate as bidirectional rings. In this case one can secure the full transmission data rate that can be achieved on the working

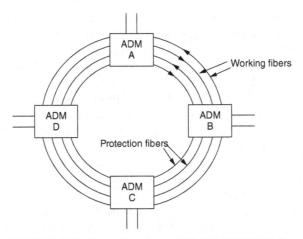

FIGURE 13.5   Four-fiber (two-pair) SONET/SDH ring.

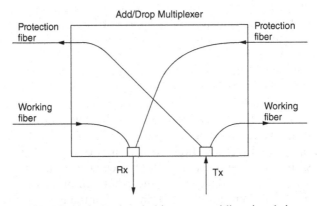

FIGURE 13.6   Path switching on a unidirectional ring.

fibers, but the protection fibers are not utilized for traffic under normal conditions. In this case it is possible to do a link recovery if one of the fibers fails
between two nodes. For example, when all the fibers between two nodes are
cut, the bidirectional ring provides restoration by routing the traffic over the
protection fibers, in the opposite direction around the ring.

There are two backup systems used on fiber: path backup and line backup.
Path backup, also known as *path switching*, is typically implemented on a unidirectional ring. Such a ring is known as a *unidirectional path-switched ring*
(UPSR). In such a system, all of the traffic is transmitted in both directions
around the ring, in one direction on the working fiber and in the other direction on the protection fiber. The SONET/SDH equipment monitors the path
traffic on both fibers and selects the traffic that is the "best" (Figure 13.6).

Because the receiver of each channel is monitoring both paths on both fibers, switching between fibers is instantaneous, with no loss of information. The restoration is handled by the receiver, without any coordination with the transmitter (i.e., no APS communication channel is needed). If multiple fiber cuts occur, additional actions are required to restore communication. The loss of an ADM NE is detected by the two neighboring ADMs; these ADMs will then place an alarm indication signal on each path that originated and terminated on the failed ADM. This is an indication to the devices on the remote end of the path that the path is no longer usable.

Path switching on a unidirectional ring implies that no coordination is needed between the receiver and the transmitter. It provides an automatic transparent (lossless) restoral. However, this approach is resource intensive (and hence, not inexpensive) because two sets of path equipment are needed for a circuit to be dropped at an ADM. Furthermore, unidirectional rings have asymmetric delay because the transmitting and propagation time required to propagate the information one way around the ring is usually different than the time it takes to propagate the information the other way around the ring. To address this issue, buffering must be supported at the path-terminating site (the amount of buffering increases as rings get larger and supports higher speeds). This is one of the reasons why unidirectional rings are utilized primarily in metropolitan networks and at lower speeds.

In the long-haul carrier backbone, bidirectional rings are much more common than unidirectional rings. As noted, in bidirectional rings the traffic between two nodes flows in two directions; for example, in the case of two nodes (e.g., C and D), traffic from node C to node D will flow in one direction around the ring, say counterclockwise, while traffic from node D to node C flows in the opposite direction. Bidirectional rings can be either two- or four-fiber systems. In a two-fiber design, each fiber can carry only half its carrying capacity because of the need to reserve the other half of the carrying capacity for backup. In a four-fiber bidirectional ring, two of the fibers are reserved for protection.

The recovery mechanism in a two-fiber bidirectional ring is called a *bidirectional line-switched ring*. When a fiber cut occurs, the only recovery possible is a ring switch, resulting in the sending of data in the opposite direction over the two fibers (the ring switch is actually called a *line switch*, hence the nomenclature). In a four-fiber bidirectional ring, a single fiber failure can be bypassed by doing a span switch, one simply to the protection fiber over that single link (failure of multiple fibers will usually require a ring switch).

Figure 13.7 shows how ADMs reroute traffic in a four-fiber bidirectional ring. The figure illustrates a fiber failure somewhere in the fiber link between ADM A and ADM B. When the fiber break occurs, both ADM A and ADM B detect the loss of signal on the fibers. At this juncture, the two ADMs will send a signal in the opposite direction over the ring in a specific set of header octets (the K1, K2 octets to be exact), with the neighboring ADM's address

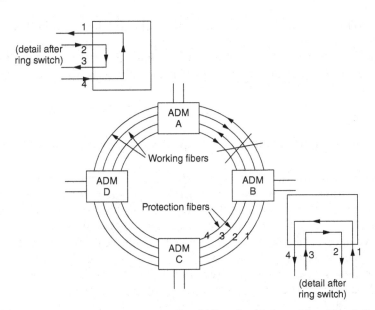

FIGURE 13.7   Example of a four-fiber bidirectional ring with a fiber failure.

in the K1 octet. ADMs C and D pass the K1, K2 octets on; when ADM A and ADM B receive the failure messages from each other, they effect a switchover.[†] The traffic arriving on fiber 1 would ordinarily be transmitted to ADM A via the fiber link 1 in the figure; postfailure, however, the traffic is placed on protection fiber link 4, which carries it in the reverse direction around the ring to ADM A. ADM A then takes the traffic on fiber link 4 and moves it to fiber link 1. In this way, the traffic on fiber link 1 on ADM B still gets to fiber 1 on ADM A, but the long way around. Traffic on fiber link 2 on ADM A is moved to fiber link 2 on ADM B in the same fashion by fiber link 3. Note that the rest of the ADMs on the ring need not be concerned about the fiber cut between ADM A and ADM B because those two ADMs handle the fault.

Recovery on a two-fiber bidirectional ring is similar to recovery on a four-fiber bidirectional ring except that a line switch is not possible—only a ring switch. Traffic is routed back around the ring in the unused/allocate capacity of the other fiber (which carries traffic in the opposite direction), not on a separate fiber pair. The protection channels carries the traffic the long way around the ring until it gets to the ADM it would have arrived at had the fiber cut not occurred.

[†]When all the fibers are cut, both ADMs detect the failure at the same time and take the same action. If only one fiber is lost, only one ADM will detect the failure. Here the ADM detecting the failure notifies the other ADM through the K1, K2 octets, and the neighboring ADM then deals with the issue as appropriate.

It is to be understood that the protection schemes discussed in this section work to protect services end to end only because the manufacturers of the various NEs, ADMs in particular, have all implemented the same SONET standards in the user and management planes (there is no control plane in SONET).

13.3 OPTICAL TRANSPORT NETWORK

In this section we discuss recent advances in all-optical networks, in particular the optical transport network (OTN). The OTN is intended to be an evolution of the carrier's network plant. OTN is designed for the transmission of multiple wavelengths per fiber, which is characteristic of DWDM systems[†]; SONET was developed to handle one optical signal per physical channel. At the same time, a need has emerged for networks capable of transporting a variety of heterogeneous types of signals directly over wavelengths carried on the same optical backbone. The OTN addresses the requirements of next-generation networks that have a goal of efficiently transporting data-oriented traffic. The OTN also aims at supporting network scalability and manageability.

As is the case for SONET/SDH, OTN is standards-based, so that interoperability among various equipment manufacturers and interfaces is ensured. In the early 2000s, the ITU reached agreement on several new standards for next-generation optical networks capable of transporting transparent wavelength services, SONET/SDH services, and data streams (Ethernet, fiber channel, ATM, frame relay, and IP). The new standards provide the ability to combine multiple client signals within a wavelength, to facilitate optimal utilization of transport capacity, and to realize improved cost-effectiveness of transport capacity while allowing switching at service rates of 2.5, 10, and 40 Gbps. The new standards also specify optical equipment functions such as performance monitoring, fault isolation, and alarming. ITU Recommendations G.709 and G.872 define an OTN consisting of optical channels within an optical multiplex section layer within an optical transmission section layer network; the optical channels are the individual light paths (beams). In this section we look at standards related to the OTN, along with architectural considerations. Motivations and positioning of all-optical networks are also discussed.

13.3.1 Motivations, Goals, and Approaches

Existing TDM SONET/SDH networks were originally designed for voice and medium-speed leased line (point-to-point dedicated) services. In some instances (e.g., outermost edges of the network), the growing trend of data

[†]Some vendors call the OTN the next-generation WDM network.

traffic could pose some technical challenges to TDM SONET/SDH networks, especially in relation to the bursty and asymmetrical nature of such traffic (however, one needs to keep in mind that in the core of a large data/IP network such as the Internet or a corporate intranet backbone, contrary to trade press assertions, *data traffic is neither bursty nor asymmetric*†). At the same time, DWDM technology, which was initially introduced to increase transport capacity, is currently available for implementing advanced optical networking functionality. The existing carrier transport networks are expected to evolve to next-generation optical networks; the expectation is that these networks ought to fulfill new emerging requirements such as fast and auto-matic end-to-end provisioning, optical rerouting and restoration, support of multiple clients and client types, and interworking of IP-based and optical networks. A mechanism that supports an IP-over-OTN protocol stack (and the underlying communications infrastructure) is a candidate for providing high bandwidth on demand and flexible, scalable support for quality of service (QoS) for transmission of multimedia services with low jitter, latency, and loss (NCS, 2003).

OTN is composed of a set of optical NEs, specifically OXCs, intercon-nected by optical fiber links, that are able to provide functionality of trans-port, multiplexing, routing, management, supervision, and survivability of optical channels carrying client signals. OXCs switch wavelength channels between their input and output fibers and are used to establish optical paths. The OTN provides the ability to route wavelengths, and therefore, when deployed in a ring topology, it has the same survivability capabilities as SONET/SDH rings. Furthermore, the OTN has the ability to improve surviv-ability by reducing the number of electrooptical network elements; these elements are, in principle, prone to failure. OTN makes use of a data plane (transport layer) and a control plane (signaling and management layer), as illustrated in Figure 13.8.

The OTN is designed to provide a cost-effective, high-capacity, survivable, and flexible transport infrastructure. The elimination of multiple service network overlays and the elimination of fine-granularity sublayers implies a reduction in the number and types of NEs and a concomitant reduction in capital and operating costs for the network provider. Figure 13.9 depicts the service consolidation that is possible with OTN. The optical transport functions include:

---

†Proponents make the pitch that with the emergence of ever-growing Internet and intranet data-oriented traffic on the network, a new network mechanism is needed to meet the network demands for scalability and manageability. This is fine, except to note that the traffic at the core of a network is rarely, if ever, bursty, since Probability 101 principles clearly demonstrate that the sum of 15 to 20 independently, identically distributed random variables leads to a distribution that is Gaussian in nature; furthermore, when the link has a service time that is small compared to the arrival rate (e.g., a 1000-person company being run on a gateway router with a single T1 uplink), the distribution is actually D/D/1 at busy hour.

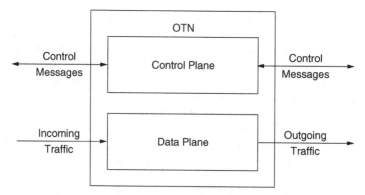

FIGURE 13.8   OTN infrastructure.

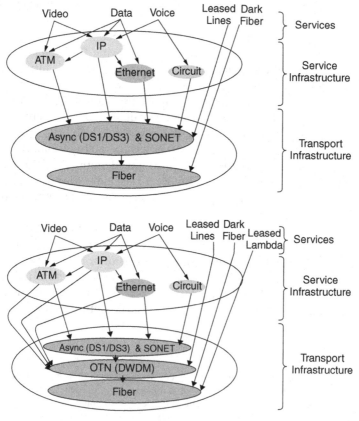

FIGURE 13.9   Pre- and post-OTN service support.

- Multiplexing functions
- Cross-connect function, including grooming and configuration
- Management functions
- Physical media functions

Some of the benefits of the OTN include:

- Maintenance signals (management plane) per wavelength
- Fault isolation capabilities (management plane)
- Small overhead
- Forward error correction (FEC) capabilities
- Protocol-agnostic nature

As just noted, a distinguishing characteristic of the OTN is its ability to transport any digital signal, independent of client-specific aspects (making it protocol-agnostic). The flexibility of OTN is built on the protocol and bit-rate independence of the information-carrying optical beams in the fiber waveguide. This transparency enables the OTN to carry many different types of traffic over an optical channel regardless of the protocol (Gigabit Ethernet, 10-Gigabit Ethernet, ATM, SONET, etc.) or bit rate (155 Mbps, 1.25 Gbps, 2.5 Gbps, etc.). The generic framing protocol (GFP) provides a means for packing nonvoice traffic into a SONET and/or OTN frame. A virtual concatenation mechanism called the link capacity adjustment scheme (LCAS) provides additional flexibility. With conformant implementation, interworking between pieces of equipment from different vendors is achievable.

In OTN there are two networks layers: an optical network layer and the user's network layer (Figure 13.10). Routers, switches, and other equipment that comprise the user's network establish connections over the underlying

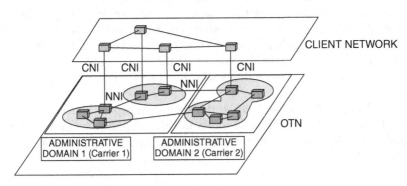

CNI - CLIENT NODE INTERFACE
NNI - NETWORK NODE INTERFACE

FIGURE 13.10   OTN environment.

network through user network interfaces (UNIs) (also known as client network interfaces). In the ITU-T model the network consists of clients; clients may be multiplexers, DWDM systems, Ethernet switches, or other devices. Clients connect into the network through one of three types of network interfaces: UNIs, external-network-to-network interfaces (E-NNIs), and internal-network-to-network interfaces (I-NNIs).

- The UNI specifies how users can access the providers' networks; only the basic information is transferred over this interface: the name and address of the endpoint, authentication and admission control of the client, and connection service messages.
- Service providers are required to share more information within their domains and among the various providers. E-NNIs support the exchange of reachability information, authentication and admission control information, and connection service information.
- The I-NNIs enable devices to get the topology or routing information for the carrier's network along with connection service information necessary to control network resources optionally. Devices within the optical network rely on I-NNIs to access network information.

As is the case in SONET, OTN defines a network hierarchy, here known as the optical transport hierarchy (OTH). OTH's basic unit, the optical transport module (OTM), supports higher transport data rates by bundling together wavelengths; OTMs can span multiple wavelengths of different carrying capacities (by contrast, to support higher speeds, SONET's STS-1's are concatenated at the electrical level). OTMs utilize the nomenclature *OTM-n.m*, where n refers to the maximum number of wavelengths supported at the lowest bit rate on the wavelength and m indicates the bit rate supported on the interface. Three rates are currently supported: 2.5 Gbps (indicated by a 1), 10 Gbps (indicated by a 2), and 40 Gbps (indicated by a 3). For example, OTM-3.2 indicates an OTM that spans three wavelengths, each operating at least at 10 Gbps. An interface could support some combination of these: for example, a 2.5-Gbps and 10-Gbps combination (1 and 2); a 10-Gbps and 40-Gbps combination (2 and 3), or a combination of all three (1, 2, and 3). For example, OTM-5.12 indicates a channel that spans five wavelengths and can operate at either 2.5 or 10 Gbps.

OTN defines of two hierarchies: the digital transport hierarchy and the OTH (Figure 13.11). The OTH supports intrinsic connections at three levels: optical channels, optical multiplex sections, and optical transmission sections (NCS, 2003):

- *Optical channel layer (OCh):* transports client signals between two endpoints on the OTN; these are conceptually similar to the SONET path. The OCh is needed to support end-to-end network functionality of the

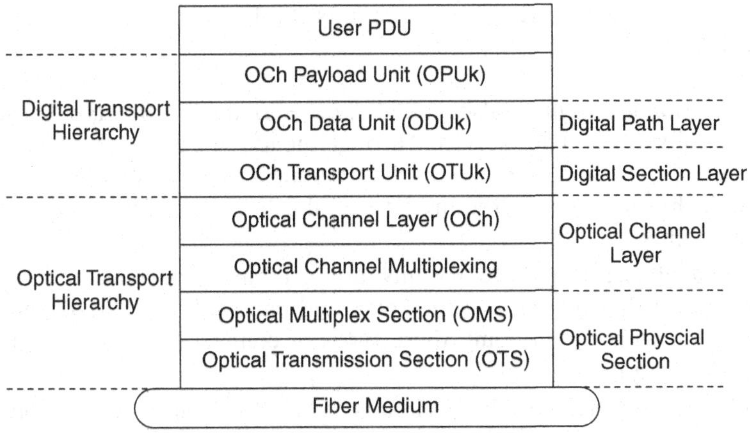

	User PDU	
Digital Transport Hierarchy	OCh Payload Unit (OPUk)	
	OCh Data Unit (ODUk)	Digital Path Layer
	OCh Transport Unit (OTUk)	Digital Section Layer
Optical Transport Hierarchy	Optical Channel Layer (OCh)	Optical Channel Layer
	Optical Channel Multiplexing	
	Optical Multiplex Section (OMS)	Optical Physcial Section
	Optical Transmission Section (OTS)	
	Fiber Medium	

FIGURE 13.11 OTN optical hierarchy.

optical channel. This layer allows transparent conveying of client information of varied formats. The signals contained in an optical channel can be transmitted using one or more wavelengths. This is the point at which the multiplexing of the channels takes place. Optical channel multiplexing combines the incoming optical wavelengths to a single optical signal.

- *Optical multiplex sections (OMSs):* describe the WDM aspects that support the optical channels. OMS data streams consist of many optical channels aggregated together. They are similar to SONET lines, but accommodate multiple wavelengths. The OMS provides the functionality between multiplexer–demultiplexer and add–drop sites in the network. This function is taken care of by the OMS overhead. This sublayer provides functionality for networking of the multiple wavelength optical signals between add–drop multiplexers and other types of multiplexers–demultiplexers in the optical network.

- *Optical transmission section (OTS):* the lowest level of the OTN link; enables transmission of signals over individual fiber spans, similar to SONET. The OTS defines a physical interface that details optical parameters such as wavelength and power level. The OTS provides for transmission of signals over individual fiber spans. OTS defines a physical interface detailing optical parameters such as frequency and power lever. This layer defines the functionality between multiplexer–demultiplexer and add–drop sites in the network. This function is taken care of by the OMS overhead. This sublayer provides functionality for networking of multiple-wavelength optical signals between add–drop multiplexers and other types of multiplexers–demultiplexers in the optical network.

Clients can request three types of circuits via the interfaces defined earlier, as follows:

1. *Provisioned circuits* (also called *hard permanent circuits*). These are basically a leased line service. Through the use of a network management station or a manual process, each NE along a required path is configured with the information required to establish a connection between two endpoints.
2. *Signaled circuits.* These circuits are established dynamically by the endpoint that requests connectivity and bandwidth. To establish a connection with an endpoint, these types of connections require network-addressing information.
3. *Hybrid connections.* These circuits have provisioned connections into the automatic switched transport network (ASTN), but then rely on switched connections within the ASTN to connect with other end nodes [these connections are also known as soft provisioned connections (SPCs)]. To the end node, an SPC and a regular permanent circuit appear the same.

One of the major advantages of OTN is its backward-compatible formulation: specifically, support for existing SONET/SDH protocols without changing the format, bit rate, or timing. The OTN allows transmission of different data packet types using the new GFP mapping; this mapping reduces the layers between the fiber and the IP layer and thus much efficient use of bandwidth. The mapping capability advantages put the OTN as protocol-agnostic carrier allowing service transparency for SDH/SONET, Ethernet, ATM, IP, MPLS, and other protocols/clients.

13.3.2 OTN Standards Support

To achieve the design goals of seamless restorability, uniform grade of service, and bandwidth optimization, standards are needed. ITU-T Recommendations G.872, G.709, and G.959.1, the initial set of standards in the OTN series approved at the turn of the decade, address the OTN architecture, interface frame format, and physical layer interfaces, respectively. This work is carried out by Study Group 15 of the ITU. The ITU-T G.709, "Network Node Interface for the Optical Transport Network," provides the basic standard specification for OTN. OTN elements' transport of a client signal over an optical channel is based on the digital signal wrapping technique defined in ITU-T G.709 recommmendation; the wrapper provides transmission protection using FEC. The OTN architecture is defined further under ITU-T G.872, "Architecture of Optical Transport Networks." By complying with ITU-T G.709, network nodes from various vendors can interoperate, although in actual practice carriers rarely, if ever, mix transport equipment from two vendors. OTN provides carrier-class robustness.

TABLE 13.1 Key OTN-Related ITU-T Standards (Partial List)

G.7041/Y.1303	Generic framing procedure specifies interface mapping and equipment functions for carrying packet-oriented payloads
G.7042	Link capacity adjustment scheme
G.709	Network node interface for the optical transport network (OTN)
G.7710/Y.1701	Common equipment management function requirements, a generic equipment management recommendation
G.7712/Y.1703	Architecture and specification for telecommunications management networks
G.7713/Y.1704	Distributed call and connection management
G.7714/Y.1705	Generalized automatic discovery techniques
G.798	Characteristics of OTN hierarchy equipment functional blocks
G.807	Requirements for the automatic switched transport network
G.8080/Y.1304	Architecture for the automatic switched optical network
G.8251	Control of jitter and wander within the OTN
G.841 and G.842	OTN protection
G.871/Y.1301	Framework of OTN recommendations
G.872	Architecture of the OTN
G.874 and G.875	Management aspects of the OTN element
G.874.1	OTN protocol-neutral management information model for the network element view
G.957	Optical interfaces for equipment and systems relating to the SDH
G.958	Digital line systems based on the synchronous digital hierarchy for use on optical fiber cables
G.959.1	OTN physical layer interfaces

In addition to basic transport there is also interest in real-time provisioning of λ's. ITU-T has grouped these new signaling architectures into two protocol-independent framework models: the general ASTN and the more specific ASON. Bandwidth on demand (called *instant provisioning* by some) requires a signaling protocol to set up the paths or connections. Work on signaling is being defined in the ASTN and ASON specifications. The Optical Internetworking Forum has taken the ITU-T ASTN and ASON models and extended the signaling capabilities of the IETF's generalized multiprotocol label switching (GMPLS). This topic is discussed in Section 13.4.2.

In what follows a more detailed survey of key recommendations is provided. OTN-related recommendations developed by Study Group 15 include the following (see also Table 13.1):

- G.709, "Network Node Interface for the Optical Transport Network (OTN)," February 2000; Amendment 1; November 2001. (This recommendation is also referred to as Y.1331 and is being renumbered is

G.7090). This recommendation provides the requirements for the OTH signals at the network node interface in terms of:

- Definition of an optical transport module of order n (OTM-n)
- Structures for an OTM-n
- Formats for mapping and multiplexing client signals
- Functionality of the overheads

An amendment to Recommendation G.709 entitled "Interfaces for Optical Transport Networks" describes the mappings for TDM-multiplexed signals in the OTN as well as extensions to allow even higher-bit-rate signals to be carried using virtual concatenation.

- ITU-T Recommendation G.7041/Y.1303, "Generic Framing Procedure (GFP)," specifies interface mapping and equipment functions for carrying packet-oriented payloads, including IP/PPP, Ethernet, Fiber channel, and ESCON (enterprise systems connection)/FICON (fiber connection) over optical and other transport networks. This recommendation, together with ITU-T Recommendation G.709 on interfaces for optical transport networks, provides the full set of mappings necessary to carry IP traffic over DWDM systems.

- ITU-T Recommendation G.7710/Y.1701, "Common Equipment Management Function Requirements," is a generic equipment management recommendation derived from knowledge gained through the development of SONET and SDH equipment management recommendations. This recommendation provides the basis for management of equipment for new transport network technologies, including the optical transport network.

- New ITU-T Recommendation G.7712/Y.1703, "Architecture and Specification of Data Communication Network," extends capabilities originally built for telecommunications management networks. It allows the use of IP protocols as well as OSIRM protocols and supports new services, such as ASTN, through communication among the transport, control, and management planes.

- G.871, "Framework of Optical Transport Network Recommendations," October 2000. This recommendation provides a framework for coordination among the various activities in ITU-T on OTN, to ensure that recommendations covering the various aspects of OTN be developed in a consistent manner. As such, this recommendation provides references for definitions of high-level characteristics of OTN, along with a description of the relevant ITU-T recommendations that were expected to be developed, together with a time frame for their development. This is also numbered Y.1301.

- G.872, "Architecture of Optical Transport Network," February 1999; revision November 2001. This recommendation describes the functional architecture of OTNs using the modeling methodology described in ITU-T G.805. The OTN functionality is described from a network-level viewpoint, taking into account an optical network layered structure, client characteristic information, client–server layer associations, networking topology, and layer network functionality providing optical signal transmission, multiplexing, routing, supervision, performance assessment, and network survivability. This recommendation is limited to the functional description of OTNs that support digital signals. The support of analog or mixed digital/analog signals is not included. A revision includes (among other features), the ability to support time-division-multiplexing signals over the OTN.

- ITU-T Recommendation G.798, "Characteristics of Optical Transport Network Hierarchy Equipment Functional Blocks," specifies the characteristics of OTN equipment, including supervision, information flow, processes, and functions to be performed by this equipment.

- ITU-T Recommendation G.8251, "The Control of Jitter and Wander Within the Optical Transport Network," includes the network limits for jitter and wander, as well as jitter and wander tolerances required of equipment for OTNs. These parameters relate to the variability of the bit rate of signals carried over the optical transport network.

- ITU-T Recommendation G.874, "Management Aspects of the Optical Transport Network Element," specifies applications, functions, and requirements for managing optical networking equipment utilizing operations support systems. It covers the areas of configuration management, fault management, and performance management for client optical network elements.

- ITU-T Recommendation G.874.1, "Optical Transport Network Protocol-Neutral Management Information Model for the Network Element View," provides a means to ensure consistency among the models for OTN equipment for specific management protocols, including CMISE (common management information service element), CORBA (common object request broker architecture), and SNMP (simple network management protocol).

- G.8080 (formerly G.ason). This recommendation describes the reference architecture for the control plane of the ASON. The reference architecture is described in terms of the key functional components and the interactions among them. This recommendation describes the set of control plane components that are used to manipulate transport network resources in order to provide the functionality of setting up, maintaining, and releasing connections. The use of components allows for the separation of call control from connection control and the separation of routing and signaling. G.8080 takes path- and call-level views of an optical connection and applies a distributed call model to the operation of these connections. In the area of optical signaling, the ITU-T has been involved in the creation of a standardized architecture for optically signaled networks. A number of different approaches have been suggested for signaling, including GMPLS and possibly other protocols. However, 35 years of history has shown that neither X.25 SVCs, nor frame relay SVCs, nor ATM SVCs have seen commercial success; we venture to guess that in the short term, switched light paths are also not going to see major commercial success for end-user enterprise applications. ASON is discussed in Section 13.4.

13.3.3   Basic OTN Technical Concepts

OTN makes use of an optical channel layer: Each wavelength λ is wrapped in an envelope that consists of a header (for overhead bytes) and a trailer (for FEC functions). The payload section allows for existing network protocols to be mapped (wrapped), thus making OTN protocol independent. OTN NEs can receive, generate, and transmit management and control information throughout the network, making performance monitoring and other network management possible on a per-λ basis. The FEC supports error detection and correction (unlike SONET's simple error monitoring, FEC has intrinsic mechanisms to correct errors and achieve BER levels of 10^{-12} to 10^{-15}).

The OTN G.709 client mapping and frame structure is shown in Figure 13.12. There are three overhead sections (also known as *areas*): an optical channel transport unit specific overhead area, an optical channel payload unit specific overhead area, and an optical channel data unit specific overhead area. The OTN frame structure is defined for three optical channel transport unit bit rates, with $k = 1, 2, 3$ corresponding to 2.5, 10, and 40 Gbps, respectively. The OTN frame structure allows end-to-end operations, administration, and maintenance (OA&M) across the network, among multivendor

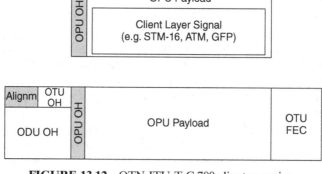

FIGURE 13.12   OTN ITU-T G.709 client mapping.

equipment; the OA&M overhead is small compared to that of SONET/ SDH.

The OTUk frame is composed of the following entities:

- *Optical channel payload unit order of k* (OPUk): includes payload and overhead. The payload contains the client information with its specific mapping technique. The overhead includes the information to support the adaptation of the specific client. Each type of client has its specific overhead structure.
- *Optical channel transport unit order of k* (OTUk): includes FEC and overhead for management and performance monitoring (section monitoring). The FEC is based on Reed–Solomon coding (ITU-T G.975).
- *Optical channel data unit order of k* (ODUk): composed of several overheads for path performance monitoring, tandem connection monitoring, communication channels, and protection control.

The OTUk overheads include the section monitoring and general communication channel fields. The section monitoring fields are used for the trail trace identifier, bit interleaved parity, backward defect indication, backward error indication and backward incoming alignment error, and incoming alignment error. The ODUk overhead includes six levels of tandem connection monitoring, path monitoring, communication channels, protection control, tandem connection monitoring activation, and fault localization. The path monitoring overhead is composed of trail trace identifier, bit interleaved parity, backward defect indication, backward error indication and status bits indicating the presence of a maintenance signal. The tandem connection monitoring overhead is composed of trail trace identifier, bit interleaved parity, backward defect indication, backward error indication and backward incoming alignment error, and status bits indicating the presence of a maintenance signal.

When migrating to higher-rate optical links, FEC mechanisms are a necessity. FEC is a signal coding/decoding technique adding redundancy data to a signal. These redundancy data identify and correct corrupted data, thus reducing the bit error rate. Per ITU-T G.975, the OTN uses the Reed–Solomon RS(255,239) error correction code; this code utilizes about 7% of the OTN bandwidth. The addition of FEC improves BER to 10^{-15}. The use of FEC also adds performance monitoring and early warning for optical link performance degradation.

In summary, the OTN overhead consumes small bandwidth while providing key operations, administration, and maintenance information. The embedded FEC mechanism ensures a better BER and/or longer reach. The client information mapping allows for support of various protocols, including SONET/ SDH. The OTN standards support efficient transport of popular data protocols such as Ethernet and fiber channel, as well as other wideband and broadband services, such as SDH/SONET, ATM, frame relay, audio/video, and IP-based services. They also specify detailed equipment functions to support performance monitoring, fault isolation, and alarming, including support for optically transparent subnetworks.

13.3.4 OTN Deployments

Currently, there are several types of carrier network infrastructures in place, as follows:

- Transport
  - Async DS1/DS3 systems
  - SONET-based systems
  - (Vendor-based) DWDM systems
- Layer 2 packet
  - Frame networks overlaid on ATM
  - ATM overlaid on async or sync optical transport
  - Metro Ethernet
- Layer 3 packet (data)
  - Private lines overlaid on async or sync optical transport
  - IP overlaid on ATM
  - IP overlaid on async or sync optical transport
  - MPLS

At face value it is desirable to migrate to a single infrastructure to reduce OAM&P costs. This is the goal of OTN. The move to this next target, however, will not happen overnight. Field trials of OTN supporting over 1 Tbps (40 Gbps × 25 light beams) were already under way at press time. It is not clear when OTN-based technology is deployed in the United States on a broad scale, but a 2010-or-beyond expectation would not be totally unreasonable.

While waiting for deployment of the OTN, enterprise users will continue to have access to SONET-based services. While carrying data traffic across an access ring, SONET link is not bandwidth efficient on that initial edge ring, it is all relative. This is no different from when a user gets a T1 tail (1.544 Mbps) to support a 384-kbps fractional T1 line or a frame relay circuit with a 64-kbps committed information rate. This is why there is grooming, where outer-edge traffic is repacked into multiplexed links of higher payload concentration. Carriers build networks using hierarchical sets of rings, not one giant ring that spans a metropolitan area. Ignoring the fact that bandwidth may, in fact, be a commodity at this juncture, virtual concatenation mechanisms have been developed of late to enable channels in SONET to be combined to support improved efficiencies. At a grooming point in the network, Ethernet streams, for example, could be multiplexed and carried across VT1.5-6v (10.368 Mbps) links instead of an STS-1 link; similarly, 100-Mbps Ethernet links could be multiplexed and carried across an STS-2c (103.68 Mbps) link instead of an STS-3c (155.520 Mbps) link.

13.4 AUTOMATICALLY SWITCHED OPTICAL NETWORKS

13.4.1 Overview

In Section 13.3 we discussed the standards baseline for a new generation of optical network, the OTN. In planning a new network, it would, in principle, be useful to have the ability for the end user to set up optical connections from and to various points in the network, in real time and as needed. In this section we extend the discussion begun in Section 13.3 by looking at the issue of signaling in more detail. The ASTN/ASON mechanisms discussed here aim at providing the OTN with an intelligent optical control plane for dynamic network provisioning; subtending capabilities include network survivability, protection, and restoration. ASTN specifications have been under development in the ITU in the recent past and were being positioned as the global framework for intelligent optical networking. The architectural choices for the interaction between IP and optical network layers, particularly the routing and signaling aspects, are key to the successful deployment of next-generation networks (Aboul-Magd et al., 2002).

With today's DWDM technology, the capacity of a single fiber can be increased 160-fold compared with a typical approach. DWDM, however, does not hold the ultimate solution to address bandwidth demand. As covered in Section 13.3, carrier-grade standards-supported work for a next-target (next-generation) intelligent optical network has been undertaken under the ASTN and ASON specifications of the ITU-T. The ASTN/ASON construct aims at providing the OTN with an intelligent optical control plane incorporating dynamic network provisioning, along with mechanisms for network survivability, protection, and restoration. In effect, a hierarchy has been built with fibers at the bottom, followed by groups of λ's, individual λ's, SONET/SDH tributaries, and packet-switch-capable connections at the top (Figure 13.13). As connections are established at each level of the hierarchy, they need to be propagated to other NEs in the system to establish higher-level connections. Dynamic routing necessitates information exchange between neighboring nodes to support the discovery of topology information; to achieve this, the out-of-band control plane is augmented with a link layer node-to-node protocol. The ASTN/ASON model is based on mesh network architectures; mesh topologies are now, in fact, being deployed by service providers. The ASTN/ASON specifications are framework models and they are protocol independent, but they do require a choice of protocols to operate, such as GMPLS.

The ASON family of standards from ITU-T Study Group 15 builds on OTN standards completed earlier in the decade. The deployment of an

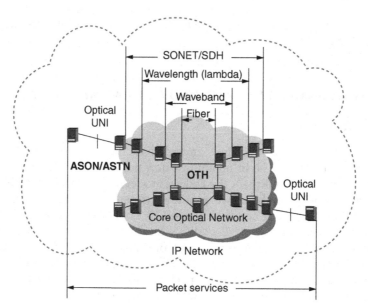

FIGURE 13.13   Example of the OTN/ASON/ASTN environment.

ASTN/ASON-enabled infrastructure allows carriers to support circuit provisioning in relatively short times (e.g., minutes); to have dynamic restoration and resiliency; to achieve flexible service selection and dynamic resource allocation; and to provide interdomain, intercarrier QoS. In certain ways, the ASTN/ASON model is relatively traditional (evolutionary) and incorporates aspects of other network technologies, such as the PSTN and ATM, (although GMPLS is based on IP specifications.) Proponents claim that these new standards can create business opportunities for network operators and service providers, giving them the means to deliver end-to-end, managed bandwidth services efficiently, expediently, and at reduced operational cost. ASON standards can also be implemented to add dynamic capabilities to new optical networks or established SONET/SDH networks. Service provider benefits could include:

- Increased revenue-generating capabilities through fast turn-up and rapid provisioning; as well as wavelength-on-demand services to increase capacity and flexibility
- Increased return on capital from cost-effective and survivable architectures that help protect current and future network investments from forecast uncertainties
- Reduced operations cost through more accurate inventory and topology information, resource optimization, and automated processes that eliminate manual steps

ASON control mechanisms provide support for both switched wavelength and subwavelength connection services in OTNs to provide bandwidth on demand. Wavelength connection services make use of an entire optical wavelength (e.g., at 1550 nm), while subwavelength services use a channel within a wavelength. The ASON control mechanisms also enable fast optical *restoration*. Traditionally, transport networks have used protection rather than restoration to provide reliability for connections; with protection, connections are moved to dedicated or shared routes in the event of failure of a fiber or of network equipment. With restoration, the endpoints can reestablish the connection through an alternative route as soon as a loss of the original connection is detected. Restoration provides a functional advantage for carriers since it makes better use of the network capacity, and with a control plane signaling standard, it can be performed faster than with restoration systems available today.

13.4.2 Standardization Efforts

An ASON infrastructure is fairly complex, with major NE-to-NE real-time interactions required. Therefore, it is critical that standards be available. ASON architecture defines a set of reference points (interfaces) that allow ASON clients to request network services across those reference points. The protocols that run over ASON interfaces are not specified in the basic ASON specifications (Mayer, 2001a,b). IP-based protocols such as GMPLS can be considered so that the ASON/ASTN work can benefit from the protocols design work done by the IETF (Aboul-Magd, 2002). The ASON model (Figure 13.14) distinguishes reference points (representing points of protocol information exchange) defined (1) between an administrative domain and a user, also known as a user-network interface; (2) between (and when needed within) administrative domains, also known as an external network–network interface; and (3) between areas of the same administrative domain and when needed

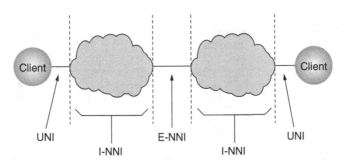

FIGURE 13.14 ASON reference model.

between control components (or simply controllers) within areas, also known as an internal network–network interface (Alanqar et al., 2003). Clients (e.g., end-user routers) can request three different types of circuits over the UNI using a signaling protocol: provisioned, signaled, and hybrid. Figures 13.15 and 13.16 (building on Figures 13.9 and 13.10) provide pictorial views of the ASTN/ASON environment.

ASON provides service providers with an evolution from early vendor products that were sold in the early 2000s as proprietary intelligent switching platforms. The ASON architectural standard design assures that service pro-

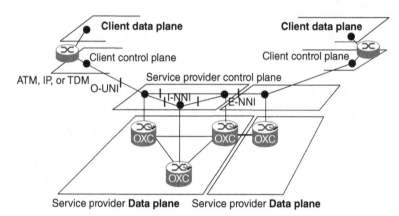

FIGURE 13.15 Pictorial view of ASTN/ASON. O-UNI: optical UNI.

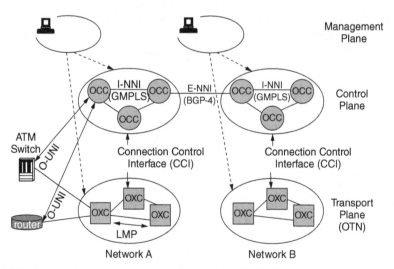

FIGURE 13.16 Another pictorial view of ASTN/ASON. LMP: link management protocol.

viders can deliver global, end-to-end λ-services. Whether λ-services are needed on a broad commercial scale between remote locations (at least in the near to midterm) remains to be established, as noted in passing earlier. Standard activities have included key efforts such as those of the ITU-T ASTN/SG15 and ASON/SG13, IETF GMPLS, and OIF's UNI 1.0. Utilizing standards, interoperability across the multivendor network environment, and across multiple service provider networks can be expected; this, in turn, ensures future-proofing the carrier and/or end-user investment.

To realize the ASTN/ASON, signaling standards are needed. GMPLS and the OIF optical user-to-network interface were considered by the ITU-T in the early part of the decade; the ITU-T has, however, also considered alternative proposals to these, including a modified form of the signaling and routing mechanisms used by ATM's private network-to-network interface. The GMPLS suite of protocol provides support for controlling different switching technologies and/or applications. These include support for requesting TDM connections, including SONET/SDH (see ANSI T1.105 and ITU-T G.707, respectively) as well as OTN (see ITU-T G.709); furthermore, it can be used to support the ASON control plane (as defined in G.8080) and ASON routing requirements (as identified in G.7715) (Alanqar et al., 2003). GMPLS is a suite of protocol extensions to MPLS to make it applicable to the control of non-packet-based switching, and particularly, optical switching. One proposal is to use GMPLS protocols to upgrade the control plane of optical transport networks. The following ASTN/ASON standardization bodies, forums, and consortia are currently involved in the standardization process:

- ITU-T SG 15 on optical and other transport network infrastructures
- ITU-T SG 4 on the telecommunications management network
- ITU-T SG 13 on next-generation network operation and maintenance
- ITU-T SG 16 for the transport of multimedia
- ITU-T SG 17 on data networks
- ITU-R WP 9B for a radio relay system
- OMG on CORBA technology
- IETF working groups in operations and management, transport, and routing (see Table 13.2 for a representative set of requests for comments
- OIF (signaling, architecture, carrier, and OAM&P working groups)
- ATIS Committee T1X1 on transport management aspects
- ATIS Committee T1M1 on generic management aspects
- TeleManagement Forum (MTNM and IPNM teams)
- W3C on XML
- IEEE 802 on Ethernet management
- Metro Ethernet Forum (MEF) on Ethernet management

TABLE 13.2   Some Applicable RFCs

RFC 4257	Framework for Generalized Multi-protocol Label Switching (GMPLS)-Based Control of Synchronous Digital Hierarchy/Synchronous Optical Networking (SDH/SONET) Networks
RFC 4258	Requirements for Generalized Multi-protocol Label Switching (GMPLS) Routing for the Automatically Switched Optical Network (ASON)
RFC 4327	Link Management Protocol (LMP) Management Information Base (MIB)
RFC 4328	Generalized Multi-protocol Label Switching (GMPLS) Signaling Extensions for G.709 Optical Transport Networks Control
RFC 4394	A Transport Network View of the Link Management Protocol (LMP)
RFC 4397	A Lexicography for the Interpretation of Generalized Multi-protocol Label Switching (GMPLS) Terminology within the Context of the ITU-T's Automatically Switched Optical Network (ASON) Architecture
RFC 4426	Generalized Multi-protocol Label Switching (GMPLS) Recovery Functional Specification
RFC 4427	Recovery (Protection and Restoration) Terminology for Generalized Multi-protocol Label Switching (GMPLS)
RFC 4428	Analysis of Generalized Multi-protocol Label Switching (GMPLS)-based Recovery Mechanisms (including Protection and Restoration)

ITU-T Recommendation G.807, the first standard in the ASTN series approved, addressed network-level architecture and requirements for the control plane of ASTN, independent of specific transport technologies. Agreement was reached in the recent past on other standards in the ASON, including (see Figure 13.17):

- ITU-T Recommendation G.8080/Y.1304, "Architecture for the Automatically Switched Optical Network (ASON)," specifies the architecture and requirements for the automatic switched transport network as applicable to SDH transport networks, defined in Recommendation G.803, and for optical transport networks, defined in Recommendation G.872. This new recommendation is based on requirements specified in Recommendation G.807.

- ITU-T Recommendation G.7712/Y.1703, "Architecture and Specification of Data Communication Network (DCN)," specifies the architecture and requirements for a data network to support the exchange of ASON messages in addition to traditional TMN communication. These communications take place among the transport, control, and management planes for ASON signaling and network management.

- ITU-T Recommendation G.7713/Y.1704, "Distributed Call and Connection Management," gives the requirements for distributed connection

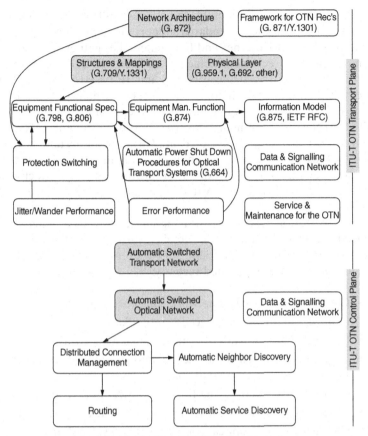

FIGURE 13.17 Key ITU-T ASON standards.

management (DCM) for both the UNI and the NNI. The requirements in this recommendation specify the signaling communications between functional components to perform automated connection operations, such as setup and release of connections. It describes DCM messages, attributes, and state transitions in a protocol-neutral fashion.

- ITU-T Recommendation G.7714/Y.1705, "Generalized Automatic Discovery Techniques," describes automatic discovery processes that support distribution connection management. Applications of automatic discovery addressed include neighbor discovery and adjacency discovery. The requirements, attributes, and discovery methods are described in a protocol-neutral fashion.

The Study Group 15 (Optical and Other Transport Network Infrastructures) recommendations (ITU, 2001) provide telecommunication equipment manufacturers with some of the necessary tools to develop interoperable products, allowing carriers to build and manage ultrahigh-capacity optical networks. In particular, these standards support efficient transport of popular data protocols such as Ethernet and fiber channel, together with other wideband and broadband services, including SDH/SONET, ATM, frame relay, IPTV (for high-definition 1080p service), and IP-based services. They also specify detailed equipment functions to support performance monitoring, fault isolation, and alarming, including support for optically transparent subnetworks. The OTN series of standards facilitate end-to-end connectivity between optical transport elements in a global network. An important new feature provided by these standards is the ability to combine multiple client signals within a wavelength to allow maximum utilization and cost-effectiveness of transport capacity while allowing switching at the OTN service rates of 2.5, 10, and 40 Gb/s.

The standardization process was still ongoing at press time. Standardization work ahead included the addition of detailed protocol specifications and the expansion of features for interoperable network restoration. Following are the "open questions" that were studied recently:

1/15. Coordination of access network transport standards

2/15. Optical systems for fiber access networks

3/15. General characteristics of optical transport networks

4/15. Transceivers for customer access and in-premises phone line networking systems on metallic pairs

5/15. Characteristics and test methods of optical fibers and cables

6/15. Characteristics of optical systems for terrestrial transport networks

7/15. Characteristics of optical components and subsystems

8/15. Characteristics of optical fiber submarine cable systems

9/15. Transport equipment and network protection/restoration

10/15. Optical fibers and cables for the access network to and in buildings and homes

11/15. Signal structures, interfaces, and interworking for transport networks

12/15. Transport network architectures

13/15. Network synchronization and time distribution performance

14/15. Management and control of transport systems and equipment

In particular, questions 12/15 and 14/15 are of direct interest to ASON.

Question 12/15: Transport network architectures. Transport network architecture Recommendations (G.805, G.809) and technology specific network architecture Recommendations (G.803, G.872, G.8010 and I.326) have been established. As operating experience is gained with employing current transport network technologies and new technologies evolve (e.g. variable size packets, high-speed transport networks), new Recommendations need to be developed, in close cooperation with the standardization activities on transport network systems and equipment. Moreover, requirements for new transport network interfaces need to be studied. The Requirements and Architecture of the Automatically Switched Optical Network (G.807, G.8080) have also been developed to enhance the capability to manage connections in the transport networks. This architecture requires enhancement to support more advanced applications. The following major Recommendations, in force at the time of approval of this Question, fall under its responsibility: G.803, G.805, G.807, G.809, G.872, G.8010/Y.1306, G.8080/Y.1304, I.326. The question is: What new or modifications to existing Recommendations are required to:

- refine and enhance the specification of Transport Network Architecture, including enhancements to G.803, G.805, G.809, G.872, G.8010, I.326?
- refine and enhance Automatically Switched Optical Network (ASON) Architecture and Requirements, including enhancements to Recommendations G.807 and G.8080?
- specify the support of packet client layers over a transport network?
- specify the client server relationships between the various network modes CO-CS, CO-PS and CLPS?

Study items to be considered include, but are not limited to:

- Next generation transport network architecture
- Converged transport networks
- Packet over transport architecture
- Transport networks using an Optical Packet server layer network
- Core/access convergence technologies
- Architecture of service management
- Transport via satellite networks

Tasks include, but are not limited to:

- Maintenance of Recommendations I.326, G.803, G.805, G.807, G.809, and G.872
- Refinement and enhancement of Recommendations G.8080, G.8010
- Functional architecture of MPLS networks

Question 14/15: Management and control of transport systems and equipment. As the level of functionality and intelligence in transport networks increases, the management and control requirements and supporting information models for these networks become more important. Management/Control Recommendations are required for all types of transport equipment, e.g., optical, wireless, terrestrial, submarine, satellite, etc., defined in ITU-T Recommendations. Management/Control requirements and information models for the interworking of standards-based transport systems/equipment and other equipment are essential to realize the full value of modern transport networks. ITU-T has established the Telecommunications Management Network (TMN) concept as a generalized framework for the definition of management requirements and supporting information models. Recommendations are needed to integrate into the

TMN framework the management functions of modern, standards-based transport equipment. Management requirements and information models should be based on the current ITU-T specified principles, functions, and techniques, for example as contained in Recommendations M.3010, M.3013, M.3020, and M.3400. ITU-T is creating standards on Automatically Switched Optical Networks (ASON) and their control mechanisms. Specifically, the ASON control mechanisms support both the activation of efficient connection configurations in a transport layer network and fast optical restoration. An architecture framework has been defined in Recommendations G.807/Y.1302 (requirements) and G.8080/Y.1304 (architecture) for ASON. Recommendations for the protocol-neutral requirements and protocol solutions are needed for ASON control, discovery, and management. The following major Recommendations, in force at the time of approval of this Question, fall under its responsibility: G.774-series, G.784, G.874-series, G.875, G.876, G.7710/Y.1701, G.7712/Y.1703, G.7713/Y.1704-series, G.7714/Y.1705-series, G.7715/Y.1706-series, G.7718/Y.1709-series, and I.752. The question is/are:

- What management requirements and information models must be specified to enable efficient and effective management of transport equipment in interoffice and long distance networks, including evolution to the Optical Transport Network (OTN), IP networks and Next Generation Networks (NGN)?
- What management requirements and information models must be specified to support the interworking among equipment and systems of different transport technologies, e.g., various interworking combinations among IP, ATM, SDH/SONET, OTN, MPLS, and Ethernet?
- What factors should be considered to identify management methodologies best suited to the characteristics of network equipment? Based on the Q-interface characterization in M.3010 and M.3013, what management protocols are appropriate for different classes of equipment? For example what management protocols are best suited to equipment that has limited management capabilities.
- What control requirements and protocol solutions must be specified to enable efficient and effective signaling, routing, automatic discovery, and management of ASON?

Study items to be considered include, but are not limited to:

- ASON control architecture based protocol-neutral requirements and associated protocol solutions
- Management aspect of control planes, including interaction between a control plane and a management plane
- Management and control aspects of Ethernet over Transport
- Management aspects of:
  - SDH equipment
  - Optical Network equipment
  - Flexible multiplexing equipment
  - PDH multiplexing equipment

- ATM equipment
- NGN equipment
- Access equipment
- GFP to include client signals such as Fiber Channel clients, FC_BB_GFPT
- Link Capacity Adjustment Scheme (LCAS)
- Virtual Concatenation
- MPLS
- Management data communication capability

Tasks include, but are not limited to:

- Revise Recommendations in the G.774 series
- Revise Recommendation G.784
- Complete draft new Recommendations I.752, G.875, G.876
- Revise Recommendations G.874 and G.874.1
- Revise Recommendation G.7710/Y.1701, Common Management Requirements
- Revise Recommendation G.7712/Y.1703, Data Communication Network
- Revise Recommendations G.7713/Y.1704 and G.7713.x/Y.1704.x series
- Revise Recommendations G.7714/Y.1705 and G.7714.1/Y.1705.1
- Revise Recommendations G.7715/Y.1706 and G.7715.x series
- Complete new Recommendations for ASON protocol-specific routing
- Complete new Recommendations G.7716/Y.1707 (on control plane initial establishment, reconfiguration and recovery) and G.7717/Y.1708 (on connection admissions control)
- Complete/Revise new Recommendation G.7718/Y.1709 on Framework for ASON Management
- Draft new Recommendation on protocol-neutral information model for LCAS and Virtual conCATenation (VCAT) mgmt
- Draft new Recommendation on protocol-neutral information model for Ethernet over Transport (EoT) mgmt
- Draft new Recommendation on protocol-neutral information model for MPLS management

The reality of the fact, however, is that today very few planners intermix equipment from various vendors within a planned network unless they are forced by some merger and acquisition or an external circumstance. Planners do not typically, by design, use SONET equipment from different vendors on the same ring, or routers from different vendors on the same peer-level backbone, and so on. Hence, the idea of interworking cannot be oversold.

The dynamic aspects of the ASTN/ASON (e.g., provisioning and restoration) require complex interactions between the optical control channels and the transport plane. In turn, this implies an interaction between the signaling and routing protocols. The result is an out-of-band control mechanism where the signaling and data paths could make use of different paths through the network.

GMPLS has been getting a lot of attention in this context. GMPLS extends MPLS to encompass time division (e.g., SONET/SDH, PDH, G.709), wavelength (λs), and spatial switching (e.g., incoming port or fiber to outgoing port or fiber). The focus of GMPLS is on the control plane of these various layers since each of them can use physically diverse data or forwarding planes (Mannie, 2004). GMPLS largely consists of (Atos, 2001; Papadimitriou et al., 2005).

- Extensions to the dynamic routing protocols intermediate system–intermediate system (IS-IS) and open shortest path first (OSPF) in order to provide management applications with the topology of the optical network, optical NE capabilities, resource availability, and so on
- Extensions to the signaling protocols constraint-based routing label distribution protocol and reservation protocol with traffic engineering (TE) in order to establish, maintain, and tear down optical paths
- A new link management protocol (LMP) for control channel management and link information exchange between neighboring optical NEs

See Figure 13.18 for an example.

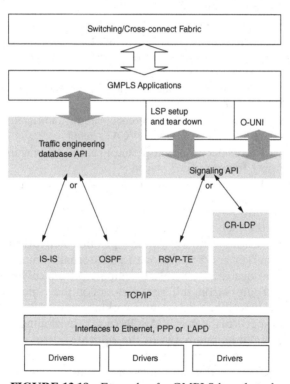

FIGURE 13.18   Example of a GMPLS-based stack.

LMP is designed to support the delivery of necessary information to the routing protocol; it also supports traditional link layer OAM&P functions (e.g., those performed by the SONET data communication channel). A modified constrained shortest path first (CSPF) engine is used to calculate a path through the network that meets required QoS criteria such as bandwidth, delay, jitter, the characteristics of fibers, route diversity, and security. The TE specific parameters are stored at each network node (as attributes of a link-state database), and this information is used by CSPF to determine appropriate routing. Once generated by a CSPF engine, this constrained route information is used by a signaling protocol, such as GMPLS, to establish the connection with each node, verifying that the resources are available as the signaling message transits the network (Shahane, 2002). Extensions to some of the existing protocols may be required. For example, with the hierarchy and multiple connections between two OXCs, each connection may carry its own or multiple control plane channels. This has the effect of possibly increasing the control plane state and computational burden, which in turn can affect both the cost and overall performance. A solution is to bundle light paths so that a single control plane channel can support multiple physical connections.

The interconnection of the IP clients (routers) and the optical control plane can be designed as being either loosely or tightly coupled. *Loose coupling* refers to an overlay model, and *tight coupling* refers to a peer model. (There is also a hybrid model, which is related to the peer model but shares attributes of both models.) The overlay signaling mechanism enables the client (the service requester) to add, modify, or delete connections over the carrier's network without providing the client with visibility to the carrier's network topology. Therefore, the overlay model (although it may use the same IP layer protocols as utilized in the client network for its routing and signaling) maintains a separation at the client-to-network interface by keeping the IP client routing, signaling protocols, topology distribution, and addressing scheme independent from the ones used by the optical layer of the carrier network (Shahane, 2002). This approach is likely to be adopted by carriers because they want to be able to control the resources of the transport network. This has traditionally been their view, going back to at least the mid-1970s (with X.25 network services), if not earlier.

GMPLS is a proposed peer-to-peer signaling protocol. It closely follows the peer-to-peer network model of IP technology. GMPLS extends MPLS with the necessary mechanisms to control routers, DWDM systems, ADMs, photonic cross-connects, and possibly other optical NEs. In a GMPLS environment, every participating label switching router sees the entire network state. Originally, there was no definition of a UNI or NNI in GMPLS but later, various standards group worked on defining a UNI for the GMPLS specification; MPLS/GMPLS was extended, particularly in the area of addressing (GMPLS extends the addressing mechanism to nodes that may not be IP capable, including support for network service access point mechanisms). GMPLS does add link-bundling and hierarchy capabilities; it also supports a bidirectional view of optical connections, as opposed to the unidirectional labeled switch paths (LSPs) that are established in MPLS.

The UNI specifies the mechanism by which a user network interface client is able to invoke transport network services with a user network interface network, which is an NE in the carrier's network. In this environment, edge (customer) routers can make their route calculations based on information collected through a GMPLS-enhanced routing protocol, such as OSPF or IS-IS. When the route is calculated, GMPLS utilizes RSVP-TE (traffic engineering) or constraint-based routing label distribution protocol to set up the LSPs. While MPLS deals only with what GMPLS identifies as packet switch capable interfaces, GMPLS adds four other types of interfaces:

1. Layer 2 switch capable interfaces recognize frames and cells.
2. Time-division-multiplexing capable interfaces forward data using time slots.
3. Lambda switch-capable interfaces such as photonic cross-connects work on individual wavelengths.
4. Fiber switch cable (FSC) interfaces work on individual or multiple fibers.

The GMPLS UNI functionality (at the logical level) is limited to three activities: connection creation, connection deletion, and connection status inquiry. Connection creation allows a connection of specified attributes to be activated between two points; these connections may be subject to policies that the operator defines, such as user group restrictions or security procedures. Connection deletion takes down an established connection. Connection status inquiry allows nodes to retrieve certain connection parameters by querying the network.

A considerable amount of work has been done by the OIF for UNI and NNI interfaces (see Table 13.3 for some of the OIF specifications). Initially, the connections that can be established are limited to SONET or SDH; additional interdomain management functions may also need to be developed (Greenfield, 2001). A number of router manufacturers have implemented the OIF UNI interface. Figure 13.19 depicts an example of use. This peer model supports the I-NNI interface of the ASTN/ASON model and assumes that all devices in the network have a complete topological view and that they participate in routing. This can be supported, for example, in the collection of OCCs for a single network, especially if this network is privately owned by the client (rather than being a carrier's network). In principle, in a hierarchical network comprised of optical switches, SONET/SDH ADMs, and IP layer 2/layer 3 devices, the entire optical core can be visible to an edge router using the same IGP routing instance over the network, such as OSPF or IS-IS (Shahane, 2002). As we noted, this approach is unlikely to be adopted by carriers because they do not want to lose control of their network by disseminating critical OTN information (such as network bandwidth, capacity, and

TABLE 13.3 OIF UNI and NNI Specifications

UNI 1.0 Signaling Specification: OIF-UNI-01.0	User Network Interface (UNI) 1.0 Signaling Specification, October 1, 2001
UNI 1.0 Signaling Specification, Release 2: OIF-UNI-01.0-R2	Common-User Network Interface (UNI) 1.0 Signaling Specification, Release 2: Common Part, February 27, 2004
OIF-UNI-01.0-R2-RSVP	RSVP Extensions for User Network Interface (UNI) 1.0 Signaling, Release 2, February 27, 2004
CDR-01: OIF-CDR-01.0	Call Detail Records for OIF UNI 1.0 Billing, April 2, 2002
SEP-01.0: OIF-SEP-01.0	Security Extension for UNI and NNI, May 8, 2003
SEP-02.1: OIF-SEP-02.1	Addendum to the Security Extension for UNI and NNI, March 31, 2006
SMI-01.0: OIF-SMI-01.0	Security for Management Interfaces to Network Elements, September 4, 2003
SMI-02.1: OIF-SMI-02.1	Addendum to the Security for Management Interfaces to Network Elements, March 31, 2006
E-NNI-01.0: OIF-E-NNI-Sig-01.0	Intra-Carrier E-NNI Signaling Specification, February 27, 2004

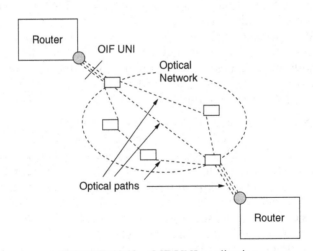

FIGURE 13.19 OIF UNI application.

topology).

A partial comparison of MPLS and GMPLS follows, based on Berger (2003).

- GMPLS differs from traditional MPLS in that it supports multiple types of switching [i.e., the addition of support for TDM, lambda, and fiber (port) switching]. The support for the additional types of switching has driven GMPLS to extend certain base functions of traditional MPLS and, in some cases, to add functionality. These changes and additions affect basic LSP properties, how labels are requested and communicated, the unidirectional nature of LSPs, how errors are propagated, and information provided for synchronizing the ingress and egress. In traditional MPLS TE, links traversed by an LSP can include an intermix of links with heterogeneous label encodings. GMPLS extends this by including links where the label is encoded as a time slot, or a wavelength, or a position in real-world physical space. Just as with traditional MPLS TE, where not all LSRs are capable of recognizing IP packet boundaries (e.g., an ATM-LSR) in their forwarding plane, GMPLS includes support for LSRs that cannot recognize IP packet boundaries in their forwarding plane. In traditional MPLS TE, an LSP that carries IP has to start and end on a router. GMPLS extends this by requiring an LSP to start and end on a similar type of LSR. Also, in GMPLS the type of payload that can be carried by an LSP is extended to allow such payloads as SONET/SDH, or 1- or 10-Gb Ethernet. These changes from traditional MPLS are reflected in how labels are requested and communicated in GMPLS.

- Another basic difference between traditional and non-PSC types of GMPLS LSP is that bandwidth allocation for an LSP can be performed only in discrete units. There are also likely to be (much) fewer labels on non-PSC links than on PSC links. The use of forwarding adjacencies provides a mechanism that may improve bandwidth utilization when bandwidth allocation can be performed only in discrete units, as well as a mechanism to an aggregate forwarding state, thus allowing the number of required labels to be reduced.

- GMPLS allows for a label to be suggested by an upstream node. This suggestion may be overridden by a downstream node, but in some cases, at the cost of higher LSP setup time. The suggested label is valuable when establishing LSPs through certain types of optical equipment, where there may be a lengthy (in electrical terms) delay in configuring the switching fabric. For example, micro mirrors may have to be elevated or moved, and this physical motion and subsequent damping takes time. If the labels and hence switching fabric are configured in the reverse direction (the norm), the MAPPING/Resv message may need to be delayed by tens of milliseconds per hop in order to establish a usable forwarding path. The label suggested is also valuable when recovering from nodal faults.

- GMPLS extends the notion of restricting the range of labels that may be selected by a downstream node. In generalized MPLS, an ingress or other upstream node may restrict the labels that may be used by an LSP along either a single hop or along the entire LSP path. This feature is driven from the optical domain, where there are cases where wavelengths used by the path must be restricted either to a small subset of possible wavelengths, or to one specific wavelength. This requirement occurs because some equipment may only be able to generate a small set of the wavelengths that intermediate equipment may be able to switch, or because intermediate equipment may not be able to switch a wavelength at all, only being able to redirect it to a different fiber.

- While traditional traffic engineered MPLS (and even LDP) are unidirectional, GMPLS supports the establishment of bidirectional LSPs.

- GMPLS supports the communication of a specific label to use on a specific interface.

- GMPLS formalizes possible separation of control and data channels. Such support is particularly important to support technologies where control traffic cannot be sent in-band with the data traffic.

- GMPLS also allows for the inclusion of technology-specific parameters in signaling.

13.4.3 Architectural Principles for ASON[†]

In this section we describe some architectural principles of the automatic switched optical network work that has recently been approved by the ITU-T. The existing transport networks provide SONET/SDH and WDM services whose connections are provisioned via network management. This process is both slow (weeks to months) relative to the switching speed and costly to network providers. An ASON is an optical/transport network that has dynamic connection capability. It encompasses SONET/SDH, wavelength, and potentially, fiber connection services in both OEO and all-optical networks. There are a number of added values related to such a capability:

- *Traffic engineering of optical channels.* Bandwidth assignment is based on actual demand patterns.

- *Mesh network topologies and restoration.* Mesh network topologies can in general be engineered for better utilization for a given demand matrix. Ring topologies might not be as efficient, due to the asymmetry of traffic patterns.

- *Managed bandwidth to core IP network connectivity.* A switched optical network can provide bandwidth and connectivity to an IP network in a dynamic manner compared to the relatively static service available today.

- *Introduction of new optical services.* The availability of switched optical networks will facilitate the introduction of new services at the optical layer. Those services include bandwidth on demand and optical virtual private networks.

ASON Architecture Principles ASON defines a control plan architecture that allows the setup and teardown of calls (and the connections that support a call) as a result of a user request. To achieve global coverage and the support of multiple client types, the architecture is described in terms of components and a set of reference points, and rules must be applied at the interface points between clients and the network, and between networks (Mayer, 2001).

ASON Reference Points In ASON architecture there is the recognition that the optical network control plan will be subdivided into domains that match the administrative domains of the network, as we alluded to earlier. The transport plane is also partitioned to match the administrative domains. Within an administrative domain the control plane may be subdivided further (e.g., by actions from the management plane). This allows the separation of resources into, for example, domains for geographic regions, which can be further divided into domains that contain different types of equipment. Within each domain, the control plane may be further subdivided into routing areas for scalability, which may also be further subdivided into sets of control components. The transport plane resources used by ASON will be partitioned to match the subdivisions created within the control plane.

The interconnection between domains, routing areas, and where required, sets of control components is described in terms of reference points. The exchange of information across these reference points is described by the multiple abstract interfaces between control components. The physical interconnection is provided by one or more of these interfaces. A physical interface is provided by mapping an abstract interface to a protocol. The reference point between an administrative domain and an end user is the UNI. The reference point between domains is the E-NNI. The reference point within a domain between routing areas and, where required, between sets of control components within routing areas is the I-NNI. Figure 13.20 shows a possible domain subdivision and the reference points between them. The difference between I-NNI and E-NNI is significant. I-NNI is applied in a single routing area where all equipment supports the same routing protocol, and detailed routing information could be exchanged between the various nodes. On the other hand, E-NNI is concerned mainly with reachability between domains that employ different routing and protection methodologies.

Call and Connection Control Separation Call and connection control are treated separately in ASON architecture. Call control is a signaling association between one or more user applications and the network to control the setup,

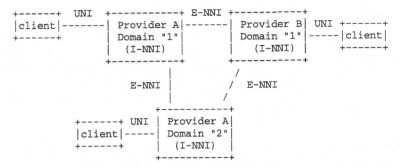

FIGURE 13.20   ASON/ASTN global architecture.

release, modification, and maintenance of sets of connections. Call control is used to maintain the association between parties, and a call may embody any number of underlying connections, including zero, at any instance of time.

Call control is provided at the ingress/egress of the network or at domain boundaries and is applicable at the E-NNI and UNI reference points. Call and connection control separation allows intermediate (relay) network elements to support only procedures needed for the support of switching connections. Access to call information at domain boundaries allows domains that use different protection or restoration mechanisms to interwork (e.g., a metro network using UPSR with a backbone network using mesh restoration) without the need for all domains to understand all possible protection/ restoration schemes.

With call and connection control separation, a single call may embody a number of connections (more than one) between user applications. This allows for the introduction of enhanced services, where a single call is composed of more than one application (e.g., voice and video). There are other situations where this separation between call and connection control is beneficial to the service provider, especially in the areas of restoration and maintenance. In those situations it is cost saving to maintain the call state while restoration actions are under way.

Policy and Security   According to the ASON, architecture policy is defined as the set of rules applied at a system boundary and implemented by port controller components. System boundaries may be nested to allow for correct modeling of shared policies with any scope. A system is defined as any (arbitrary) collection of components. In general, a system boundary will coincide with a domain boundary; this allows the application of a common policy for all interfaces that cross the domain boundary. The nesting of system boundaries allows the application of additional (more stringent) policies if the domain boundaries are between cost centers within a single network (administration) or between different networks (administrations).

Federation   Connection control across multiple domains requires cooperation between controllers in the various domains. A *federation* is defined as a
community of domains that cooperate for the purpose of connection management. Two types of federations are defined: a *joint federation model*, where
one connection controller has authority over connection controllers that reside
in different domains, and a *cooperative model*, where there is no concept of a
parent connection controller.

ASON Control Plane Requirements   A well-designed control plane architecture should give service providers better control of their network while
providing faster and improved accuracy of circuit setup. The control plane
itself should be reliable, scalable, and efficient. It should also be sufficiently
generic to support different technologies and differing business needs and
different partitions of functions by vendors (i.e., different packaging of the
control plane components). In summary, the control plane architecture
should:

- Be applicable to a variety of transport network technologies (e.g., SONET/
  SDH, OTN, PXC). To achieve this goal, it is essential that the architecture
  isolates technology-dependent aspects from technology-independent
  aspects, and address them separately.
- Be sufficiently flexible to accommodate a range of different network
  scenarios. This goal may be achieved by partitioning the control plane
  into distinct components. This allows vendors and service providers to
  decide the location of these components and allows the service provider
  to decide the security and policy control of these components.

The control plane should support either switched connections or soft permanent connections of basic connection capability in transport networks.
These connection capability types are:

- Unidirectional point-to-point connections
- Bidirectional point-to-point connections
- Unidirectional point-to-multipoint connections

The control of connectivity is essential to the operation of a transport
network. The transport network itself can be described as a set of layered
networks, each acting as a connecting function whereby associations are
created and removed between the inputs and outputs of the function. These
associations are referred to as *connections*. Three types of connection establishment are defined: provisioned, signaled, and hybrid.

Establishment of a *provisioned connection* is triggered by a management
system and is referred to as *hard permanent connection*. *Signaled connections*
are established on demand by the communicating endpoints using a dynamic

protocol message exchange in the form of signaling messages. In a *hybrid connection* a network provides a permanent connection at the edge of the network and utilizes a switched connection within the network to provide end-to-end connections between the permanent connections at the network edges.

The most significant difference between the three methods is the party that sets up the connection. In the case of provisioning, connection setup is the responsibility of the network operator, whereas in the signaled case, connection setup may also be the responsibility of the end user. Additionally, third-party signaling should be supported across a UNI.

ASON Functional Architecture The components of the control plane architecture are:

1. *Connection controller function* (CC). The connection controller is responsible for coordination among the link resource manager, routing controller for the purpose of the management, and supervision of connection setup, release, and modification.

2. *Routing controller* (RC). The role of the RC is to respond to requests from the CC for route information needed to set up a connection and to respond to requests for topology information for network management purposes.

3. *Link resource management* (LRM). The LRM component is responsible for the management of subnetwork links, including the allocation and deallocation of resources, providing topology and status information.

4. *Traffic policing* (TP). The role of the TP is to check that the incoming user connection is sending traffic according to the parameters agreed upon.

5. *Call controller.* There are two types of call controller, a calling/called party call controller and a network call controller. The role of the call control is the generation and processing of call requests.

6. *Protocol controller* (PC). The PC provides the function mapping of the parameters of the abstract interfaces of the control components into messages that are carried by a protocol to support interconnection via an interface.

ASON Reference Points and GMPLS Protocols The ASON CP as shown in Figure 13.20 defines a set of interfaces or reference points, already discussed:

- UNI runs between the optical client and the network.
- I-NNI defines the interface between the signaling network elements within the same domain or between routing areas.

- E-NNI defines the interface between ASON control planes belonging to different domains.

The various ASON interfaces are described in the next few sections. Candidate GMPLS-based protocols for use at the interfaces are also discussed.

ASON User–Network Interface   ASON UNI allows ASON clients to perform a number of functions, including:

- *Connection create:* allows clients to signal to the network to create a new connection with specified attributes. Those attributes might include bandwidth, protection, restoration, and diversity.
- *Connection delete:* allows ASON clients to signal to the network the need to delete an already existing connection.
- *Connection modify:* allows ASON clients to signal to the network the need to modify one or more attributes for an already existing connection.
- *Status enquiry:* allows ASON clients to inquire into the status of an already existing connection.

Other functions that might be performed at the ASON UNI are client registration, address resolution, and neighbor and service discovery. Those functions could be automated or manually configured between the network and its clients. Client registration and address resolution are tightly coupled to the optical network address scheme. Requirements for optical network addresses and client names are outlined by Lazar (2001). In general, the client name (or identification) domain and optical address domain are decoupled. The client ID should be globally unique to allow for the establishment of end-to-end connections that encompass multiple administration domains. For security it is required that the nodal addresses used for routing within an optical domain not cross network boundaries. The notion of closed user groups should also be included in ASON addressing to allow for the offering of OVPN services.

ASON UNI realization requires the implementation of a signaling protocol with sufficient capabilities to satisfy UNI functions. Both LDP and RSVP-TE have been extended to be used as the signaling protocol across the ASON UNI. The extensions involve the definition of the necessary TLVs (type-length values) or objects to be used for signaling connection attributes specific to the optical layer. New messages are also defined to allow for connection status enquiry. The OIF adopted both protocols in its UNI specification (Rajagopalan, 2001). It should be noted, however, that there have been some efforts to move CR-LDP off the standards track, with only RSVP-TE as the recommended solution.

ASON Internal Node-to-Node Interface The I-NNI defines the interface between adjacent connection controls in the same domain or between routing areas. There are two main aspects of I-NNI: signaling and routing. Path selection and setup through the optical network requires a signaling protocol. Transport networks typically utilize explicit routing, where path selection can be done either by operator or by software scheduling tools in management systems. In ASON, end-to-end optical channels (connections) are requested with certain constraints. Path selection for a connection request should employ constrained routing algorithms that balance multiple objectives:

- To conform to constraints such as physical diversity
- To load balancing of network traffic to achieve the best utilization of network resources
- To follow policy decisions on routing, such as preferred routes

To facilitate the automation of the optical connection setup, nodes in the optical network must have an updated view of its adjacencies and of the utilization levels at the various links of the network. This updated view is sometime referred to as *state information*. State information dissemination is defined as the manner in which local physical resource information is disseminated throughout the network. First, the local physical resource map is summarized into logical link information according to link attributes. This information can then be distributed to the various nodes in the network using the control plane transport network IGP. ASON I-NNI could be based on two key protocols, IP and MPLS. Since MPLS employs the principle of separation between the control and the forward planes, its extension to support I-NNI signaling is feasible.

GMPLS defines MPLS extensions to suit types of label switching other than the in-packet label. Those other types include time slot switching, wavelength and waveband switching, and position switching between fibers. CR-LDP and RSVP-TE have been extended to allow for the request and binding of generalized labels. With generalized MPLS, an LSP is established with the appropriate encoding type (e.g., SONET, wavelength). LSP establishment takes into account specific characteristics that belong to a particular technology. MPLS traffic engineering requires the availability of routing protocols that are capable of summarizing link-state information in their databases. Extensions to IP routing protocols, OSPF and IS-IS, in support of link-state information for generalized MPLS are described in various IETF documents.

ASON External Node-to-Node Interface   E-NNI is the external NNI between domains. Those domains may belong to the same network administration or to different administrations. In some sense, E-NNI could be viewed as similar to the UNI interface, with some routing functions to allow for the exchange of reachability information between different domains. Border gateway protocol (BGP) is the IP-based protocol that is commonly deployed between domains. It could be used to summarize reachability information between ASON domains in the same manner as it has been in use today for IP networks. BGP is rich in policy that makes a good candidate to satisfy service requirements such as diversity. where policies could be used in choosing diverse routes. (Note, however, that whereas BGP is well suited for the transport of richly policy-labeled information between routing domains, it is not well optimized for rapid convergence times; this may have implications when considering restoration times vs. initial path selection.)

ASON/ASTN CP Transport Network (Signaling Network)   In this section, some architectural considerations for the makeup of the transport network that is used to transport the control plane information are discussed. For circuit-based networks, the ability to have an independent transport network for message transportation is an important requirement. The control network represents the transport infrastructure for control traffic and can be either in-band or out-of-band. An implication of this is that the control plane may be supported by a different physical topology from that of the underlying ASON. There are fundamental requirements that control networks must satisfy to assure that control plane data can be transported in a reliable and efficient manner. In the event of control plane failure (e.g., communications channel or control entity failure), while new connection operations will not be accepted, existing connections will not be dropped. Control network failure would still allow dissemination of the failure event to a management system for maintenance purposes. This implies a need for separate notifications and status codes for the control plane and ASON. Additional procedures may also be required for control plane failure recovery.

It is recognized that the interworking of the control networks is the first step toward control plane interworking. To maintain a certain level of ease, it is desirable to have a common control network for different domains/subnetworks or types of network. Typically, control plane and transport functions may coexist in a network element. However, this may not be true in the case of a third-party control. This situation needs further study. Furthermore, addressing issues in the control plane vis-à-vis the transport network is also for further study. ASON CP transport network requirements include the following:

- Control plane message transport should be secure. This requirement stems from the fact that the information exchanged over the control plane is service-provider specific and security is of utmost importance.
- Control message transport reliability has to be guaranteed in almost all situations, even during what might be considered catastrophic failure scenarios of the controlled network.

- The control traffic transport performance affects connection management performance. Connection service performance depends largely on its message transport. Time-sensitive operations such as protection switching may need certain QoS guarantees. Furthermore, a certain level of survivability of the message transport should be provided in case of control network failure.
- The control network needs to be both upward and downward scalable in order for the control plane to be scalable. Downward scalability may be envisioned where the ASON network offers significant static connections, reducing the need for an extended control network.
- The control plane protocols should not assume that the signaling network topology is identical to that of the transport network. The control plane protocols must operate over a variety of signaling network topologies.

Given the foregoing requirements, it is critical that the maintenance of the control network itself not pose a problem to service providers. As a corollary, this means that configuration-intensive operations should be avoided for the control network.

Common channel signaling links are associated with user channels in the following ways:

- *Associated*, whereby signaling messages related to traffic between two network elements are transferred over signaling links that connect the two network elements directly.
- *Nonassociated*, whereby signaling messages between two network elements A and B are routed over several signaling links, whereas traffic signals are routed directly between A and B. The signaling links used may vary with time and network conditions.
- *Quasi-associated*, whereby signaling messages between nodes A and B follow a predetermined routing path over several signaling links, whereas the traffic channels are routed directly between A and B.

Associated signaling may be used where the number of traffic channels between two network elements is large, thereby allowing a single signaling channel to be shared among a large number of traffic channels. Quasi-associated signaling may be used to improve resiliency. For example, consider a signaling channel that has failure mechanisms independent of the traffic channels. Failure of the signaling channel will result in loss of signaling capability for all traffic channels, even if all the traffic channels are still functional. Quasi-associated signaling mitigates against this by employing alternative signaling routes. In other words, the signaling network must be designed such that failure of a signaling link should not affect the traffic channels associated with that signaling channel.

Transport Network Survivability and Protection   In this section we describe the strategies that can be used to maintain the integrity of an existing call in the event of failures within the transport network. The terms *protection* (replacement of a failed resource with a preassigned standby) and *restoration* (replacement of a failed resource by rerouting using spare capacity) are used to classify these techniques. In general, protection actions are completed in the tens of millisecond range, while restoration actions are normally completed in times ranging from hundreds of milliseconds to up to a few seconds.

The ASON control plane provides a network operator with the ability to offer a user calls with a selectable class of service (CoS) (e.g., availability, duration of interruptions, errored seconds). Protection and restoration are mechanisms (used by the network) to support the CoS requested by the user. The selection of the survivability mechanism (protection, restoration, or none) for a particular connection that supports a call will be based on the policy of the network operator, the topology of the network, and the capability of the equipment deployed. Different survivability mechanisms may be used on the connections that are concatenated to provide a call. If a call transits the network of more than one operator, each network should be responsible for the survivability of the transit connections. Connection requests at the UNI or E-NNI will contain only the requested CoS, not an explicit protection or restoration type.

The protection or restoration of a connection may be invoked or temporarily disabled by a command from the management plane. These commands may be used to allow scheduled maintenance activities to be performed. They may also be used to override the automatic operations under some exceptional failure conditions. The protection or restoration mechanism should:

- Be independent of, and support any, client type (e.g., IP, ATM, SDH, Ethernet).
- Provide scalability to accommodate a catastrophic failure in a server layer, such as a fiber cable cut, which affects a large number of client layer connections that need to be restored simultaneously and rapidly.
- Utilize a robust and efficient signaling mechanism, which remains functional even after a failure in the transport or signaling network.
- Not rely on functions that are non-time critical to initiate protection or restoration actions. Therefore, consideration should be given to protection or restoration schemes that do not depend on fault localization.

Relationship to GMPLS Architecture   The relationship between ASON/ASTN control plane architecture and GMPLS-based protocols was discussed earlier; different GMPLS protocol can be utilized for realization of the different ASON/ASTN external interfaces. It is important to note that there is no conflict between GMPLS architecture and the ASON network architecture.

ASON/ASTN provides a functional architecture of a control plane that allows the establishment of switched paths in optical networks. It provides the set of external interfaces that are necessary for the ASTN/ASON network to have a global reach. It does that, however, in a protocol-independent fashion that can be realized in different ways provided that its requirements are satisfied. The GMPLS architecture focuses more on the applications of GMPLS-defined protocols [e.g., CR-LDP for the setup of generalized LSP (GLSP) at the various interfaces of the network: e.g., I-NNI, UNI, etc.].

13.5 STATUS AND DEPLOYMENTS

Work on the ASTN/ASON, GMPLS, and O-UNI standards was progressing at a steady pace middecade and vendors claim to have field-deployed standard implementations. The challenges to full-scale adoption of these models lie in several areas: (1) the remaining standards development work, (2) persuading carriers to adopt IP technology in the control plane, and (3) deriving the economic benefits carriers will accrue from deploying optical signaling technology (Shahane, 2002). New investments are required by carriers to realize the OTN/ASTN/ASON. The long-term benefits include rapid circuit provisioning and other capabilities listed above. Many of these features, however, are available in one form or another in today's SONET/SDH-based networks.

It needs to be noted at the outset that *while* switched-λ services are within technical grasp in the core of the network, these services are a *real challenge* in practical end-to-end applications. For example, in the United States there are approximately 4.5 million commercial buildings, with over 735,000 being office-space building. Yet, only approximately 30,000 buildings were expected to use fiber loops as of 2007. Of the 30,000 buildings cited, many are "carrier hotels," central office–like real estate housing the telecommunication equipment of competitive carriers. That leaves the actual end-user-penetrated building at an even smaller number.

The issue is that it generally takes about $0.25 million to penetrate a building, often even more. Although it is fashionable to assume that there will be a large number of customers that require gigabit-level connectivity in a building, the realities are quite different. As a quick anecdotal calculation, one can assume that (at least for the foreseeable future) only the Fortune 5000 companies[†] are candidates for switched-λ services. Assuming that these companies had on average 10 locations that are tier 1 network locations requiring gigabit-level connectivity, there would be 50,000 data sources/sinks. Assuming that these companies had all chosen the "intelligent building" locations that are on fiber (30,000 cited above), this would translate to about two customers per

[†]To provide a view, note that there are about 16,000 publicly traded companies in the United States; this is not to imply that only publicly traded companies need gigabit networks, but these companies tend to be the largest companies around.

building, if that. This implies that the penetration of fiber and related services will take a relatively long time.[‡]

Network features of interest to enterprise users almost invariably have to be end to end in order to be of value: reliability has to be end to end; QoS has to be end to end; security has to be end to end; and similarly, provisionability has to be end to end. It is immaterial to the user that the core can be provisioned in 2 seconds when the user has to wait nine months to get fiber installed and deployed in a building. Today, we can already provision an OC-3 in seconds using DCS technology, but if a building in the outskirts of town does not have a fiber loop, it could take months to obtain an end-to-end OC-3. Herewith the assumption is made that the issues just highlighted are well understood by the developers of next-target (next-generation) optical networks, and hence that the features discussed in the chapter are not just "pie in the sky" for most potential enterprise users. Also, the instantaneous provisionability frankly will not occur until a certain large majority of buildings of interest are on fiber facilities. Today in the United States, only 5% of office buildings are on fiber; for instantaneous provisionability to be a reality, perhaps 50, 65, 80, or 95% of the buildings have to be on fiber networks.

With reduced capital expenditures in the early part of the decade, service providers have shown a degree of reluctance to invest in new technologies unless there is careful economic and rate-of-return analysis regarding what needs to be an increased revenue opportunity; there also has to be an assurance that the embedded base of hundreds of billions of dollars can continue to be utilized in a cohesive manner with the new technology proposed. Proponents make the pitch that the ability to provision bandwidth more rapidly enhances revenue by enabling service providers to support new value-added services and realize substantial cost savings. Provisioning capabilities, however, have to be end to end to be of real value. Once the capabilities are available end to end, or at least edge to edge, the dynamic bandwidth allocation that can be accomplished with optical signaling (particularly in the core of the network) can be leveraged to maximize the reuse of existing transmission facilities, to reduce the need to overprovision, and to optimize service reliability.

KEY POINTS

- Networks consists of various network elements (NEs), such as ADMs, switches, and routers, that use signaling protocols to provision resources

---

[‡]One should not confuse the deployment of fiber to the home (FTTH) with OTN/SONET. The former typically supports bandwidths in the range 10 to 100 Mbps and is designed for a price point of $1000 to 2000 per end; this technology is *not* designed to transport symmetric end-to-end multigigabit/second connections.

dynamically and to provide network survivability using protection and restoration techniques

- Protection switching exists at layer 1 (physical layer protection switching), layer 2 (LAN, frame relay, or ATM switching), and layer 3 (internal gateway protocols, exterior gateway protocols, or multiprotocol label switching). Typically, layer 1 protection switching operates at the 50-ms level, layer 2 in the second-to-multisecond level, and layer 3 operates on the tens-of-seconds level, as it requires topology information from significant portions of the network

- Network management requires existence of open industry standards for the user, control, and management planes for all NEs in order for various pieces of networking equipment to function together properly

- The user plane refers to data forwarding in the transmission path; the industry standards deal with transmission parameters (electrical, optical, mechanical, frame/packet formats)

- The control plane refers to a session setup and teardown mechanism; the industry standards allow switches from two different vendors to talk to each other

- The management plane deals with procedures and mechanisms to deal with fault, configuration, accounting, and performance management

- SONET/SDH is the de facto infrastructure technology for communication services at the present time, as effectively all communication in the world is taking place over SONET or SDH systems ranging from DS3 to OC-768

- Automatic protection switching (APS) is the mechanism used in SONET/SDH to restore service in the case of an optical fiber failure or a NE failure. SONET standards require restoration of service within 50 ms

- Fiber cuts are the most common failures in SONET rings; NEs typically have mean time between failures of about 20 years

- The most common SONET topology for carrier networks is the ring. Rings are preferred, as they provide an alternative path to support communication between any two nodes

- A two-fiber ring can operate as a unidirectional or bidirectional ring. In a unidirectional ring, traffic is limited to one fiber and it always flows the same direction around the ring; the second fiber is the protection path. With bidirectional designs, information is sent on both fibers; when data are sent between two nodes, they flow over the two fibers connecting them. To provide backup, each fiber in a bidirectional ring can only be utilized to half its capacity.

- Four-fiber rings always operate as bidirectional rings. Full transmission capability is achieved on the working fibers, while the protection fibers are not utilized for traffic under normal conditions

- The optical transport network (OTN) addresses the requirements of next-generation networks that have a goal of efficiently transporting data-oriented traffic. OTN is based on industrial standards (ITU) to ensure interoperability among various NE manufacturers. A distinguished characteristics of the OTN is its ability to transport any digital signal, independent of client-specific aspects, making it a protocol-agnostic technology

- OTN makes use of a data plane (transport layer) and a control plane (signaling and measurement layer)

- The optical transport functions include multiplexing, cross-connecting including grooming and configuration, management functions, and physical media functions)

- OTN benefits include maintenance signals per wavelength, fault-isolation capabilities, forward error correction (FEC) capabilities, protocol-agnostic nature

- OTN defines a network hierarchy known as the optical transport hierarchy (OTH). OTH's basic unit is the optical transport module (OTM). OTMs utilize the nomenclature OTM-$n.m$, where n refers to the maximum number of wavelengths supported at the lowest bit rate on the wavelength and m indicates the bit rate supported on the interface. For example, OTM-3.2 indicates an OTM that spans three wavelengths, each operating at least at 10 Gb/s.

- OTH supports intrinsic connections at three levels: optical channels, optical multiplex sections, and optical transmission section.

- The optical channel layer transports client signals between two endpoints on the OTN; this is conceptually similar to the SONET path

- Optical multiplex sections (OMSs) describe the WDM aspects that support the optical channel; they are conceptually similar to SONET lines but accommodate multiple wavelengths

- Optical transmission sections (OTS) enable transmission of signals over individual fiber spans; this is conceptually similar to the SONET section concept

- OTN makes use of an optical channel layer: each wavelength is wrapped in an envelope that consists of a header (for overhead bytes) and a trailer (for FEC functions). The payload section allows for existing network protocols to be mapped (wrapped), making OTN protocol independent

- The OTNk frame structure is defined for three optical channel transport unit bit rates, with $k = 1, 2, 3$ corresponding to 2.5, 10, and 40 Gb/s, respectively. The OTUk frame is composed of an optical channel payload unit, an optical channel transport unit, and an optical channel data unit

- Automatically switched optical networks (ASONs) aim at providing the OTN with an intelligent optical control plane for dynamic network provisioning. The ASON model is based on mesh network architectures

- The dynamic aspects of ASONs (e.g., provisioning and restoration) require complex interactions between the optical control channels and the transport plane. ASON uses out-of-band control mechanism where signaling and data paths could make use of different paths through the network
- Generalized MPLS (GMPLS) extends MPLS to encompass time-division (used in SONET/SDH, G.709), wavelength (λ's), and spatial switching. The focus of GMPLS is on the control plane of various OSIRM layers since each of them can use physically diverse data or forwarding planes
- GMPLS can be understood as a peer-to-peer signaling protocol as it extends MPLS with necessary mechanisms to control routers, DWDM systems, ADMs, and photonic cross-connects
- ASON is an optical/transport network that has dynamic connection capability, leading to the following network benefits: traffic engineering of optical channels (bandwidth issues based on actual demand), mesh network topologies and restoration, managed bandwidth to core IP network connectivity, and the ability to introduce new optical services
- ASON interfaces can be either an internal node-to-node interface (I-NNI) or an external node-to-node interface (E-NNI). ASON I-NNI could be based on two key protocols: IP and MPLS. ASON E-NNI could be viewed as the UNI interface with some routing functions to allow for the exchange of reachability information between domains

REFERENCES

Aboul-Magd, O., B. Jamoussi, S. Shew, G. Grammel, S. Belotti, and D. Papadimitriou, Automatic switched optical network architecture and its related protocols, IPO WG, Internet Draft, draft-ietf-ipo-ason-02.txt, March 2002. Copyright © The Internet society. All rights reserved. This document and translations of it may be copied and furnished to others, and derivative works that comment on or otherwise explain it or assist in its implementation may be prepared, copied, published and distributed, in whole or in part, without restriction of any kind, provided that the above copyright notice and this paragraph are included on all such copies and derivative works.

Alanqar, W., et al., Requirements for generalized MPLS (GMPLS) routing for automatically switched optical network (ASON), draft-ietf-ccamp-gmpls-ason-routing-reqts-00.txt, December 2003.

Atos, Promotional material from Atos Origin—Systems Integration Telecom Technologies Business Unit, Paris, http://www.marben-products.com.

Berger, L., Ed., *Generalized Multi-Protocol Label Switching (GMPLS) Signaling Functional Description*, IETF, RFC 3471, January 2003.

Greenfield, D., Optical standards: a blueprint for the future, *Network Magazine*, October 5, 2001.

ITU, 2001, Recommendations adopted by Study Group 15 in October 2001, http://www.itu.int/newsroom/Recs/SG15Recs.html.

Lazar, M., et al., *Alternate Addressing Proposal*, OIF Contribution OIF2001.21, January 2001.

Mannie, E., Ed., *Generalized Multi-protocol Label Switching (GMPLS) Architecture*, IETF, RFC 3945, October 2004.

Mayer, M., Ed., *Requirements for Automatic Switched Transport Networks (ASTN)*, ITU G.8070/Y.1301, V1.0, May 2001a.

Mayer, M., Ed., *Architecture for Automatic Switched Optical Networks (ASON)*, ITU G.8080/Y1304, V1.0, October 2001b.

Minoli, D., and A. Alles, *LAN, ATM, and LAN Emulation Technologies*, Artech House, Norwood, MA, 1996.

Minoli, D., *Telecommunications Technology Handbook*, 2nd ed., Artech House, Norwood, MA, 2003.

Minoli, D., Metro Ethernet: Where's the beef? *Networkworld*, December 11, 2006.

NCS (National Communications System), *Internet Protocol over Optical Transport Networks*, Technical Information Bulletin 03-3 Ncs Tib 03-3, National Communications System, Arlington, VA, December 2003.

Papadimitriou, D., J. Drake, et al., *Requirements for Generalized MPLS (GMPLS) Signaling Usage and Extensions for Automatically Switched Optical Network (ASON)*, RFC 4139, July 2005.

Rajagopalan, B., Editor, *User Network Interface (UNI) 1.0 Signaling Specifications*, OIF Contribution OIF2000.125.7, October 2001.

Shahane, D., Building optical control planes: challenges and solutions, *Communication Systems Design*, January 7, 2002.

INDEX

Network Infrastructure and Architecture: Designing High-Availability Networks,
By Krzysztof Iniewski, Carl McCrosky, and Daniel Minoli
Copyright © 2008 John Wiley & Sons, Inc.